Artificial Intelligence in Design

Artificial Intelligence in Industry Series

Series Editor: A. Kusiak

Titles in the series:

ARTIFICIAL INTELLIGENCE IN INDUSTRY

Artificial Intelligence in Design

Edited by D. T. Pham

With 229 Figures

Springer-Verlag
London Berlin Heidelberg New York
Paris Tokyo Hong Kong

DT Pham, PhD
School of Electrical, Electronic and Systems Engineering
University of Wales College of Cardiff
PO Box 904, Cardiff CF1 3YH, UK

ISBN 3-540-50634-9 Springer-Verlag Berlin Heidelberg New York
ISBN 0-387-50634-9 Springer-Verlag New York Berlin Heidelberg

British Library Cataloguing in Publication Data
Artificial intelligence in design.
1. Engineering. Applications of artificial intelligence
I. Pham DT (Duc Truong) 1952- II. Series
620.0028563
ISBN 3-540-50634-9

Library of Congress Cataloging-in-Publication Data
Artificial intelligence in design/ edited by DT Pham.
p. cm. – (Artificial intelligence in industry series)
Includes index.
ISBN 0–387–50634–9 (U.S.)
1. Engineering design. 2. Artificial intelligence. I. Pham,
DT II. Series. TA174.A79 1991
620'.00425'02863--dc20   90-39029
                                                       CIP

Typeset by Photo·graphics, Honiton, Devon
Printed by Page Brothers, Norwich, Norfolk
69/3830-543210   Printed on acid-free paper

Series Editor's Foreword

The rapid development of manufacturing and computer technologies has generated new problems. To solve these problems modern tools and techniques are required. Artificial intelligence (AI) offers one of the most appropriate sets of tools for solving complex industrial problems.

This series represents an effort to disseminate valuable information on applications of AI in industry which have become well-utilized throughout the world.

AI has been recognized in many industrial countries as a means of solving complex problems arising in modern industry. All books in this series highlight issues that arise in the development and application of AI systems.

The series is intended for production and industrial engineers, managers, system designers, and programmers.

Andrew Kusiak
The University of Iowa,
Iowa City, Iowa

Preface

Computers have been employed for some time in engineering design mainly as numerical or graphical tools to assist analysis and draughting. The advent of the technology of artificial intelligence (AI) and expert systems has enabled computers to be applied to less deterministic design tasks which require symbolic manipulation and reasoning, instead of only routine number processing. This book presents recent examples of such applications, focusing on mechanical and manufacturing design. The term "design" is interpreted here in its wider sense to include creative activities such as planning. The book covers a wide spectrum of design operations ranging from component and product design through to process, tooling and systems design. Its aim is to expose researchers, engineers and engineering designers to several developments in the emerging field of intelligent computer-aided design (CAD) and to alert them of the possibilities and opportunities in this exciting field.

The book comprises five sections. Section A deals with concepts and techniques of intelligent CAD. The first chapter, by Pham and Tacgin, provides an overview of techniques underlying the field of intelligent knowledge-based systems, particularly those which can be used to develop intelligent design programs. The chapter introduces and defines many of the terms encountered in the rest of the book. The second chapter, by Aldefeld et al., describes the concept of variational geometry in design and a method for implementing this concept based upon geometric reasoning. The latter technique is further examined in the third chapter, by Martin, which reviews its various aspects and relationships with topics such as computer algebra, automated theorem proving and expert systems. The final chapter in this section, by Akagi, discusses a powerful concept for representing design knowledge, the object-oriented concept, and presents an object-oriented expert system for engineering design.

Section B is concerned with the design of both individual components and complete products. The first chapter, by Dong and Soom, treats the problem of component dimensional tolerancing and suggests AI techniques for automating the tedious tasks of tolerance analysis and synthesis. The second chapter, by Kroll et al., is about product design for assembly and describes an intelligent knowledge-based system for this purpose. The third chapter, by ElMaraghy, also deals with intelligent product design. The emphasis of the chapter, however, is on integrating design and manufacture. The chapter presents a framework for supporting intelligent product design and manufacture. The framework comprises a feature-based modeller for design, high-level design languages and an assortment of expert process planning programs. Section B concludes with the chapter by Huang and Brandon which focuses on the knowledge-based design of machine tools after considering wider issues of managing knowledge bases for machine design.

Section C is devoted to process design. The first chapter, by Inui and Kimura, describes an intelligent system incorporating product modelling techniques for designing process plans for machining prismatic parts and bending sheet metal components. The second chapter, by Milacic, proposes the use of the theories of formal grammars and automata to acquire and represent knowledge for expert process design systems. The chapter also outlines two such systems, one for designing process plans for machining rotational parts and the other, for producing plans for prismatic components. The third chapter, by Ito and Shinno, discusses the philosophy to be adopted in the development of the next generation of AI-based systems for machining process design. The last chapter in this section, by Wright et al., reports on the progress achieved by the authors' team in building an AI system that integrates machining process design and workholding element configuration.

Section D covers tooling design. The first chapter, by Pillinger et al., describes an intelligent knowledge-based system for designing metal-forming dies. The system combines rule-based techniques with finite-element simulation and is able to improve its design rules automatically as it gains experience. The second chapter, by Nee and Poo, reviews the state of the art in expert CAD systems for jigs and fixtures. The third chapter, by Pham and de Sam Lazaro, presents two knowledge-based programs for jig and fixture design, one to provide design advice and the other to carry out design operations automatically.

Section E deals with systems design. The first chapter, by Li et al., describes part of an intelligent software package for designing hydraulic systems. The second chapter, by McGuire and Wee, discusses the design of control systems and the use of knowledge-base technology in a tool for generating control programs for programmable logic controllers. The third chapter, by Kusiak and Heragu, presents a knowledge-based system for selecting equipment for a manufacturing plant. The system, which also performs optimization, is to be used in the design of the plant. The fourth and final chapter, by Parthasarathy and Kim, considers fundamental issues in designing intelligent manufacturing systems and discusses the representation and utilization of design knowledge by formal decision rules and performance measures.

Acknowledgements

My thanks are extended to the chapter authors for their contributions and to Dr Brian Rooks (IFS), Mr Nicholas Pinfield (Springer-Verlag) and Miss Linda Schofield (Springer-Verlag) for their expert help with the production of this book.

DT Pham

Contents

SECTION E. Systems Design

Contributors

S. Akagi
Department of Mechanical Engineering, Osaka University,
2-1 Yamadaoka, Suita, Osaka 565, Japan

B. Aldefeld
Philips GmbH, Forschungslaboratorium Hamburg, Vogt-
Kölln-Str. 30, Postfach 54 08 40, 2000 Hamburg 54, Federal
Republic of Germany

J.A. Brandon
School of Engineering, University of Wales College of
Cardiff, PO Box 917, Cardiff CF2 1XH, UK

T.A. Dean
School of Manufacturing and Mechanical Engineering,
University of Birmingham, Edgbaston, Birmingham B15 2TT,
UK

Z. Dong
Department of Mechanical Engineering, University of
Victoria, PO Box 1700, Victoria, British Columbia VBW 2Y2,
Canada

H.A. ElMaraghy
Centre for Flexible Manufacturing Research and
Development, McMaster University, 1280 Main Street West,
Hamilton, Ontario L8S 4L7, Canada

P.J. Englert
AT&T Bell Laboratories, Systems Packaging Development
Department, Rm 4C-251 244A, 1 Whippany Road, Whippany,
NJ 07981-0913, USA

P. Hartley
School of Manufacturing and Mechanical Engineering,
University of Birmingham, Edgbaston, Birmingham B15 2TT,
UK

C.C. Hayes
Robotics Institute, Carnegie Mellon University, Pittsburgh,
PA 15213, USA

S.S. Heragu
Department of Management and Marketing, School of
Business and Economics, The State University of New York,
Plattsburgh, NY 12901, USA

G.Q. Huang
School of Engineering, University of Wales College of
Cardiff, PO Box 917, Cardiff CF2 1XH, UK

S.H. Huang
Department of Mechanical Engineering 2, Huazhong
University of Science and Technology, Wuhan, Hubei, China

M. Inui
Factory Automation Laboratory, RCAST: Research Center for
Advanced Science and Technology, The University of Tokyo,
Komaba 4-6-1, Meguro-ku, Tokyo 153, Japan

Y. Ito
Department of Mechanical Engineering for Production,
Tokyo Institute of Technology, 2-12-1 Ookayama, Meguro-ku,
Tokyo, Japan

S.H. Kim
Laboratory for Manufacturing and Productivity,
Massachusetts Institute of Technology, 77 Massachusetts
Avenue, Cambridge, MA 02139, USA

F. Kimura
Factory Automation Laboratory, RCAST: Research Center for
Advanced Science and Technology, The University of Tokyo,
Komaba 4-6-1, Meguro-ku, Tokyo 153, Japan

E. Kroll
Department of Mechanical Engineering, Texas A&M
University, College Station, TX 77843, USA

A. Kusiak
Department of Industrial and Management Engineering, The
University of Iowa, Iowa City, Iowa 52242, USA

E. Lenz
Faculty of Mechanical Engineering Technion, Israel Institute
of Technology, Haifa 32000, Israel

C.X. Li
Department of Mechanical Engineering 2, Huazhong
University of Science and Technology, Wuhan, Hubei, China

B.R. McGuire
Cincinnati Milacron, 4701 Marburg Avenue, Cincinnati, Ohio
45209, USA

H. Malberg
Philips GmbH, Forschungslaboratorium Hamburg, Vogt-
Kölln-Str. 30, Postfach 54 08 40, 2000 Hamburg 54, Federal
Republic of Germany

R.R. Martin
Department of Computing Mathematics, University of Wales
College of Cardiff, PO Box 916, Cardiff, CF2 4YN, UK

V.R. Milacic
Faculty of Mechanical Engineering, University of Beograd, 27
Marta 80, 11000 Beograd, Yugoslavia

A.Y.C. Nee
Department of Mechanical and Production Engineering,
National University of Singapore, 10 Kent Ridge Crescent,
Singapore 0511, Republic of Singapore

S. Parthasarathy
Laboratory for Manufacturing and Productivity,
Massachusetts Institute of Technology, 77 Massachusetts
Avenue, Cambridge, MA 02139, USA

D.T. Pham
School of Electrical, Electronic and Systems Engineering,
University of Wales College of Cardiff, PO Box 904, Cardiff
CF1 3YH, UK

I. Pillinger
School of Manufacturing and Mechanical Engineering,
University of Birmingham, Edgbaston, Birmingham B15 2TT,
UK

A.N. Poo
Department of Mechanical and Production Engineering,
National University of Singapore, 10 Kent Ridge Crescent,
Singapore 0511, Republic of Singapore

H. Richter
Philips Medical Systems, Roentgenstr. 24, Postfach 63 05 60,
D 2000 Hamburg 63, Federal Republic of Germany

A. de Sam Lazaro
Department of Mechanical and Materials Engineering,
Washington State University, Pullman, Washington 99164,
USA

H. Shinno
Department of Mechanical Engineering for Production,
Tokyo Institute of Technology, 2-12-1 Ookayama, Meguro-ku,
Tokyo, Japan

A. Soom
Department of Mechanical and Aerospace Engineering, State
University of New York at Buffalo, Amherst, NY 14260, USA

C.E.N. Sturgess
School of Manufacturing and Mechanical Engineering,
University of Birmingham, Edgbaston, Birmingham B15 2TT,
UK

E. Tacgin
School of Electrical, Electronic and Systems Engineering,
University of Wales College of Cardiff, PO Box 904, Cardiff
CF1 3YH, UK

K. Voss
Philips GmbH, Forschungslaboratorium Hamburg, Vogt-
Kölln-Str. 30, Postfach 54 08 40, 2000 Hamburg 54, Federal
Republic of Germany

Y.G. Wang
Department of Mechanical Engineering 2, Huazhong
University of Science and Technology, Wuhan, Hubei, China

W.G. Wee
Department of Electrical and Computer Engineering,
University of Cincinnati, Cincinnati, Ohio 45221, USA

J.R. Wolberg
Faculty of Mechanical Engineering Technion, Israel Institute
of Technology, Haifa 32000, Israel

P.K. Wright
Robotics and Manufacturing Research Laboratory,
Department of Computer Science, Courant Institute of
Mathematical Sciences, New York University, 719 Broadway,
NY 10003, New York, USA

Concepts and Techniques

This section contains four chapters. The first chapter, by Pham and Tacgin, provides an overview of techniques underlying the field of intelligent knowledge-based systems, particularly those which can be used to develop intelligent design programs. The chapter introduces and defines many of the terms encountered in the rest of the book. The second chapter, by Aldefeld et al., describes the concept of variational geometry in design and a method for implementing this concept based upon geometric reasoning. The latter technique is further examined in the third chapter, by Martin, which reviews its various aspects and relationships with topics such as computer algebra, automated theorem proving and expert systems. The final chapter in this section, by Akagi, discusses a powerful concept for representing design knowledge, the object-oriented concept, and presents an object-oriented expert system for engineering design.

## Chapter 1

Techniques for Intelligent Computer-Aided Design

D.T. Pham and E. Tacgin

Introduction

The majority of the computer-aided design (CAD) systems developed to date are not true design systems. They are in most cases mere draughting or analysis packages lacking the intelligence and creative faculty of the human designer. Due to the recent availability of massive computing power at relatively low cost, opportunities have arisen for building CAD systems with more genuine design abilities [1–12]. These systems apply techniques drawn from the branch of computer science known as artificial intelligence (AI). The most promising techniques are those of expert systems or intelligent knowledge-based systems. Several of these techniques will be discussed in different parts of the book. They, together with others, will be assembled and overviewed in this chapter. The chapter contains two main sections. The first deals with techniques underlying intelligent knowledge-based systems in general. The second is devoted to techniques applicable to intelligent knowledge-based systems for design.

The Technology of Intelligent Knowledge-Based Systems

An intelligent knowledge-based system (IKBS) [1] generally consists of the following components:

1. A knowledge base containing knowledge (facts, information, rules of judgement) about a problem domain.

2. An inference mechanism (also known as inference engine, control structure, or reasoning mechanism) for manipulating the stored knowledge to produce solutions to problems.
3. A user interface (or explanation module) to handle communication with the user in natural language.
4. A knowledge acquisition module to assist with the development of the knowledge base.

The first two components constitute the kernel of the IKBS and will be examined in more detail below.

Knowledge Representation

The three most popular ways of representing knowledge are rules, frames and semantic nets. Rule-based representation is a shallow representation, whereas schemes using frames and semantic nets are deep representations.

Rule-based representation

In a rule-based system, knowledge is represented in terms of facts pertinent to a problem area and rules for manipulating the facts. Many systems also incorporate information about when or how to apply the rules (that is, meta-knowledge, or knowledge about knowledge).

Facts are asserted in statements which explicitly classify objects or specify the relationships between them, such as "A Flywheel Engine is-an Engine", "An Engine is-a Prime Mover", "A Prime Mover has Moving Parts", "A Prime Mover is-a Machine", and "A Machine is-an Artefact".

Rules are modular "chunks" of knowledge of the form "IF antecedent THEN consequent", or "IF situation THEN action", meaning "If the situation described in the antecedent part of the rule is true, then produce the action specified in the consequent part", hence the names "IF-THEN rules", "situation–action rules" or "production rules". Examples of rules are:

1. IF "more energy storage capacity is required" THEN "increase the size of the engine's flywheel".
2. IF "the engine has been exposed to high levels of moisture" AND "the engine parts are metallic" THEN "there is a strong possibility ⟨0.85⟩ that the energy dissipation rate of the engine will be high".

Rule 2 illustrates two points. First, the consequent part of a rule does not have to specify an action. Instead it can be an assertion, hypothesis or conclusion which will be added to the knowledge base if the antecedent part is satisfied. Second, certainty factors can be used in a rule (or a fact) to indicate the degree of confidence attached to it. This enables the expert system to deal with information which is inexact or not completely reliable. The handling of uncertain information will be discussed in a later section.

It has been mentioned that a system may also possess meta-knowledge to guide its own reasoning. In a rule-based system this is expressed as meta-rules. An example of a meta-rule is:

IF "the inertias of the flywheel and the load have been determined" THEN "select the rule for calculating the gear ratio"

When diverse types of knowledge have to be handled, the rules (including meta-rules) are sometimes grouped into specialized independent sets, each corresponding to one type of knowledge. These so-called knowledge sources all operate on a common central database, the blackboard, and communicate their results to one another via this blackboard.

Frame and semantic net-based representations

A frame (or, its near equivalents, concept, schema, and unit) is a record-like structure, a form for encoding information on a stereotyped situation, a class of objects, a general concept, or a specific instance of any of these. For example, one frame might represent a particular type of machine (a flywheel engine), another, a whole range of machines (engines, gearboxes, robots, computers etc.) and yet another, the more general class of artefacts (bridges, machines, buildings etc.) (see Fig. 1.1).

Associated with each frame is a set of attributes, the descriptions or values of which are contained in slots. For instance, the frame for the "Flywheel Engine" subclass of engines might have slots entitled "flywheel size", "moving part material", "input-wheel-to-flywheel gear ratio" (see Fig. 1.2). In addition to attribute values, slots can also store other information such as constraints on the nature and range of these values ("value class", "cardinality min", "cardinality max") and procedures (arbitrary pieces of computer code) which are executed when the values are changed. Frames are usually organized into a hierarchy, which enables frames to inherit attributes from other frames located above them in the hierarchy.

Knowledge representation schemes using semantic nets are similar to those based on frames. A semantic net is a network of nodes linked together by arcs. A simple semantic net expressing the concept of a machine is illustrated in Fig. 1.3. There, arcs represent "is-a", and "has-part" relations which establish the same kind of property inheritance hierarchy as in a frame-based system.

The Inference Mechanism

Although, as mentioned previously, the knowledge base is the most important component in an expert system, the latter will not be useful

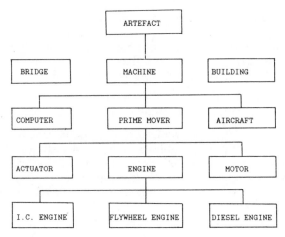

Fig. 1.1 Frame representation of part of the hierarchy of artefacts.

```
FRAME : Engine

      Superclass : Prime Mover
      Subclass   :
      Member of  : Class of Physical Objects

          Member Slot : Flywheel size
              Value Class      : Real numbers
              Cardinality Min : 1
              Comment          : "Diameter in mm."
              Values           : Unknown

          Member Slot : Moving part material
              Value Class      : Metal
              Cardinality Min : 1
              Values           : Unknown

          Member Slot : Input wheel-to-flywheel gear ratio
              Value Class      : Real numbers
              Cardinality Min : 1
              Values           : Unknown
```

(a)

```
FRAME : Flywheel Engine

      Member of : Engine
          Own slot : Flywheel size
              Value : 10 mm.

          Own slot : Moving part material
              Value : Steel.

          Own slot : Input wheel-to-flywheel gear ratio
              Value : 3
```

(b)

Fig. 1.2 Frames representing the "Engine" subclass of "Prime Mover" and the object "Flywheel Engine" in that subclass.

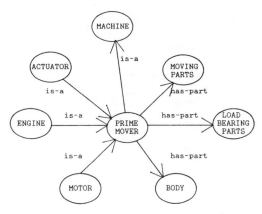

Fig. 1.3 A simple net partially representing the concept of a machine.

unless it has a good inference mechanism to enable it to apply the stored knowledge.

Different inference mechanisms are possible, depending on the type of knowledge representation adopted. In a rule-based system the inference mechanism, also called rule interpreter, examines facts and executes rules contained in the knowledge base according to some logical inference and control procedure.

Reasoning by the exercising of inference rules can proceed in different ways depending on different control procedures. One strategy is to start with a set of facts or given data and to look for rules in the knowledge base the "If" portion of which matches the data. When such rules are found, one of them is selected based upon an appropriate "conflict-resolution" criterion and executed or "fired". This generates new facts and data in the knowledge base which in turn causes other rules to fire. The reasoning operation stops when no more new rules can fire. This kind of reasoning is known as "forward chaining" or "data-driven inferencing". It is illustrated in Fig. 1.4.

An alternative approach is to begin with the goal to be proved and try to establish the facts needed to prove it by examining the rules having the desired goal as the "Then" portion. If such facts are not available in the knowledge base, they are set up as subgoals. The process continues until all the required facts are found, whereupon the original goal is proved, or the situation is reached when one of the subgoals cannot be satisfied, in which case the original goal is disproved. This method of reasoning is called "backward chaining" or "goal-directed inferencing". An example of backward chaining is given in Fig. 1.5. In practice, forward and backward chaining are sometimes integrated and an iterative convergence process is used to join these opposite lines of reasoning together at some intermediate point to yield a problem solution.

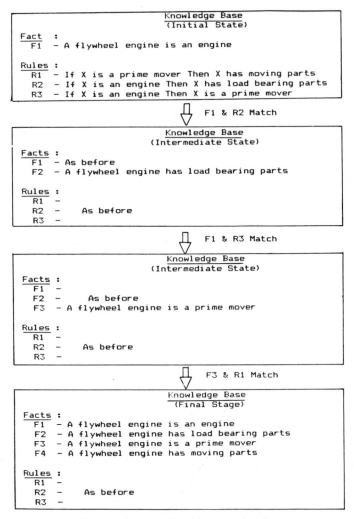

Fig. 1.4 An example of forward chaining.

Regardless of the control procedure, from any given goal state or start state there are usually alternative paths leading to different possible solutions. These paths constitute the branches of a "search tree" rooted at the goal state or start state in question. The search tree may be explored "depth first" or "breadth first". A depth-first search plunges from the root deep into the tree, considering a sequence of successors to a state until the path (line of reasoning) is exhausted. The search then proceeds to the next branch of the tree, exploring it in depth. With breadth-first search, all possible alternatives at the root are generated, then the

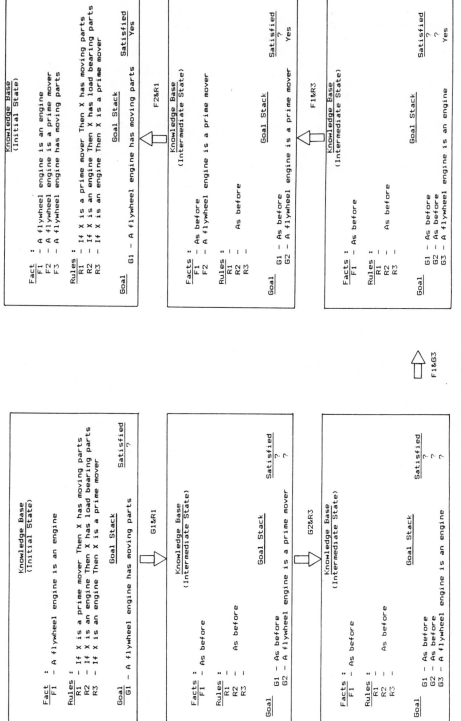

Fig. 1.5 An example of backward chaining (the goal stack stores the goals to be satisfied).

alternatives at the next level are produced, and so on. The search is thus
conducted in breadth across the tree.

In a frame-based or semantic-net-based system, inferencing is usually
achieved by exploiting the inheritance characteristic of frame and net
structures mentioned previously. For instance, since the "Flywheel
Engine" frame in Fig. 1.1, has a subclass link to the "Engine" frame and
the latter has a subclass link to the "Prime Mover" frame, the inference
engine will automatically "retrieve" the belief that "Flywheel Engine" is
a subclass of "Prime Mover" and as a consequence inherits attributes
from the latter. Other inference methods, readily implemented in the
frame formalism, use constraints, such as specifications of the nature
(value-class) and range (cardinality) of attribute values, to determine
whether a given item could be a value of a given slot. For example, when
a value is being added to the "Power Source" slot in the flywheel engine
frame (Fig. 1.2), the value is automatically rejected if it does not represent
a recognized power source. As will be seen in the next section, reasoning
by propagation of constraints has been used to limit search and refine
plans in some advanced design expert systems.

Some AI Techniques for Design Problems

This section reviews some of the main techniques in the field of AI
which are applicable to intelligent knowledge-based systems for design.
The techniques discussed include those for handling design goals and
constraints, for creating and validating design solutions, for better
representing fundamental design principles, for reasoning with qualitative
and uncertain design information and multiple contexts (nonmonotonic
reasoning).

Goal Handling

A design is generally specified in terms of the goals to be achieved.
Although there is usually only one final main goal, the latter may be
divided into several subgoals. This "divide-and-conquer" technique
simplifies the structure of the main goal and allows it to be handled (or
conquered). However, augmenting the number of subgoals and the
relationships between the dependent goals can regenerate more complexity. It is very important to avoid this early in the design process by
proper determination and ordering of subgoals. Subgoal determination
and redesign will be treated in more detail later.

There are different kinds of goals and subgoals. Consider, for example,
a flywheel engine for a toy car, which employs the kinetic energy of a

flywheel as its power source (see Fig. 1.6). One subgoal might be to add an object to the system to increase the inertia of the engine. A condition (subgoal) is that the mass of the object be distributed so that the motion of the whole system is as smooth as possible. It is evident that a circular object would satisfy the above condition. Another subgoal might be to connect this object to the system in such a way that the connection will increase the apparent inertia. For instance, if the object to be added and the rest of the system (wheels, shafts etc.) are coupled to each other by means of a pair of gears, then the object should be on the side of the smaller gear (the pinion).

A common problem with the design process is the lack of well-defined goals or subgoals [12]. In certain cases, although some of the characteristics are known, the solution may not be unique so that different designers may reach different solutions, each of which satisfies the specifications. In addition to this, in most situations, subgoals are also in conflict with one another.

A clear sign of conflict is the impossibility of achieving the design with the given goals or subgoals. Another sign can be spotted when some requirements are incompatible with others during the design process. Avoiding conflicts between subgoals is a major part of the process of goal satisfaction because in the absence of conflicts each subgoal may be independently and therefore more readily satisfied. There are several methods for dealing with conflict between subgoals. One method is backtracking to earlier stages of the design problem. The subgoals may then be modified and the problem retried. Another method to resolve conflict is to change subgoals. If one subgoal is violated by a second, starting with the latter may solve the problem in some cases. Consider, for instance, the following two subgoals in an assembly process plan: "Locate the shaft which carries the flywheel onto the body of the engine" and "Locate the pinion onto the shaft on which the flywheel is mounted". If the former subgoal was the first to be pursued, the latter

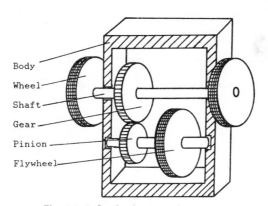

Fig. 1.6 A flywheel engine for a toy car.

could not be satisfied but this can be prevented by starting the assembly with the second subgoal.

Constraint Handling

Constraints may be generally defined as relationships between two or more attributes or variables in any stage of a design process. For example a constraint regarding the design in Fig. 1.6 may be "The object to be added must have no relative motion with respect to the shaft on which it is mounted". Two different types of constraints are "hard" and "soft" constraints. Hard constraints are definite and can only be either satisfied or violated whereas soft constraints can be met to different degrees varying from "nil" to "perfect" [13–14].

One aim of using constraints is to limit the set of possible solutions, or search space, to prevent unnecessary searches so that the given goals or subgoals can be reached. This is because an unnecessary search branch created by a missing constraint might have many sub-branches, all of which will lead to dead ends.

So that constraints may be handled, they must first be represented in a suitable format. Constraints can be represented as rules, frames etc. and given as a set of upper and lower boundaries which could be fixed (e.g. all lengths should be greater than zero) or variable (e.g. the mass of the object to be added should be greater than that of the entire engine without the object). Constraint representation is thus part of the process of forming relationships between design attributes.

Design is a hierarchical activity and constraint information has to be propagated up and down the hierarchy. This involves creating new constraints from existing ones, the main task of a constraint propagator being to partition the high-level constraints into components and to disseminate them. This operation also allows for interactions between subgoals at one level by reformulating constraints at a higher level.

As the design proceeds from the top to the lowest level in the hierarchy, failure to achieve a subgoal may occur at any level. If this happens, the design information is passed back to the higher level in the form of a failure report. The constraints are then reallocated, the design style is changed, or failure is confirmed. As in the case of goal conflict resolution, this procedure allows backtracking to previous design decisions taken at a higher level in the design hierarchy. This iterative model also supports a constraint handling procedure which propagates design styles and parameters both upwards and downwards. Constraint propagation and failure reporting help the design specifications to be complete by providing communication between different design levels.

Constraint propagation is also a mechanism for satisfying constraints, that is, for determining the allowable values of the design variables. Since constraints are often representations of goals or subgoals [15], the procedure for satisfaction of constraints is very similar to that for satisfying goals.

Generate and Test

In the case of a well-defined or routine problem domain, one possible design approach may be to perform a preliminary design, evaluate it and iterate this procedure until an acceptable solution is found. This is called the generate and test method. The design generation process is carried out randomly with the designs either produced one by one after each failure or all in advance, more likely by the forward-chaining strategy.

After a proposal design has been generated, it is evaluated against the design constraints or requirements to check whether it is acceptable. If not, the redesign process will be carried out and the proposal re-evaluated [16–17]. An acceptable solution may not be found. A possible reason might be exhaustion of the available time. In this case, either the acceptability standards can be modified or some of the specifications changed.

The efficiency of this method depends heavily on the determination of the initial design and how close it is to being acceptable. In practice an initial design is often given as an old design to be modified or developed to meet new requirements. In such cases, the generate and test method may perform well. For example, the rules for generating and testing the final design of the flywheel engine, using the old design shown in Fig. 1.7 as the starting point, might be:

1. If an additional weight is mounted on the shaft, then the inertia of the system will increase.
2. If a weight is located on the shaft on the pinion side, then it will rotate faster than the wheels which are mounted on the gear side.

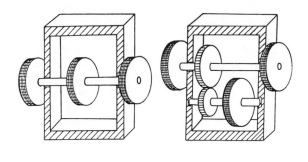

a   b

Fig. 1.7 Flywheel engines: **a** a possibile initial design; **b** an acceptable final design.

3. If the speed of rotation of any part of the engine rises, then the kinetic energy of the system will increase.

4. If either the total inertia or the kinetic energy of the engine increases, then the toy car will travel further.

Note that the first two rules are for generating a design proposal to achieve the goal which is to make the toy car travel further. The last two rules are for testing whether or not the goal is satisfied.

Object-Oriented Representation

Object-oriented representation [18–21] is a way of representing knowledge based on three concepts: classes, objects and messages. Classes which include objects and messages are fundamental units for organizing a set of knowledge. They may have many subclasses each of which can also be divided into other subclasses in a hierarchy.

In an object-oriented design system, design variables are modularized as design knowledge elements and represented as "objects". These are primitive frames accessible to the user and structured as instance variables, that is, instances of one or more classes. A possible object-oriented structure for representing a flywheel engine is illustrated in Fig. 1.8. Information transfer between objects is carried out through messages. In order to determine the value of a design variable, a message-passing procedure is activated and a message is passed to the corresponding object. During this procedure if other variables are needed to determine the required design variable, more messages are communicated to appropriate places. A class in an object-oriented design system delineates a concept which is common to its instances or design variables. In other words, a class describes a general strategy which governs the basic design procedures.

Figure 1.8 illustrates how the above mentioned flywheel engine may be modelled as a relational network consisting of a set of design variables represented as objects. These design variables are manipulated by message-passing procedures as shown in the figure. The message-passing feature characterizes object-oriented representation and differentiates it from frame-related formalisms which otherwise share basic similarities with it, such as hierarchical organization and property inheritance.

To design systems, object-oriented representation offers several advantages including modularity and hence ease of addition or modification of design knowledge. Another advantage, linked to the ability of this type of representation to express general concepts, is the possibility of handling "deep knowledge".

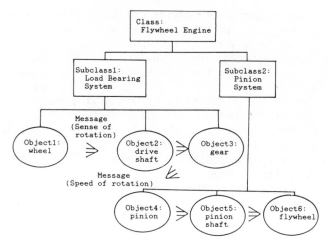

Fig. 1.8 Object-oriented representation of a flywheel engine.

Deep Knowledge

For design, traditional rule-based knowledge representation has several drawbacks. First, the knowledge has to be very detailed and specific because every possible combination of situations has to be considered. Second, it is difficult to organize heuristic rules in practice, even though the basic theory of a domain may be well structured. Third, extraction of knowledge from an existing design problem for use in another problem, even if the latter is very similar to the former, is not straightforward.

Although a general way of eliminating the above mentioned difficulties does not exist, there are some common methods of alleviating them based on the use of "deep knowledge". These involve:

1. Explanation of the structure of the problem domain so that the properties of each artefact are explicitly defined together with all their inter-relationships.
2. Separation of functions of artefacts from their structures. This is because different structures can result in different functions and generality could be lost if functions and structures are intertwined.
3. Detailed representation of all cause-and-effect relationships.
4. Use of symbolic representation in preference to numerical representation.

Deep knowledge representation is a technique for representing domain knowledge such that the above conditions are satisfied. A possible deep knowledge application is device simulation for design diagnosis. For example the flywheel engine may be simulated using a deep knowledge system, which could discover from its "understanding" of kinematic principles that the output shaft carrying the the flywheel should rotate faster than the input shaft on which the toy car's wheels are mounted, given the nature of the gearing.

The use of deep knowledge has a number of benefits. These include more systematic and simpler knowledge acquisition (particularly in a design process, due to the complete knowledge of artefacts and their inter-relationships), ease of extraction from similar domains, readiness of a deep knowledge system to be extended, and ability of such a system to solve rare problems (or even unpredicted problems) and to explain its behaviour accurately, clearly and in detail [22].

Deep knowledge systems also have disadvantages. These are related to the large amount of effort to build them mainly because of the need for knowledge engineers to master fundamental principles in a domain prior to the building process.

Qualitative Reasoning

In a general sense qualitative reasoning [23–25] deals with fundamentals rather than intuitive heuristic rules so that deep knowledge about a system is inevitably required. Qualitative reasoning also involves general relationships between elementary objects in a given domain such as "adjacent to", "mounted on", "overlapping with", "inside", "collinear with", etc. rather than their precise coordinates. This type of reasoning is particularly useful in mechanical design for explaining the relationships between parts. For example, the two gears in Fig. 1.6 may be conveniently described as "in mesh", without specifying their relative location exactly.

Another aspect of qualitative reasoning is the handling of symbolic information to carry out nonnumeric simulation. Although this approach has the disadvantage of possibly giving rise to ambiguity where absolute values are needed for a definite result, it also possesses many advantages. For example, it can handle incomplete information as well as providing more explanation and justification facilities. Also, nonnumeric simulation can be performed much more quickly than numerical simulation, and therefore requires much less computation time.

Geometric Reasoning

Representation of geometrical data and knowledge is one of the main issues in mechanical CAD. Traditional representation techniques make CAD systems inflexible and time consuming to use. An alternative method for representing and manipulating geometric data and knowledge developed over the last decade or so is known as geometric reasoning, a form of qualitative reasoning [26]. This process employs the features of geometric entities as well as the relationships between them. Geometric features can be divided into low- and high-level categories. High-level features can be extracted from those at a lower level, for instance, points, lines, arcs, surfaces, primitive solids etc. The identification of high-level features such as suitability of a workpiece for a machining or casting process varies according to different extraction systems. For example, a system that designs a machining process will recognize a part differently from one that plans a casting process for it [27–28].

Geometric features could be manipulated for two purposes, one of which is to deduce implicit information from the data given, for example, the centre of gravity of a part from its dimensions. Another purpose is to obtain information regarding a part by using the geometrical relationships between it and other related parts. For instance, the shape of a fixture required to hold different workpieces is found by geometric reasoning from the geometry of the workpieces to be gripped.

Uncertainty Handling

Uncertainty may arise in a design expert system due to conflicting, redundant or missing rules or incomplete and probabilistic data. Conventional logic can be extended to include probabilistic statements in order to deal with uncertain facts and imperfect rules. Probabilistic reasoning in engineering design may be used to calculate or report on the degree of certainty that the design will meet the given criteria. It also serves to control decisions on what design option to try in order to decrease the uncertainty in the next step.

Two common methods for handling uncertainties are the Bayesian and fuzzy logic methods. In the former, the degree of relationship between evidence and hypothesis is described by three numbers: logical sufficiency (LS), logical necessity (LN) and a-priori probability of the hypothesis (PP) [29]. PP is the probability of the hypothesis being true prior to observing the evidence. If the evidence is observed to be true, PP is increased by the LS factor. Conversely, if the evidence is found to be false, PP is decreased by the LN factor. The values of LN and LS in an expert system need to be adjusted according to the degree of uncertainty. Although the proponents of uncertainty-handling systems have used several different approaches to adjust these values, there is still no formal method for doing this.

One of the major disadvantages of the Bayesian approach lies with the technique of evaluating hypothesis from evidence. This is based on assumptions regarding the hypothesis such as exclusivity and exhaustivity which are invalid in many design problem domains. Another disadvantage concerns the computation of the LS, LN and PP factors. In some design domains these can be so critical that even small variations in their values can result in large changes in the way an expert system behaves.

The basic difference between the Bayesian and fuzzy logic approaches is that the former deals with the randomness of future events whereas the latter is concerned with the imprecision of current and past events [29]. For example, a conclusion derived using the Bayesian approach might be "There is 90% certainty that the gears will fail in bending rather than fatigue". On the other hand, a fuzzy statement might be: "The gear surface is very hard". In the former example, uncertainty is related to the future event "failure" whereas in the latter the uncertainty or fuzziness pertains to the state of the item "gear" (how hard its surface is).

Although the fuzzy logic approach being linguistically biased has advantages over the Bayesian approach such as providing a better explanation facility, it also has difficulties. For example, the functions which define the grades of membership to the different fuzzy sets for each of the variables used in a model are in general chosen subjectively so that controversy can arise.

Apart from these two well-known methods, some uncertainties occurring during a design could also be handled by techniques such as nonmonotonic reasoning which were originally conceived for other purposes. As discussed below, the latter is a technique that allows previously derived facts or conclusions to change when new data are introduced.

Nonmonotonic Reasoning

During a design process, a conventional deductive inference mechanism has the ability to manipulate rules and facts to reach a given conclusion. Any conclusion drawn by such an inference mechanism is considered as certain and does not change during the reasoning. This feature is known as monotonicity. For example, let two rules regarding the engine in Fig. 6 be "If the rotational speed of the shaft on which the flywheel is mounted is increased, then the kinetic energy of the system will increase" and "If a gear is mounted on the input shaft in mesh with a pinion on the shaft carrying the flywheel, then the rotational speed of the latter will increase for a constant input speed". In this example the conclusion will always be "A pair of gears, the smaller of which is connected to the shaft of the flywheel, will increase the kinetic energy of the engine system for a constant speed of the input".

Alternatively, the design conclusion drawn by the inference engine does not always have to be certain. For example, consider the following rules: "If an object with a large mass is fixed to the output shaft, then motion will be smoother" and "If the diameter of a cylindrical object is increased, then its mass will increase". A conclusion obtained from these rules could be: "Increasing the diameter of a cylindrical object fixed to the output shaft makes the engine run more smoothly". It is obvious that this conclusion is not always true; in fact, it is only valid for a particular way of assembling the object to the output shaft. Therefore, this conclusion must be changed when new information is received. For example, it is found that the object must be symmetrically assembled with respect to the shaft axis. The new conclusion will then be "Increasing the diameter of a cylindrical object which is fixed to the output shaft such that their axes coincide makes the engine run more smoothly". The ability of an inference engine to change old conclusions or facts with the introduction of new data is known as nonmonotonicity. The above example illustrates how uncertain knowledge can be handled by nonmonotonic reasoning.

For a design problem nonmonotonic inference could be thought of in two ways. In one, a default conclusion is drawn in the absence of information to the contrary. In the above example, the rule for assigning the default value might be "If the type of connection between two elements is not defined, then they are fixed to each other". When new information stating the type of connection is received, the previous conclusion will be discarded and a new one will be assigned. The concept of default reasoning can be readily extended to hierarchies of defaults. This explains how, except when otherwise specified, the values of design attributes can be inherited from other attributes in a hierarchical frame-based knowledge representation scheme.

With the second approach, it is assumed that the design database is complete and attempts are made to minimize the extent of some predicate or relation in the database. This assumption is called the "closed-world" assumption. "Minimizing" means assuming the number of arguments for which the predicate is true to be as small as possible. If no argument is found to be true, the predicate is taken as false. Once again the conclusion will be defeatable so that the previous conclusion will be valid until new information to the contrary is added to the database.

Consider an example relating to an engine design database. One of the predicates of interest is possible_gear_material. The extent of this predicate is {steel, brass}. That is, steel and brass are believed to be possible gear materials and the database contains only the following information with respect to the predicate possible_gear_material:

possible_gear_material (steel)
possible_gear_material (brass).

According to the closed-world assumption and the minimization concept, the above set of arguments {steel, brass} is taken to be complete and cannot be any larger. Thus, because the database does not include it, possible_gear_material (plastic) will be regarded as false until it is asserted explicitly.

It should be noted that nonmonotonic reasoning can also provide answers to questions requiring implicit knowledge without violating the minimization principle. For example, suppose there is an entry in the database stating that gear_A is in contact with gear_B. Using the symmetry of the "contact" predicate, namely contact (a,b) ⇔ contact (b,a), it is possible to derive the implicit fact gear_B is in contact with gear_A even though this fact does not exist explicitly in the database.

Minimization is one of the key approaches for obtaining default values in nonmonotonic reasoning and underlies techniques such as circumscription [30]. The idea of circumscription is based on the description of default rules in terms of "abnormalities". According to this, for example, "two elements are connected" normally means that they are brought together to an adjacent position and then fixed to each other. This is the default conclusion.

There are a number of objections to nonmonotonic reasoning [31–33]. One is that the notion of "nonmonotonic logic" is inconsistent with that of logic, which is by definition monotonic [31]. Another concerns its inferior handling of uncertainty compared with the better understood Bayesian and fuzzy logic methods. Problems in nonmonotonic reasoning systems also require repeated consistency checking. Consequently, these systems tend to be very slow. In addition, they are liable occasionally to make mistakes or even loop indefinitely, through circular dependencies between items in the knowledge base [31, 33].

Truth Maintenance

Closely related to the concept of nonmonotonic reasoning is that of "truth maintenance" or "consistency management". A truth maintenance system enables a nonmonotonic reasoning program to revise its beliefs when new discoveries contradict previously held assumptions. It does this by recording and maintaining the reasons for each belief.

Reasoning mechanisms incorporating truth maintenance systems (TMS) consist of two components: a conventional problem solver and a TMS [34]. The problem solver includes a design knowledge base and an inference mechanism, both able to communicate with the TMS. The TMS determines which data are to be believed and which are not, given the reasons sent by the problem solver. Arguments for beliefs are recorded and maintained by the TMS in order to demarcate the current set of beliefs continuously. The TMS in a design process manipulates two types of data structures: nodes and justifications which represent design beliefs and design reasons for beliefs, respectively. It can also create new nodes to which a new statement of beliefs can be attached. A node may have more than one justification forming a justification set, each member of which represents a different reason for the given node to be "believed". A node is believed or "in" if at least one of its justifications is valid. Conversely, a node which has no valid justification is disbelieved or "out". The TMS is also able to perform dependency-directed backtracking in case a contradictory node is "in" during the truth maintenance process. Dependency backtracking is carried out to find and remove at least one of the current assumptions to make this contradictory node "out". The following example adapted from [34] illustrates the functioning of a TMS. Consider an expert system program for designing flywheel engines. During the design process let the program assume that battery-powered flywheel engines provide smoother motion and should be adopted. The rulebase of the program includes two rules, the first of which draws the conclusion that flywheel engines with batteries are considered "expensive" and the latter concludes that expensive flywheel engines are not to be chosen. The nodes and rule-constructed justifications in this example can be written as follows:

Node	Statement	Justification
I1	TYPE=BATTERY	[SL() (I2)]
I2	TYPE≠BATTERY	
I3	COST=HIGH	[SL(R98 I1)()]

Note that the justifications belong to the class of "support-list" (SL) justifications. A support-list justification comprises two lists of nodes and is a valid reason for belief if and only if each of the nodes in the first list is believed and each of the nodes in the second list is not believed [34]. In the above example, the two justifications for I1 and I3 are the only existing justifications; I2 is not a current belief since it has no justifications at all. I1 is assumed true since the justification for I1 specifies that this node depends on the lack of belief in I2, which is the case so far. The justification for I3 shows that I3 depends on a presumably believed node R98. In this case, R98 represents a rule acting on node I1.

Subsequently another rule, represented by node R61, acts on beliefs about the cost of batteries (represented by node I84) to reject the assumption I1.

I2	TYPE≠BATTERY	(SL(R61 I84)())

To allow for this new justification, the TMS will modify the current set of beliefs so that I2 is believed, and I1 and I3 are not. It does this by tracing "upwards" from the node to be changed (I2), to see that I1 and ultimately I3 depend on I2. It then examines the justifications for each of these nodes to see that the justification for I2 is valid, so that I2 is "in". From this it follows that the justification for I1 is invalid, so I1 is "out", and hence that the justification for I3 is invalid and I3 is also "out".

Justification-based TMSs have a number of limitations. For example, they can consider only one design solution at a time, so that comparison of two solutions is very difficult. Another limitation arises in the case of a contradiction between two nodes: a justification-based TMS can only work on one of the nodes. Also, a justification-based TMS cannot change design assumptions temporarily. This is because in order to change an assumption, a contradiction should be introduced. After this is carried out the contradiction can no longer be removed. Some of these problems may be overcome by using another strategy called assumption-based truth maintenance.

An assumption-based truth maintenance system (ATMS) [35] is a variant of a TMS which is able to manipulate assumption sets as well as justification sets, whereas a justification-based TMS can only manipulate justifications. It is, therefore, more effective in dealing with inconsistent information in a design process. Justification information is stored by labelling each datum with the set of assumptions associated with it. The idea is to derive data from the assumptions which are considered to be the primitive form of all data. This is because manipulating the assumption sets is easier than manipulating the data sets that they represent. An ATMS provides more facilities and avoids many of the limitations faced by a justification-based TMS. For example, it is not restricted to searching only one point of the search space at a time and does not require much dependency-directed backtracking [35].

Conclusion

This chapter has reviewed techniques underlying the field of knowledge-based systems and artificial intelligence, with particular reference to those which can be applied to the building of intelligent knowledge-based systems for design. It has been seen that there is a wide range of techniques which are capable of enhancing traditional CAD systems with advanced reasoning abilities thus increasing their prospect of being tools for intelligent design.

References

1. Pham DT, Pham PTN. Expert systems in mechanical and manufacturing engineering. Int J Adv Manuf Technol 1988; 3: 3–21
2. Dym CL. Expert systems: new approaches to computer aided engineering. In: Proc. 25th AIAA-ASME-ASCE-AHS, structures, structural dynamics and materials, Palm Springs, CA, May 1984, pp 99–115
3. Dixon JR. Artificial intelligence and design: a mechanical engineering review. In: Proceedings AAAI-86, 5th National Conference on AI, University of Pennsylvania, vol 2, August 1986, pp 872–877
4. Rychener MD. Expert systems for engineering design: experiments with basic techniques. In: Proceedings IEEE Conference, trends and applications on automating intelligent behaviour: applications and frontiers, Gaithsburg, MD, May 1983, pp 21–27
5. Rychener MD. Expert systems for engineering design. Expert Syst 1985; 2: 30–44
6. Simmons MK. Artificial intelligence for engineering design. Comput Aided Eng, 1984; April: 75–83
7. Heragu SS, Kusiak A. Analysis of expert systems in manufacturing. IEEE Trans Syst Man Cybernet, 1987; SMC-17: 898–912
8. Dimitrov II. Knowledge representation for mechanical systems design. In: Sriram D et al. (eds) Knowledge-based expert systems in engineering: planning and design. Computational Mechanics Publications, Southampton, 1987, pp 367–376

9. Joskowicz L. Shape and function in mechanical devices. In: Proceedings 6th national conference on AI, Lawrence, KS, July 1987, pp 611–615
10. Tomiyama T, Yoshikawa H. Knowledge engineering and CAD. In: International symposium on design and synthesis, Tokyo, July 1984, pp 11–13
11. Rasdorf WJ. Perspectives on knowledge in engineering design. In: Proceedings ASME international computers in engineering conference, Boston, MA, vol 1, 1985, pp 249–253
12. Dixon JR, Simmons MK. Computers that design: expert systems for mechanical engineers. In: Proceedings ASME international computers in mechanical engineering conference, November 1983, pp 10–18
13. Sriram D. ALL-RISE: a case study in constraint-based design. Artif Intell Eng, 1987; 2: 186–203
14. Taylor NK, Corlett EN. An expert system which constrains design. Artif Intell Eng, 1987; 2: 72–75
15. Coyne R. Knowledge-based planning systems and design: a review. Architect Sci Rev, 1985; 28: 95–103
16. Dixon JR, Simmons MK. Expert systems for engineering design: standard V-belt drive design as an example of the design-evaluate-redesign architecture. In: Proceedings ASME international computers in engineering conference, Las Vegas, Nevada, 1984, pp 332–337
17. Dixon JR, Simmons MK, Cohen PR. An architecture for application of artificial intelligence to design. In: IEEE, 21st design automation conference, Albuquerque, NM, June 1984, pp 634–640
18. Sheu P, Kashyap RL. Object-based process planning in automatic manufacturing environments. In: IEEE International conference on robotics and automation, ROBOT 87, Raleigh, NC, vol 1, 1987, pp 435–440
19. Barbuceanu M. An object-centred framework for expert systems in computer-aided design. In: Gero JS (ed) Knowledge engineering in computer-aided design. Elsevier Science Publishers, IFIP, North-Holland, 1985, pp 223–253
20. Akagi S, Fujita K. Building an expert system for the preliminary design of ships. AI EDAM 1987; 1: 191–205
21. Bose G, Krishnamoorthy CS. GEAREX – a Unix based gearbox selection expert system. In: Sriram D and Adey RA (eds) KBES for engineering classification, education and control. Computational Mechanics Publications, Southampton, Boston, 1987, pp 345–358.
22. Price C, Lee M. Applications of deep knowledge. Artif Intell Eng, 1988; 3: 12–17
23. Green DS, Brown DC. Qualitative reasoning during design about shape and fit: a preliminary report. In: Gero J (ed) Expert systems in computer-aided design. Elsevier Science Publishers, IFIP, North Holland, 1987, pp 93–117
24. Price CJ. Augmenting qualitative diagnosis. Technical report, Department of Computer Science, University College of Wales, Aberystwyth, UK, February 1989 (No. RRG-TR-143-89)
25. Dyer MG, Flowers M. Automating design invention. In: Proceedings AUTOFACT-6 Conference Computer Automated Systems Association of SME, Anaheim, CA, October 1984, pp 25/1–25/21
26. Martin RR. Geometric reasoning for computer-aided design. In: Pham DT (ed) AI in design. IFS and Springer-Verlag, Bedford and London, 1990
27. Dixon JR, Libardi Jr EC, Luby SC, Vaghul M, Simmons MK. Expert systems for mechanical design: examples of symbolic representations of design geometries. Applications of knowledge-based systems to engineering analysis and design. ASME Publications, New York, 1985, pp 29–46
28. Arbab F, Wing JM. Geometric reasoning: a new paradigm for processing geometric information. In: Yoshikawa H, Warman EA (eds) Design theory for CAD. Elsevier Science Publishers, IFIP, North Holland, 1987, pp 145–159
29. Mills P, Jones R, Sumiga J. Evaluation of fuzzy and probabilistic reasoning in a design quotation expert system. In: Proceedings 3rd international conference on expert systems. Learned Information (Oxford), London, 1987, pp 145–158
30. McCarthy J. Circumcription – a form of nonmonotonic reasoning. Artif Intell, 1980; 13: 27–39

31. Israel D. What's wrong with nonmonotonic logic? In: Ginsberg ML (ed) Readings in nonmonotonic reasoning. Morgan Kaufmann Publishers, Los Altos, CA, 1987, pp 53–55
32. Reiter R. A logic for default reasoning. Artif Intell, 1980; 13: 81–132
33. Perlis D. On the consistency of commonsense reasoning. In: Ginsberg ML (ed) Readings in nonmonotonic reasoning. Morgan Kaufmann Publishers, Los Altos, CA, 1987, pp 56–66
34. Doyle J. A truth maintenance system. Artif Intell, 1979; 12: 231–272
35. Kleer J de. Assumption-based TMS. Artif Intell, 1986; 28: 127–162

## Chapter 2

Rule-Based Variational Geometry in Computer-Aided Design

B. Aldefeld, H. Malberg, H. Richter and K. Voss

Introduction

Geometric information processing is a core topic in the computer-aided design of mechanical products, common to a variety of tasks such as draughting, geometric modelling, finite-element analysis and production planning. The development of adequate software tools that support a wide range of geometry-related applications is therefore of major interest in view of the economy of the design process and the quality of the envisaged products. This concerns the user interface as well as the manipulation of internal product descriptions.

A major requirement to be met by a tool for geometric applications is that it can adequately support the concept of variational geometry, that is, the concept of generic geometric types and their instances. Generic descriptions, because of their generality, are the suitable basis for solutions that offer flexibility and a high degree of automation. Recurring applications where this is especially relevant are the design of parametric part families, adaptive design, kinematic simulation of mechanisms, tolerance analysis and modification of design-oriented geometries for manufacturing purposes.

Specifically, it is important to meet the following objectives:

- The man–machine interface should allow for good visualization, flexibility, conciseness and precision at the same time.
- The system should supervise the validity of geometric models, checking for the consistency and sufficiency of the constraining scheme and providing hints, if required, that aid in the correction of errors.
- Variants of a given geometry should be automatically derivable according to actual model parameters.

In this chapter, we describe a method for variational geometry which combines interactive graphical techniques for man–machine communication with symbol processing for reasoning about the internal models. Geometries are described as dimensioned drawings – the graphical language that has been traditionally used to express geometric information in engineering practice. Such a graphical input is translated into a language-oriented description, in which the geometric elements and relationships are represented as atomic formulae. A construction of the geometry is then performed on a symbolic level using rule-based methods.

Special attention is given to the question of how the graphical data can be interpreted in a completely automatic way, recognition of implicit constraints included. Machine interpretation capabilities can greatly contribute to the convenience of the man–machine dialogue, allowing to use drawings in their conventional form, with no additional input required, no restrictions on the sequence of input steps, and no mandatory adherence to some guided construction procedure. A system accepting such input can also process existing drawings not previously intended as generic models.

Previous Work

The "classic" method which has been widely used for many years to make parametric design more economical, is based on a description of geometry by means of a textual language. The constructs of the language are used to program a sequence of instructions whose execution, with suitable parameter values provided as input data, generates variants of a given geometric type as desired. This method does not require complex software, but a severe shortcoming is that the description of such a procedural model is an excessively cumbersome task.

An approach that is more congenial to the user – at the expense of complex software – is based on a declarative method of description, defining a geometric object by a set of elements and a set of constraints imposed on them. This approach was pioneered by Sutherland [1], who used the idea of constraint-based geometries in his early SKETCHPAD system. Considerable research has been spent on this problem since then, and a number of solutions based on different concepts have been developed.

One approach is to describe a constrained geometry by a system of nonlinear simultaneous equations and to solve such a system using numerical techniques. Hillyard and Braid [2, 3] developed basic ideas of the concept, using the analogy of a constained geometry with a mechanical mechanism. Gossard et al. [4,5,6] developed solutions for more general types of shapes, algorithms with improved convergence and methods for checking the correctness of the dimensioning scheme. Gossard et al. [7] also described an approach based on the formalism of *constructive solid geometry* (CSG), in which dimensions are included as *relative position operators* into the CSG representation. This method allows variation of shapes with little computational cost once a CSG tree has been built up.

Another approach to the problem is oriented towards symbol-processing techniques. Based on a set of rules and inferencing, a symbolic step-by-step construction of the geometry is attempted. The goal of the procedure is to prove all geometric elements as constrained, analogous to the proof of a geometry theorem [8,9]. Variants of the geometry are generated through numerical evaluation of the symbolic construction. Solutions along these lines have been described by Kimura et al. [10–13], Brüderlin [14] and Aldefeld [15,16]. Another nonnumerical method has been described by Todd [17], who used graph-searching techniques to test constrained geometries for manufacturability of the technical objects they represent.

The known methods focus mainly on the mathematical or logical part of the problem, that is, on constraint propagation and satisfaction, given the set of constraints in a formal description. Little attention has been given to the question of how this description can be automatically derived from a graphics-oriented input. In attempting to solve this problem, additional difficulties must be overcome, which are caused by the implicitness of part of the constraints. These constraints must be recognized in a way which is both logically consistent and in conformance with the designer's intentions.

The Graphical Model

A dimensioned drawing represents a generic geometric model, defined by an outline, termed *prototype geometry* in the following, and a constraining scheme. The prototype geometry defines the number of elements and their types, e.g. whether an element is a straight line segment, circular arc or other type of line segment, and it determines, implicitly, part of the constraining scheme. Instances of the model are defined through assignment of dimension values within allowed or reasonable ranges.

Constraints

Constraints can be subdivided into two main classes. Those of the first class, which will be termed *dimensional* or *metric* constraints, determine distances, angles, radii and/or diameters. Each constraint has an associated attribute (the dimension value), which ranges over a numerical interval. Constraints of this class are explicitly described in the drawing by the graphical symbols of the dimensioning scheme, that is, by witness lines, arrows and alphanumeric expressions.

Constraints of the second class determine topological and other spatial relationships with which no variable attribute can be reasonably associated. Examples are connectivity, parallelism, tangency and symmetry of elements. We subsume these constraints under the term *structural constraints* in the following. Constraints of this class are implicit in that they (usually) do not have corresponding annotations in the drawing. However, auxiliary lines, especially lines of symmetry, play an important role in clarifying what constraints are intended.

Structural constraints can be conveyed implicitly in a drawing because they are easily inferred by a person who is familiar with the draughting rules. The prototype geometry and the distribution of explicit constraints provide sufficient information for this purpose. How this information can be captured in a computational model, however, is not obvious. The draughting rules are rather informal, mostly explained by way of examples, and do not provide a rigorous syntax that would be adequate to guide the interpretation of the graphical data. In Requicha's characterization of representation schemes [18], engineering drawings are even classified as "ambiguous".

The main difficulty in recognizing the structural constraints by an algorithm lies in the fact that a *relationship* among elements of the prototype geometry is not necessarily a *constraint*. A structural relationship exhibited in the drawing is possibly a singularity of the model representation, which holds in the prototype geometry but is not meant to be a characteristic feature of the geometric type under consideration. Thus the structural constraints cannot be safely recognized simply on the basis of a "measurement" in the drawing, as this would also capture all *incidental* relationships.

The example shown in Fig. 2.1 illustrates this point. The perpendicularity of T_2 and T_7 with each other is incidental (under common interpretation), holding in the prototype geometry (Fig. 2.1a,b) but not necessarily in the variants. If one of the dimensioned angles is varied, the perpendicularity is no longer present (Fig. 2.1c). Situations like this are encountered quite often, especially in more complex models. A designer tries to avoid them only insofar as a human reader of the drawing might be misled in his interpretation of the structure.

Compounding this problem is the fact that consistency considerations are not sufficient, either, to find out what the real constraints are. Several subsets of relationships, if viewed as constraints, may result in a consistent overall solution. For example, if the perpendicularity mentioned is assumed as a constraint while the parallelism of T_7 and T_9 with each other is discarded, the resulting model would still be consistent in a mathematical sense. Only when dimension values are changed will it be

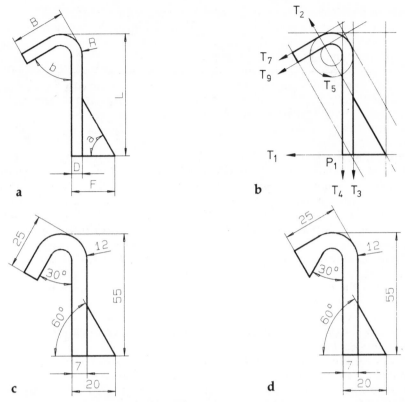

Fig. 2.1 Example of a parametric geometry: **a** prototype; **b** illustration of internal model; **c** correct; **d** incorrect variant.

recognized that the resulting geometry does not correspond to the designer's intention (Fig. 2.1d).

Data Structure

Concurrently with the creation of a drawing on the interactive display device, a corresponding internal model is built up in the database of the CAD system. Such a model can be designed in a variety of ways and may include higher-level structures and nongeometric information. For the purpose of this chapter, a simplified data structure as shown in Fig. 2.2 is sufficient for illustrating the main ideas. (A name enclosed in a rectangle denotes an entity type, a simple arrow denotes a 1:n type of relationship, two arrows pointing to a circle denote an n:m type of relationship.)

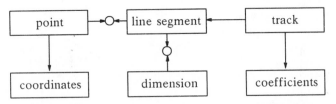

Fig. 2.2 Simplified data structure of the graphical model.

Entities of the types *point, line segment* and *track*, together with the relationships between them describe the topology of the model. The term *track* subsumes all infinite or unbounded curves (including straight lines). A *line segment* is a section of a track between two points. Entities of the types *coordinates* and *coefficients* contain the numerical information, namely, the coordinates of points and the coefficients of track equations, respectively.

The set of entities of the type *dimension* defines the explicit part of the constraining scheme. Each dimension entity contains the information about both the dimension value and the layout of the graphical symbols in the drawing. This includes the layout of the witness lines, which are generated from this information when the dimensioning scheme is displayed. (Witness lines do not fall under *line segments* as they are not part of the geometry.)

Tracks describe the unbounded geometry that corresponds with the bounded geometry displayed in the drawing. Additional tracks are generated such that each witness line is also positioned on exactly one track (Fig. 2.1b). Each track is considered directed, so that a left-hand and a right-hand side can be distinguished. Such directional information is important for avoiding ambiguous interpretations when structural relationships are expressed in a symbolic language.

An important point is that the internal graphical model corresponds closely with the drawing. If structural constraints are not expressly specified by the user – as is assumed here – they are as implicit in the internal model as they are in the drawing. Also, the internal model is fully declarative, as no information about the history of input operations is included.

Constructs of the Language-Oriented Model

The graphical model is suitable for supporting the interactive input of geometries, but it would not be an adequate basis for symbolic reasoning. For that purpose, we describe a generic model in a language-oriented form using the formalism of first-order predicate calculus. The constructs for describing such a model are given in this section. The next section then addresses the instantiation of actual models.

A model is expressed as a set of atomic formulae, whose constituent elements are taken from three sets, namely, a set of predicates, $\mathcal{R}$, denoting relationships, a set of constants, $\mathcal{E}$, denoting points, line segments and tracks and a set of constants, $\mathcal{D}$, denoting symbolic dimension values or structural attributes (see below). Each atomic formula takes one of two forms,

$$(R\ e_1\ e_2\ \ldots\ e_n)\ \text{or}\ (R\ e_1\ e_2\ \ldots\ e_n\ d)$$

where $R \in \mathcal{R}$, $e_i \in \mathcal{E}$ and $d \in \mathcal{D}$.

Each point of the model is defined by an atomic formula

$$(\text{POINT}\ P_i)$$

whose interpretation is that the element named P_i is of the type *point*. Similarly, each track is defined. For each line segment, L_i, an atomic formula

$$(\text{LINE-SEGMENT}\ L_i\ T_j\ P_k\ P_l)$$

states that L_i is a line segment lying on track T_j and ranging from the startpoint P_k to the endpoint P_l.

An example of a metric constraint is

$$(\text{ANGLE}\ T_l\ T_m\ \phi)$$

which expresses that the angle between the tracks T_l and T_m is ϕ. The predicates RADIUS and DISTANCE are used in a similar way to express radial and distance constraints. The constraints on the absolute position and orientation of a geometry are expressed by fixing the coordinates of one point, P_i, and the direction (the angle with the x-coordinate axis) of one track, T_j. We write this as

$$(\text{COORDS}\ P_i\ *)$$

and

$$(\text{DIRECTION}\ T_j\ *)$$

where a "$*$" stands for one or more numerical values that are considered to be arbitrary. (These constraints are usually not specified in a drawing.)

Structural relationships are described using the predicates PARALLEL, TANGENT, PERPENDICULAR, ON, MIDPOINT, SYMMETRIC and EQUAL. The last argument in each atomic formula, taken from the set $\mathcal{D}$, ranges over a small set of string values, which distinguish different structural arrangements that fall under the same predicate name. For example,

$$(\text{PARALLEL } T_l \ T_m \ \text{SAME})$$

states that T_l and T_m are parallel, proper. The string ANTI would indicate antiparallelism. For predicates where no multiple structures are possible, the string TRUE is chosen.

Symbolic formulae of the types described are the basis for geometric reasoning using formal rules of inference. The connection with the numerical world is established by each formula's algebraic interpretation. As an example, (PARALLEL $T_l \ T_m$ SAME) is interpreted as

$$\sin \phi_l = \sin \phi_m \text{ and } \cos \phi_l = \cos \phi_m$$

where ϕ_l and ϕ_m denote the angles of T_l and T_m, respectively, with the x-coordinate axis, and (PARALLEL $T_l \ T_m$ ANTI) is interpreted as

$$\sin \phi_l = -\sin \phi_m \text{ and } \cos \phi_l = -\cos \phi_m$$

Model Translation

The set of atomic formulae constituting the language-oriented model is generated by a number of procedures that examine the graphical model. They find all elements and relationships and output the atomic formulae in the syntax required. In the language-oriented model, not only metric but also all structural relationships are explicitly represented, including both constraints and incidental relationships, if present.

Information that is explicitly encoded in the data structure becomes immediately available through simple interrogation. For example, the algorithm for generating all line segments consists of a navigation through the data structure with iteration of the following steps:

1. Find the next line segment and get its name, L.
2. Find the track, T, associated with L.
3. Find the startpoint, P, and the endpoint, Q, of L.
4. Output the expression (LINE-SEGMENT $L \ T \ P \ Q$).

A similar straightforward procedure generates the atomic formulae for the dimensional constraints, using the information stored in the attributes of the dimension records and the topological structure of the graphical model. To make the set of metric constraints complete, the coordinates of a point and the direction of a track are declared fixed. A number of heuristics are used to select elements that are prominent in the structure.

Structural relationships are made explicit through numerical examination of the prototype geometry. Whenever the coordinates and/or coefficients of some combination of elements satisfy a symbolic relationship's algebraic interpretation, a corresponding atomic formula is generated. Some of these relationships immediately qualify as safe constraints on the basis of local information. This applies, for example, to parallelism between tracks that are at the same time related through a distance dimension. Also, point-track incidences (ON relationships) that are necessary to preserve the topological structure are safe constraints. The remaining structural relationships are candidate constraints, and are subject to verification or rejection in the global context of the solution.

Since the criterion of a consistent solution is not sufficient to determine the intended set of constraints among the candidates, additional information, which suggests some candidates as preferable over others, must be extracted from the graphical model. Such information is certainly available in some form or other because a dimensioned drawing is (usually) unambiguous and easily comprehensible for a human reader, allowing a safe discrimination between constraints, on the one side, and incidental relationships, on the other. One possibility is that the needed information is embedded in the structure and appearance of the graphical elements. Probably, general knowledge about shapes and functions of technical objects also plays a role.

As a first approach, we use only graphical information. The main idea is that a relationship must be sufficiently visible in the graphical structure if it is to communicate constraining information in a reliable and unambiguous way. Thus we estimate a visibility score (*visibility* for short) for each relationship. This score is expressed as a number in the interval [0, 1] and defines a ranking order on the set of candidate constraints.

The computation of visibilities is based on a number of heuristics, which are specific to the particular type of relationship. The main criteria are spatial proximity and lengths of line segments. The following examples give an idea:

- The perpendicularity of two tracks with each other is well visible if there are line segments on these tracks that meet at one point (forming a visible right angle). It is less visible if the line segments are separated from each other (like the line segments on tracks T_2 and T_7 in Fig. 1b, for example).

- The visibility of a point-track incidence is (among others) dependent on whether or not the point in question lies on a part of the track that is visible in the drawing (i.e. on a line segment or on a witness line).

Figure 3 illustrates the second criterion. The relationship (ON P_1 T TRUE) is given a high visibility score for both ways of dimensioning.

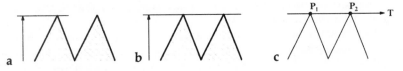

Fig. 2.3 Illustration of different visibilities of the relationship (ON *P T* TRUE).

(ON P_2 *T* TRUE) is given a low visibility score in Fig. 2.3a and a high visibility score in Fig. 2.3b. The particular value is computed as a function of the distance between the point in question to the nearest point on the witness line. The function smoothly decreases toward zero as the distance becomes large.

The heuristics used to compute visibilities are rather weak if considered separately from the global context, so the set of intended constraints certainly cannot be determined on the basis of this information alone. But the ranking of the candidates allows us to make the best choice in cases where, during the subsequent inference process, a decision has to be made between alternatives that cannot be distinguished otherwise.

Table 1 shows some of the (candidate) constraints with their visibility scores, generated from the graphical data in Fig. 2.1. For uniformity of the representation, all safe constraints are also given a visibility score, which is set equal to 1. A point to note is that generally a large proportion of the structural relationships are redundant, but each one is given its individual visibility score.

Inferencing

The goal of the inference process is to prove all points and tracks of the geometric model as constrained and to find the sequence in which they can be constructed, step by step. Moreover, in case several constructions are possible, the goal is to find a construction that is compatible with the intended constraints. For the solution of this problem, we define a

Table 1. Part of the symbolic description for the model shown in Fig. 2.1

Relationship	Visibility	Relationship	Visibility
(ANGLE T_1 T_2 *a*)	1.0	(PERPENDICULAR T_2 T_7 PLUS)	0.2
(RADIUS T_5 *R*)	1.0	(PARALLEL T_7 T_9 SAME)	0.9
(DISTANCE T_3 T_4 *D*)	1.0	(TANGENT T_5 T_3 ANTI)	0.9
(DIRECTION T_4 *)	1.0	(ON P_1 T_1 TRUE)	1.0
(COORDS P_1 *)	1.0	(MIDPOINT P_4 T_5 TRUE)	1.0

set of rules, which is then applied to the language-oriented description previously derived.

Figure 2.4 shows the global architecture of the inference component, which is modelled after a production system. The factbase, or working memory, contains the specific data of the model being processed. It is initially filled with the constructs of the language-oriented model. The rulebase contains the exhaustive set of rules that is potentially applicable in the domain of interest. The inference engine tries the rules on the facts, resolves conflicts if more than one rule is applicable at the same time and stores inferred facts. The ouput is a *construction plan* – a procedural description of how the geometry can be constructed from the set of constraints.

Rules

Each rule is written in the form

$$A_1 \wedge A_2 \wedge \ldots \wedge A_n \to B_1 \wedge B_2 \wedge \ldots \wedge B_m$$

where each A_i and each B_i is an atomic formula, "$\wedge$" and "$\to$" denote logical conjunction and implication, respectively, and universal quantification over all variables in A_i and B_i is implicitly assumed. Formulae in the condition part may be negated, whereas each formula in the conclusion part is nonnegated. Note that rules are defined on a symbolic, nonnumerical level. Their numerical evaluation is considered separately later on.

For a simple example of a rule, consider Fig. 2.5a. Assume that the coordinates of point P as well as the direction of track l are given, that is, have either been specified as constrained or been proved constrained before. Then the coefficients of l are also (indirectly) constrained. The rule is written

(ON P l TRUE) $\wedge$ (COORDS P _) $\wedge$ (DIRECTION l _)
$\to$ (COEFFS l *)

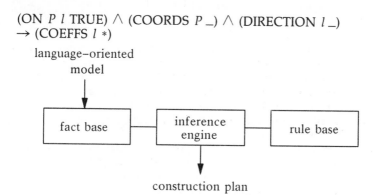

construction plan

Fig. 2.4 Block diagram of the inference component.

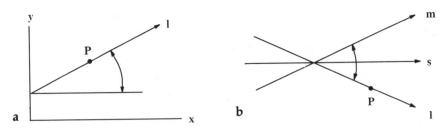

Fig. 2.5 Sketches to illustrate sample rules.

P and l are variables ranging over the set of points and tracks, respectively. The symbol "_" is an anonymous variable [19], standing for arguments that are irrelevant for the validity of a rule. The "*" stands for the triple of track coefficients, indicating that they are fixed, though not yet actually computed.

For another example of a rule, consider Fig. 2.5b. The tracks l and m are symmetric with respect to track s, and point P lies on track l. Assuming that the coefficients of s, the coordinates of P and the angle between l and m are given, the coefficients of both l and m are constrained. The rule is written

(COEFFS s _) $\wedge$ (COORDS P _) $\wedge$ (ON P l _) $\wedge$ (SYMMETRIC l m
s _) $\wedge$ (ANGLE l m _) $\rightarrow$ (COEFFS l *) $\wedge$ (COEFFS m *)

Rules where, like in the two examples above, the predicate COEFFS occurs in the conclusion, can be viewed as construction rules for tracks, and, similarly, rules with predicate COORDS can be viewed as construction rules for points, to use the analogy to geometric constructions on a draughting board. Rules for inferring other relationships may be formulated as well, for example

(PARALLEL l m _) $\wedge$ (PERPENDICULAR m n _)
$\rightarrow$ (PERPENDICULAR l n *)

As for the validity of a rule, the analogy with a known geometric construction may be accepted as a sufficient proof. However, if desired, a formal proof can be derived on the basis of the algebraic interpretations of a rule's constituent formulae. The set of interpretations of a rule's condition part form a small system of simultaneous constraint equations, from which the equations that express the interpretation of the conclusion part are easily derived [16].

As the above examples show, the rules considered here have been designed with an orientation towards two-dimensional (2D) geometries, according to the presupposition that man–machine communication takes place via dimensioned drawings. The three-dimensional (3D) case can be covered in this philosophy in combination with 2D–3D reconstruction

[20], for which an example will be shown later. The formulation of rules directly on geometric entities in 3D space is possible as well and has been described elsewhere [13, 14].

Generating a Construction Plan

Trying a given rule is basically a pattern-matching process [21], in which the factbase is searched for a set of expressions that match a rule's condition with all variables bound consistently over all formulae. If all relationships were constraints, each rule application would result in a valid conclusion. But because constraints are uncertain, the decision of what rules to select and apply is critical.

Finding the "best" solution in which some global function of all visibility scores is optimized is a search problem of large proportions, with many different alternative construction sequences to be examined and weighed according to the scores. This procedure would require unacceptable computational cost for practical applications when using common contemporary computer equipment. Besides, there is no satisfactory theory at present about how to compute meaningful visibility scores for inferred relationships, which presents an especially severe problem if the chains of inferences are long.

As a compromise, a hill-climbing type of search strategy with a local evaluation function has been implemented, in which the best choice is made at each decision point and no backtracking considered. The procedure begins with a small number of forward-inference steps, in which dimensional constraints are propagated. For example, a radius specified for circle T_i is assigned to all undimensioned circles T_j that have an EQUAL-RADIUS relationship with T_i. A sufficient set of explicit data is then available in the factbase to continue with a backward-inference process of depth one. It consists of iterating the following steps:

1. For each element not yet proved constrained, successively try all relevant rules. Compute a measure of confidence in each rule instantiation.
2. Select the best rule instantiation, insert the instance of the conclusion part into the factbase, record the details of this inference step and mark the element(s) concerned as "constrained".
3. Stop if all elements are marked as "constrained", or if no rule was applicable.

The measure of confidence assigned to a rule instantiation is computed from the visibility scores of the relevant relationships. The minimum operator, being widely employed in uncertain information processing [22], is a plausible choice as a first approximation. Let $R_1^{(i)}$, $R_2^{(i)}$, ..., $R_n^{(i)}$ be the set of relationships matched in rule instantiation i, and let $v_1^{(i)}$,

$1 \leq l \leq n$, be the visibility score associated with $R_l^{(i)}$. Then, the measure of confidence for rule instantiation i is computed as

$$m^{(i)} = \min (v_1^{(i)}, v_2^{(i)}, \ldots, v_n^{(i)})$$

The rule instantiation with the maximum confidence is considered the best construction.

Provided that the set of rules is adequate, the procedure proves each point and each track of the model as either constrained or unconstrained, depending on whether or not the constraining scheme is consistent and sufficient. Moreover, assuming that the concept of visibility is sufficiently powerful, only intended contraints will be used in the procedure if the dimensioning scheme is correct. In the example geometry of Fig. 2.1, the parallelism of T_7 and T_9 with each other is preferred over the perpendicularity of T_2 and T_7, which has a lower visibility.

Besides constituting a proof procedure, the chain of recorded inferences can also be interpreted as a construction plan that specifies a step-by-step construction of the geometric elements by elementary operations. Through execution of this plan, different variants of the prototype geometry can be generated when desired. All these variants conform exactly to the model constraints.

An example of how the construction plan may be recorded is shown in Fig. 2.6, which contains the first two steps out of 22 generated for the model shown in Fig. 1. In the first step, track T_4 is constructed using the first sample rule written down earlier. In the second step, T_3 is constructed using the result of the first step and the parallel distance between T_3 and T_4.

Checking the Construction

Erroneous dimensioning on the part of the user cannot be excluded. The resulting model is then either under- or overdimensioned. It is therefore desirable that the system can effectively check for such cases and, if there

---

(COEFFS T_4 *)   inferred   using RULE–DIR1   (ON P_1 T_4 TRUE)
                                               (COORDS P_1 *)
                                               (DIRECTION T_4 *)
(COEFFS T_3 *)   inferred   using RULE–D1   (DISTANCE T_3 T_4 D)
                                               (PARALLEL T_3 T_4 SAME)
                                               (COEFFS T_4 *)

Fig. 2.6 Two steps of the construction plan for the model shown in Fig. 2.1.

is evidence, provide information that helps the user to locate the source of the error(s).

If no complete construction plan can be generated, underdimensioning is obvious. If a complete construction plan has been generated, different types of errors are still possible. The model may still be underdimensioned if it contains one or more incidental relationships that have been taken for constraints in the construction plan. The model may also be overdimensioned, in which case one or more constraints are not accommodated in the construction plan.

For the detection of such errors, two tests are performed. The first test checks whether or not all (candidate) constraints derived from the graphical model have been satisfied by the construction. A dimensional constraint has been satisfied if it has been used in the construction plan. A structural constraint has been satisfied if it has been used in the construction plan or if it is derivable from the set of constraints used in the construction plan. This derivation can be based on inferencing, again, with a less complex procedure because visibility scores have no relevance here. (Forward inferencing is adopted in the prototype system at present.)

In case some of the (candidate) constraints have not been satisfied, they are made known to the user, in the form of a textual message together with highlighting of selected portions of the drawing. Suppressing redundant messages is important to keep the information as concise as possible. Incidental relationships not satisfied can be ignored. If one or more intended constraints have not been satisfied, the model is overdimensioned. After the dimensioning scheme has been corrected, the inference process must be run again.

The second test informs the user about all relationships that have been used in the construction plan, yet have a low visibility score. Assuming that incidental relationships have a score below a certain threshold, v_{min}, then all relationships with $v < v_{min}$ are suspicious and should be checked. If this assumption is invalid, some rare cases of underdimensioning might pass undetected.

For the example geometry shown in Fig. 2.1, the system issues the message

(PERPENDICULAR T_2 T_7) is not *considered a constraint*

which is to be understood as a warning against possible overdimensioning.

Another example is given in Fig. 2.7. The model in Fig. 2.7a is overdimensioned (radius $R15$) and also contains implicit constraints with low visibility scores (four equality constraints). The system generates a complete construction plan for this model. Radius $R15$ is used as a constraint, but a tangency condition is ignored instead, about which the user is notified as above. In addition, a message that warns against possible underdimensioning is given for each constraint used with low visibility, for example

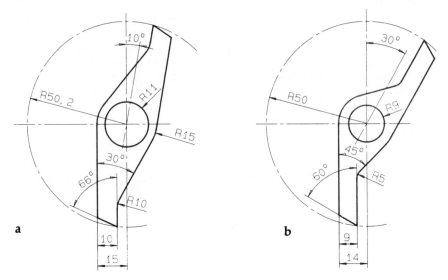

Fig. 2.7 a Overdimensioned model; **b** variant of corrected model.

(EQUAL-ANGLE T_9 T_{11} T_5 T_1) *is considered a constraint*

which relates to the equality of the 30° angles. An example of a variant for the corrected model is shown in Fig. 2.7b.

Variation of the Prototype

Once a construction plan for a geometric type has been established, instances in any number are readily generated as required. When working interactively, the user may select one or more dimensions and specify actual parameters or change current numerical values. Alternatively, parameter values may be supplied by an application program. The system then executes the construction plan and computes the explicit geometry exactly in accordance with the new dimension values.

Execution of the construction plan consists of executing a set of analytical programs, termed *execution procedures* in the following, which carry out the numerical computations that correspond to the symbolic constructions listed in the plan. Each execution procedure corresponds to exactly one rule and embodies the numerical solution of the algebraic constraint equations behind this rule. Only elementary geometric computations are needed here, so the plan–execution process is computationally inexpensive.

At execution time, the rule name listed in each step of the plan is interpreted as the name of the corresponding execution procedure. It computes the values represented by "$*$" in the inferred formula, using as input data the arguments of the formulae matched in the rule's condition. These data are safely available when needed. They have been either given as actual dimensions, have been computed in previous steps of the plan, or are default data of the prototype (fixed point and fixed direction).

For example, in the first construction step in Fig. 2.6 the program RULE-DIR1 computes the coefficients of T_4. Input data are the coordinates of P_1 (which lies on T_4) and the direction of T_4. The computed coefficients are then available to replace the "$*$" in the formula (COEFFS T_4 $*$) when needed as input in the second, and possibly subsequent, steps of the plan.

An execution procedure may return more than one solution for a given set of input data. Although structural atttributes (e.g. SAME, ANTI for parallelism) reduce the number of such ambiguities considerably, they are not avoided altogether. For example, if the coefficients of a circle are computed, given its radius and two (noncoincident) points on its circumference, two solutions are obtained. Multiple solutions, however, do not present a problem because removal of ambiguity is possible simply through selection of the alternative that keeps the structural similarity with the prototype. For more details and similarity criteria, see [16].

The computed numerical data are finally written into the records "coordinates" and "coefficients" of the graphical model (or into a copy of it if the prototype is to be kept). This completes the generation of a new variant as far as the basic geometry is concerned. Elements of the auxiliary geometry (e.g. lines of symmetry), which are not completely constrained under the dimensioning scheme, as well as the dimensioning layout may require correction if the variations have been large. User-guided and automatic corrections can be implemented, which shall not be further considered here.

Figure 2.8 shows an example of a 3D application. The model is described as a two-view drawing, into which the dimensions have been distributed as convenient. (At present, a distance dimension with arbitrary value, D, has to be included to fix the two views relative to each other.) The model is processed in the same way as a single-view description. The explicit 3D shapes displayed in the figure have been constructed from the 2D information using a user-guided 2D–3D reconstruction procedure [20].

It is obvious that the plan–execution process is absolutely robust and never incurs numerical instabilities. At the same time, variants are obtained with high precision, depending only on the precision (e.g. double) of the computations implemented in the execution procedures but unaffected by the precision (e.g. single) of the graphical data. This is especially important in applications where variations are confined to

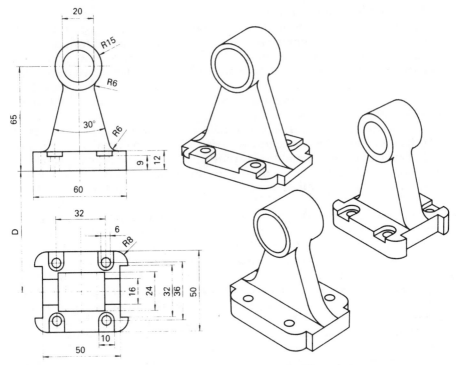

Fig. 2.8 Two-view drawing with variants in 3D rendering.

minute amplitudes. An example is tolerance analysis. It can be based on repeated execution of the construction plan, varying one or more dimension values within their tolerance limits and examining the effects on critical regions of the geometry [12].

Conclusion

The method described covers variational geometry in its various aspects, ranging from graphical description of a generic model to generation of explicit variants. Rule-based geometric reasoning is the core of the method. Rules are formulated such that they describe mathematically provable inferences, while the data – the candidate constraints – are associated with a measure of uncertainty. The necessity to handle uncertain information in the procedure is dictated by the fact that the recognition of implicit constraints from engineering drawings is not possible on only mathematical considerations, but is also influenced by informal draughting conventions.

About 100 rules, implemented in a prototype system for variational geometry (programmed in Common Lisp), were found sufficient to cover even complex constructions in their domain of geometries composed of straight-line segments and circular arcs. The visibility concept, though implemented in a rudimentary version so far, was found to be already very effective in guiding the search for the correct solution when two or more alternatives are possible. Refinement of the criteria for the assignment of visibility scores and addition of rules to cover further types of curves can improve the method within the current frame.

Further in-depth research into the problem, however, is desirable. This concerns primarily the inference procedure, which should be extended because the one-step backward strategy presently used is too restricted as a general method. Backward inferencing with unlimited depth but controlled in such a way that shorter chains of reasoning are preferred over longer chains is a promising direction. In this context, a theory on which to base the propagation of visibility scores along chains of inferences is also required.

It is finally mentioned that the method described in this chapter should be suitable also for high-level interpretation of data obtained by optical scanning of paper drawings. Automatic recognition of constraints and their contextual interpretation are essential in this application for the reconstruction of semantically correct and numerically precise models from the imprecise data.

References

1. Sutherland IE. SKETCHPAD: a man–machine graphical communication system. Proceedings AFIPS spring joint computer conference, 1963, pp 329–346
2. Hillyard RC, Braid IC. Analysis of dimensions and tolerances in computer aided mechanical design. Comput Aided Des 1978; 10: 161–166
3. Hillyard RC, Braid IC. Characterizing non-ideal shapes in terms of dimensions and tolerances. Comput Graphics 1978; 12: 234–238
4. Lin VC, Gossard DC, Light RA. Variational geometry in computer-aided design. Comput Graphics 1981; 15: 171–177
5. Light RA, Gossard DC. Modification of geometric models through variational geometry. Comput Aided Des 1982; 14: 209–214
6. Light RA, Gossard DC. Variational geometry: a new method for modifying part geometry for finite element analysis. Comput Struct 1983; 17: 903–909
7. Gossard DC, Zuffante RP, Sakurai H. Representing dimensions, tolerances, and features in MCAE systems. IEEE Comput Graphics Applic 1988; 8: 51–59
8. Gelernter H. Realization of a geometry-theorem proving machine. In: Siekmann J, Wrightson G (eds) Symbolic computation. Automation of Reasoning Series 1, Springer-Verlag, Berlin Heidelberg New York, 1983, pp 99–121
9. Gilmore PC. An examination of the geometry theorem machine. Artif Intel 1970; 1: 171–187

10. Kimura F, Suzuki H, Ando H, Sato T, Kinosada A. Variational geometry based on logical constraints and its applications to product modelling. Ann CIRP 1987; 36: 65–68
11. Kimura F, Suzuki H, Sata T. Variational product design by constraint propagation and satisfaction in product modelling. Ann CIRP 1986; 35: 75–78
12. Kimura F, Suzuki H, Wingard L. A uniform approach to dimensioning and tolerancing in product modelling. In: Proceedings CAPE '86, Elsevier Science Publishers (North Holland), 1986, pp 165–178
13. Suzuki H, Kimura F, Sata T. Treatment of dimensions on product modelling concept. In: Yoshikawa H (ed) Design and synthesis, Elsevier Science Publishers (North Holland), 1985, pp 491–496
14. Brüderlin B. Constructing three-dimensional geometric objects defined by constraints. In: Proceedings workshop on 3D interactive graphics, Durham, NC, 1986
15. Aldefeld B. Rule-based approach to variational geometry. In: Knowledge engineering and computer modelling in CAD (Proceedings CAD '86), Butterworth, London, 1986, pp 59–67
16. Aldefeld B. Variation of geometries based on a geometric-reasoning method. Comput Aided Des 1988; 20: 117–126
17. Todd P. An algorithm for determining consistency and manufacturability of dimensioned drawings. In: Knowledge engineering and computer modelling in CAD (Proceedings CAD '86). Butterworth, London, 1986, pp 36–41
18. Requicha AAG. Representations for rigid solids: theory, methods, and systems. Comput Surv 1980; 12: 437–464
19. Clocksin WF, Mellish CS. Programming in Prolog. Springer-Verlag, Berlin Heidelberg New York, 1984
20. Aldefeld B, Richter H. Semiautomatic three-dimensional interpretation of line drawings. Comput Graphics 1984, 8: 371–380
21. Bundy A. The computer modelling of mathematical reasoning. Academic Press, London, 1983
22. Bhatnagar RK, Kanal LN. Handling uncertain information: a review of numeric and non-numeric methods. In: Kanal LN, Lemmer JF (eds) Uncertainty in artificial intelligence. Machine Intelligence and Pattern Recognition Series 4, North Holland, Amsterdam, 1986, pp 3–26

Chapter 3

Geometric Reasoning for Computer-Aided Design

R.R. Martin

Introduction

Geometric computation has been widely used for over 20 years in computer-aided design (CAD), but until fairly recently, the emphasis has been on the end user deciding what geometric constructions to make. The main mode of use has been to treat a CAD system as being to geometry as a calculator is to arithmetic – the CAD system can perform various geometric manipulations and draw the results, but it has no built in knowledge of geometric theorems and concepts, just as a calculator has no knowledge of number theory.

For intelligent design, the next logical step appears to be for CAD systems to be able to perform *geometric reasoning*. What exactly geometric reasoning means, and what it is intended to do varies very much depending on who one talks to, as might be expected in a relatively new area of research. Nevertheless, there seem to be two main facets to geometric reasoning. The first of these is the ability of software to define geometric information and deduce properties of geometric objects at a high level rather than directly in terms of the details of specific geometric elements. One example of the way in which information can be processed at higher levels is feature recognition in a solid modeller, where a boss may be defined as a protrusion from the main body of a solid, irrespective of its particular geometric shape, as shown in Fig. 3.1. Various intermediate level properties of the object such as convexity, loops of connected edges and faces, and relative sizes of faces have to be deduced from the low-level geometric information, which are then combined to decide whether various highest level features are present. This must all be done in a way which is independent of the particular geometry which exists at the lowest level. A rather different example of manipulation of geometric information at a higher level is to represent all curves and surfaces as

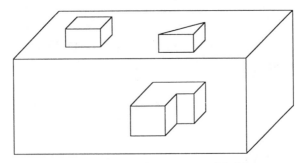

Fig. 3.1 Different bosses on a solid object.

algebraic polynomial expressions. As will be explained later, this enables generic solutions to geometric problems such as finding the intersection of any two algebraic curves, rather than specific algorithms for the intersection of two straight lines, a straight line and a circle, two circles and so on.

The second facet of geometric reasoning is the ability of software to convert one representation of a geometric object to another representation more suited to the particular task in hand. This process may involve selection and rejection of the original information, as well as deriving new information from it. For example, consider the object shown in Fig. 3.2. When trying to deduce what weight it will support, it is natural to think of the object as being composed of four rectangular blocks supporting each other. However, when trying to decide how to manufacture the object from a solid piece of metal, it is much more useful to think of it as block minus a hole. Producing a description of the latter type is exactly what feature recognition described in the previous paragraph is attempting to do.

Given these ideas, this chapter attempts to review various aspects of geometric reasoning, and to show how they might be useful for intelligent design. However, it does not try to be comprehensive in its coverage, but instead concentrates on various areas the author feels to be of major interest. One extensive area of geometric reasoning which will not be detailed in this survey is its use for image understanding. Although this is obviously of great importance for many tasks, its main uses in manufacturing are at the machining, assembly and inspection stages rather than the design stage.

Two particular publications which include papers on a wide variety of topics in geometric reasoning are [1,2].

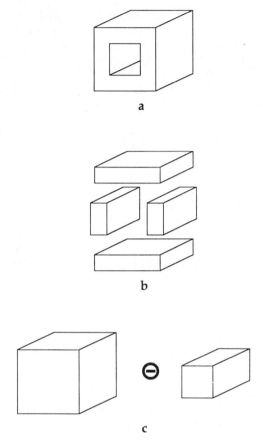

Fig. 3.2 Alternative representations of solids. **a** Original solid; **b** decomposition for evaluating load carrying capacity; **c** decomposition for deciding machining strategy.

Geometry and Computer Algebra

In the introduction, computer systems were mentioned for performing arithmetical and geometrical calculations. Not surprisingly, systems have also been created and are commercially available for performing algebraic calculations. Such systems are called computer algebra systems. Although they originally used various heuristic methods for such tasks as evaluating indefinite integrals, the more recent trend is towards more deterministic methods often based on deep mathematical results. For a useful introduction both to the use of such systems, and the theory they are based on, see Davenport et al. [3].

The relevance of computer algebra methods to geometric reasoning is twofold. Firstly, restatement of a geometric problem in algebraic terms (usually polynomial equations) may mean that it can readily be solved by use of standard algebraic methods. This is a specialized case of the more general paradigm of transforming knowledge and problem statements from one representation to another which may be easier to work with than the original representation. Note that although algebraic methods are used for these computations, the results themselves can readily be converted back into geometric terms when necessary.

Secondly, and even more important, is the fact that using algebraic methods can be rather more general than solving individual geometric problems, as already mentioned in the Introduction. The example given there will be explained in further detail. In conventional CAD systems, problems like finding the intersection of a straight line and a circle are typically solved by the software developer solving the problem geometrically on paper, deciding that it requires the solution of a particular quadratic equation, and then hard coding that solution. When using algebraic methods, the computer is now able to deal with higher level concepts such as algebraic variables and polynomials. Finding the points of intersection of *any* two curves which can be expressed in the form $f(x,y) = 0$ can be performed by the algebraic computation of the resultant of two polynomials in x and y, and using Sturm sequences to find out how many solutions exist. The particularly useful point about doing this is that should it be desired to add ellipses to the CAD system, in the former case, special code would need to be developed to find intersections of ellipses and straight lines, and ellipses and circles, whereas if algebraic methods are used, the solution to a whole class of such problems is available.

A further, but different example of the generic nature of solutions provided by algebra systems comes when drawing a solid model using ray tracing. In a conventional system, many individual rays must be tested for intersection with the object. Using computer algebra, it is possible to solve once where a generic ray with variable parameters intersects the model, and then substitute for the actual rays required in this general solution.

Two algebraic methods which are of particular relevance to geometric reasoning are cylindrical algebraic decomposition [4–6], which allows the decomposition of an arbitrary part of n-dimensional space defined by a set of polynomial equations and inequalities into connected regions, and the construction of the adjacency relations between these regions [7]. These relations form a graph representing, for example how space is occupied by a moving mechanism, or whether various objects will fit inside another object.

The second technique is Buchberger's algorithm for constructing Gröbner bases [6], which are useful for attacking various problems involving systems of algebraic polynomial equations, such as deciding whether the equations are independent, finding common roots of the equations, and eliminating variables between the equations. The last computation repeatedly occurs when solving geometric problems by algebraic means, for example in turning parametric formulations of curves and surfaces into implicit ones, in solving various intersection problems and in detecting singularities.

On the practical side, Bowyer et al. [8] are in the process of developing a subroutine library called GAS (geometric algebra system) for performing various algebraic tasks specifically required by geometry manipulating programs such as solid modellers. Their ultimate aim is to create a solid modeller which performs its computations using algebraic methods wherever possible in place of floating-point numerical methods.

A more particular use of algebraic methods has been described by Martin and Stephenson [9], for finding the swept volumes of moving solid objects. Algebraic techniques are used to find what envelope surfaces are generated by each of the surfaces in the original model as it moves, while a projection-based algorithm then decides which part or parts of each of these surfaces contributes to the final swept volume. An example of the use of this computation arises in designing suitable housings for mechanisms such as a moving radar dish. A second, and perhaps rather more widespread use is in determining the volume of metal removed by a cutter as it sweeps out a path during machining in an integrated CADCAM system.

Todd [10] describes how algebraic methods can be used for designing mechanisms. Here, high-level constraints (such as distances between points) are converted into algebraic form, and then compiled (or solved) by a computer algebra system into a conventional program. This allows various design parameters to be changed interactively while still satisfying the constraints, and the results to be displayed graphically.

A rather different use of algebraic methods is in the automatic creation of blending surfaces. Here, the sharp edge of intersection of two existing main surfaces is replaced by a small piece of blending surface joining both of the original surfaces with tangential continuity. This allows the designer to concentrate on the main features of the design, and then to smooth (fillet or chamfer) automatically the edges of these features as necessary. Various slightly different formulations of this method exist [11,12]; the basic idea is that if A and B are the algebraic equations of two surfaces, then $(1 - u)AB + uP^2$ is a new surface which is tangent to A and B and touches them along the curve where P, a third, control surface, intersects them. The parameter u can be varied to control the fullness of the blend.

Finally, it should be noted that other advantages are also to be had in using methods from computer algebra systems. For example, by using rational arithmetic it is possible to avoid some of the numerical instabilities which arise when performing real arithmetic with computer limited precision. Nevertheless, as is often the case, there is a tradeoff. The gain in precision is obtained at the expense of the size of the rational coefficients, which must be represented internally using multiple precision arithmetic, with many tens if not hundreds of decimal digits. Indeed, a well-known general problem with computer algebra is *intermediate expression swell*, where although the input expressions and the output expression may be just a few lines long, intermediate expressions can grow exponentially in length before collapsing back to the size of the final result. Current algebra systems are for this reason heavy users of memory, and computer time.

Automated Theorem Proving

When reasoning about geometry, the high-level concepts under consideration eventually need to be broken down into lower-level queries to solve problems. Sometimes this can be done directly. For example, telling whether one convex polyhedral object is inside or above another can be done by comparing their vertices. Again, however, particular code must be written to solve each such specific problem. In the spirit of geometric reasoning working at a higher level, there exist techniques of *automatic theorem proving* where a query about an object is expressed as a theorem, and code which can check *any* such geometric theorem is used to answer the query. It is not too difficult to see the parallel with expert systems – in that case a general purpose inference engine works from a set of stored facts and input data describing a particular problem towards a goal, while in this case a general purpose theorem prover works from a particular set of facts and conjectures towards a proof.

A powerful and general algebraic method for proving geometric theorems has been given by Wu [13,14], and has been implemented with considerable success by various people, including Wu himself and Chou [15]. The method basically works by writing the given facts and conjectures as polynomials, where the variables in these polynomials are coordinates of points. The goal is then to find whether the conjecture follows from the given facts. This is so if the common zeros of the given statements are also zeros of the conjectures. Various methods may be used to find these common zeros, Wu's original idea being based upon pseudo-division and polynomial factorization, while another possibility relies on the Gröbner basis method described above [16]. Wu's method may also be extended to describe results which involve inequalities, as well as equations, by introducing extra variables.

One advantage of Wu's technique is that it solves problems generically, in other words, any theorems proved will still be true in degenerate cases such as ones where distances between points become zero, lines are accidentally parallel, circles reduce to a single point, and so on. This is an important advantage, as enumerating and handling special cases with special code is a tedious and error prone concern for most current CAD systems. Nevertheless, if it is wished to rule out specific degenerate cases from the proof (X is true except when Y), this can be done by adding extra equations describing the exclusions.

It should also be noted that other theorem proving methods have been suggested, such as the method due to Kapur [17], which uses similar ideas to Wu's, but instead uses proof by contradiction, and tries to prove that the negation of the conjecture is inconsistent with the given facts. An interesting extension which Kapur's method can handle is as follows: given a set of initial facts which do not lead to the desired conclusion, then it can find additional facts, if they exist, which are consistent with the original facts, and which when added to the original set of facts, do cause the set of facts to imply the desired conclusion. These additional facts often turn out to be extra constraints, e.g. that certain points must be distinct.

It is not difficult to see that theorem proving has many potential applications for design, for example when checking that a set of geometric constraints on a design lead to other desired geometric properties. However, the complexity of the algebraic problem to be solved would seem to grow rapidly with the geometric complexity of a design, and whether Wu's method can be practically useful has yet to be determined. Although simple theorems with a small number of variables can be solved in a few seconds, the rather more complex problems posed in design may well take up to several tens of hours, or more.

A second problem with theorem proving systems is that they work in the field of complex numbers rather than the reals. Although complex numbers are more general than the reals, loosely speaking, and so any theorem which is true for complex numbers must include the same theorem for real numbers as a special case, the reverse of this is unfortunately not true. Thus, a theorem may be true for the reals, but the theorem prover may not be able to prove it, because it is not true for complex numbers. Taking a concrete example, for real numbers the given fact $x^2 + y^2 = 0$ implies that the theorem $x = 0$ is also true, but this is not the case if x and y are allowed to be complex, as shown by the pair $x = i$, $y = 1$. The implications of this observation for solving practical problems of real geometry are not yet well understood.

Reasoning about User Input

Although many current CAD systems allow geometric input, they only do so in a straightforward way, where geometric shapes drawn represent themselves. Geisow [18] contends that diagrams can be used for much wider purposes, as often they can be more readily understood than textual descriptions. While this observation has been widely used for output, its uses for input have been quite limited up to the present. Here, geometric reasoning will be an essential tool for capturing designers' intentions presented in schematic diagrams. Geisow proposes a twofold approach which first builds up a structured diagram from the primitive graphical elements (a syntactical step), followed by mapping of the structured diagram to the particular application domain (a semantic step). As he points out, this will allow modular software construction from reusable components. Related work in this area has also been carried out by Pereira [19] and Arya [20].

Another area which is attracting current interest is sketch input to CAD systems. Here, artists' sketches are used as a starting point for automatic derivation of a CAD model, as opposed to the more usual methods of precise user input of desired geometric elements. Such sketches are by their nature incomplete, and unreliable in terms of both linear and angular dimensions. Here, geometric reasoning is needed to infer the artists' intentions from the sketch. Suffell and Blount [21] note that at different phases of the design, different amounts of detail are required, allowing further information to be added to that provided by sketches at a later stage. They advocate the use of computational stereo techniques for matching information provided in multiple sketches, but note that allowing for differences in sketches due to "artistic licence" may be problematical.

Fisher and Orr [22] note that geometric consistency and merging operations need to be performed when information is obtained from several different features in the two views. They describe an extension of ACRONYM [23] in which such information is expressed as constraints, which can be manipulated algebraically. These constraints are associated with object hypotheses to form networks which must be forced into consistency. Fisher and Orr expect fast convergence of such methods, especially when executed in parallel.

Indeed, it is not difficult to see that many of the various techniques proposed for geometric reasoning in the field of computer vision are of potential use in interpreting sketched input, or even for using photographs as a basis for design.

Another use of geometric reasoning in input lies in enabling the user to identify higher levels of a solid object than just its faces, edges and vertices. He may wish, for example, to move the position of a hole or pocket, a composite item made of many simpler geometric elements, within the object. For such reasons, and also for uses such as automatic generation of process plans [24] and part codes [25], there has been much interest in the use of geometric reasoning to identify geometric features in solid objects.

In another approach to the problem of input, Martin [26] points out that most current CSG solid modelling systems use relational algebra to define shapes in terms of set operators which explicitly combine primitive shapes. An alternative is to use relational calculus instead to define a shape in terms of the points it must contain. The former is prescriptive (do this, do that), and is well suited to traditional languages like Pascal, whereas the latter is descriptive (the result required has these properties), and is much more suited to declarative languages like Prolog. Although it may be more natural for users to use the former for input, the latter is probably more suitable as an internal representation for geometric reasoning modules. Importantly, relational algebra and relational calculus are equivalent, and can readily be converted from one form into the other [27].

The problem of feature recognition may be solved in part by suitable choice of relational algebra operators, especially if they correspond to particular manufacturing operations, as some features can then be captured at the input stage. However, as Jared [28] points out, different geometrical aspects of a solid object may well be regarded as features at different stages of design and production. Designers tend to think in terms of functional features while process planners require manufacturing features (Husbands et al. [29]). A further observation made by these authors is that interrelationships between the features are often more important than the isolated features themselves. As they point out, feature input by the designer is a time-consuming process which urgently needs to be replaced by automatic feature recognition.

For an excellent discussion of the whole area of automating feature recognition, see the speculative paper by Woodwark [30]. In particular, he notes that most approaches, perhaps apart from Woo's [31], are based on boundary models, but this can lead both to storing explicitly information which is not required in such detail, and to expensive global computations to ensure that other regions of the component do not invalidate a feature (a bolt hole whose opening is blocked by another part of the object is of no use as a bolt hole). He instead proposes a method of feature recognition based on set theoretic models, despite many problems of using this approach.

Expert Systems

Because design is often based on heuristics as well as scientific reasoning, it is not surprising that the use of expert systems has often been proposed as a suitable method for storing these heuristic rules. However, although expert systems are well suited to *logical* reasoning, and possibly *statistical* reasoning when fuzzy logic and similar methods are used, in general it would seem that present expert systems probably do not have the necessary mathematical tools (such as algebraic manipulation capabilities) for the more complex *geometric* reasoning tasks. Nevertheless, several attempts have been made to perform geometric reasoning tasks using expert systems.

For example, as Murray and Miller point out [32], various relationships exist between the different components of an assembly. If geometric features of one part of an assembly are changed, it is often necessary to make changes to other parts of the assembly as a result. Thus, if the diameter of a shaft is increased, the diameter of any bearings supporting that shaft will also need to be increased. One way of allowing this type of change is to rely heavily on standardized parts whose geometric shape is specified by a set of parameters. Murray suggests the use of an expert system for keeping track of the geometric relationships between the parts of an assembly. Firstly, facts and rules about the design are entered, and then in a second step, parameters are assigned to create a specific instantiation of the assembly, where the facts and rules ensure that all of the design constraints are adhered to. Note that alternative designs may be created not only by changing the parametric values, but also by modifying the design rules themselves. Typical rules allow for "part of" and "type of" relationships, while the user may also define higher-level rules of his own, such as what "force-fitted" means in geometric terms.

Let us go back to the earlier remarks made about the general suitability of expert systems for performing geometric reasoning. If an expert system were to be interfaced to a conventional solid modeller, such a system could offer the reasoning capabilities of the expert system, while using the geometric modeller to solve any geometric problems posed. One possible area the current author has identified where this approach may be of some use is in checking the conformance of a design with standards of a geometric nature. One example might be whether the correct shape and number of brackets have been used to support a component. It is not too difficult to further envision the use of automatic finite-element mesh generation from solid models, as described by Wordenweber [33], to answer further questions asked by the expert system about strength and other properties. It is hoped that a future project will tell us whether a system where the geometry and the reasoning are only loosely coupled can be effective for solving problems of this type, or whether an integrated system is necessary for solving geometric reasoning problems.

Another type of task where expert systems and CAD systems have already been linked is in the area of ergonomic design. Bonney et al. [34] describe the use of the SAMMIE CAD system and ALFIE expert system in this field. The layout of workplaces or equipment can be checked for a variety of requirements, from simple ones of whether everything will fit, to the more complex criteria involved in deciding whether an operator can see and easily operate all of the controls. A particular requirement and feature of their system is the ability to make geometric models of human beings.

Other Possibilities

Although the main part of this chapter has concentrated on various major aspects of geometric reasoning which the author feels are particularly relevant to design, many other possibilities are also being actively pursued. In this final section, a brief summary of some of the more interesting ones will be given.

The use of networks of constraints has already been mentioned with respect to input above. Another area where very similar ideas are of use is in the automatic analysis of geometric tolerances, and Fleming [35] suggests a method of doing this based upon representing tolerances by a network of datum values and tolerance zones.

Other work for representing uncertain geometry has been carried out by Durrant-Whyte [36], principally for reasoning in the area of robotics, but it would also seem that his ideas might well be of use in representing tolerances in design. He particularly considers the problems of how different uncertainty measures may be combined. To do so, he represents uncertain points, lines and surfaces as stochastic point processes described as a probability distribution on the parameter space describing the underlying object. The manipulation of uncertainty measures can then be performed by transforming and combining these probability measures.

An interesting idea due to Canny [37] links together algebraic ideas with solid modelling ideas. In his parlance, a semi-algebraic set is what CAD practitioners would call a CSG (computational solid geometry) solid model. He gives an algebraic procedure for constructing a "Roadmap" of a semi-algebraic set, which is a one-dimensional structure, such that each connected component of the original model corresponds to a single connected component of roadmap. Furthermore, it is possible, given a point in the original set, to find rapidly a corresponding point in the roadmap, and as a corollary, decide quickly if two given points are in the same component of the original. Although he originally proposed these ideas in the context of robot path planning (where existence of a path between two points depends on whether the starting and end point are in the same component of a set in configuration space), it is obvious that this technique has much wider potential use in design. A reasoning module may wish to tell whether two points are in the same component of an assembly or mechanism when deciding how moving one point will affect the other. More exotic possibilities include deciding whether a geometric design satisfies certain basic requirements of vibrational or electrical insulation by testing whether certain points are in the same or different components of the model.

Although one advantage of algebraic systems is their immunity to numerical problems, this is offset by their relative inefficiency, and so alternative approaches are also being studied to the accuracy problem. One example is the work by Milenkovic [38] on how verifiably correct geometric algorithms can be constructed, which are still based on limited precision arithmetic. He proposes two methods, the first of which, data normalization, transforms the geometric structure of the problem into a configuration where all limited precision calculations give correct answers. The second method, the hidden variable method, constructs configurations which belong to infinite precision objects, but without explicitly representing them.

Another rapidly expanding area of great significance is theoretical computational geometry. Many efficient data structures and algorithm design techniques have been devised for reasoning about geometric structures. An excellent tutorial on this subject is provided by Guibas and Stolfi [39]. One current problem, highlighted by this chapter, is that the resulting algorithms proposed by such research are usually quite subtle and extremely complex, and relatively few of them have been implemented in practice.

As a final statement, the need for hierarchical representations of geometric information and the combination of different reasoning methods at different levels should be noted, a point made very clearly by Barry et al. [40]. To a certain extent, current boundary representation modellers do this already by separating their data into the topological and geometrical components, where restructuring problems are abstracted from particular geometric computations. Barry et al. take these ideas further. They propose a model which stores both explicit geometric (and topological) data, and logical relations about geometric entities which are used to control numerical accuracy problems. For example, a given vertex may be computed perhaps by the intersection of any two edges meeting at it, or as the intersection of any three faces. Depending exactly on which of these possible definitions is used to find the vertex, slightly different numerical results may be obtained. The logical relations are used to merge these possibilities, by perturbing the geometric entities to ensure that any geometric computation which leads to the same result gives the same geometric answer. Nevertheless, incorrect logical decisions can be made when, for example, two points are accidentally close together, and are determined to be logically the same point. A higher reasoning level is then needed to keep track of which geometric entities should and should not be logically associated.

As has been pointed out above, algebraic and theorem proving methods can consume large amounts of computer time, so multi-level reasoning systems which decide when and how to use these powerful tools to best effect will be necessary.

References

1. Kapur D, Mundy JL (eds). Geometric reasoning. MIT Press, Cambridge, MA, 1989.
2. Woodwark JR (ed). Geometric reasoning (proceedings of a conference held at the IBM UK Scientific Centre). Oxford University Press, 1989
3. Davenport JH, Siret Y, Tournier E. Computer algebra systems and algorithms for algebraic computation. Academic Press, London, 1988
4. Arnon DS, Collins GE, McCallum S. Cylindrical algebraic decomposition I: the basic algorithm. SIAM J Comput 1984; 13: 865–877

5. Arnon DS, Collins GE, McCallum S. Cylindrical algebraic decomposition II: an adjacent algorithm for the plane. SIAM J Comput 1984; 13: 878–889
6. Buchberger B, Collins GE, Kutzler B. Algebraic methods for geometric reasoning. Ann Rev Comput Sci 1988; 3: 85–119
7. Davenport JH. Robot motion planning. In: Woodwark JR (ed) Geometric reasoning (proceedings of a conference held at the IBM UK Scientific Centre). Oxford University Press, 1989
8. Bowyer A, Davenport J, Milne P, Padget J, Wallis AF. A geometric algebra system. In: Woodwark JR (ed) Geometric reasoning (proceedings of a conference held at the IBM UK Scientific Centre). Oxford University Press, 1989
9. Martin RR, Stephenson PC. Swept volumes in solid modellers. In: Handscomb DC (ed) Mathematics of surfaces III. Oxford University Press, 1989
10. Todd SJP. Programming interactions by contraints. In: Woodwark JR (ed) Geometric reasoning (proceedings of a conference held at the IBM UK Scientific Centre). Oxford University Press, 1989
11. Hoffmann C, Hopcroft J. Quadratic blending surfaces. Comput Aided Des, 1986; 18: 301–306
12. Middleditch A, Sears K. Blend surfaces for set theoretic modelling systems. ACM SIGGRAPH Comput Graphics 1985; 19: 161–170
13. Kapur D, Mundy JL. Wu's method: an informal introduction. In: Kapur D, Mundy JL (eds) Geometric reasoning. MIT Press, Cambridge, MA, 1989
14. Wu W. On the decision problem and the mechanisation of theorem proving in elementary geometry. Sci Sin 1978; 21: 150–172
15. Chou SC. Proving elementary geometry theorems using Wu's algorithm. In: Bledsoe, Loveland (eds) Theorem proving: after 25 years. Contemp Math 1984; 29: 243–286
16. Kutzler B, Sifter S. Automated geometry theorem proving using Buchberger's algorithm. 1986 symposium on symbolic and algebraic computation (SYMSAC 86), Waterloo, Canada, 1986
17. Kapur D, Mundy JL. A refutational approach to geometry theorem proving. In: Kapur D, Mundy JL (eds) Geometric reasoning. MIT Press, Cambridge, MA, 1989
18. Geisow A. Recognition and generation of symbolic diagrams. In: Woodwark JR (ed) Geometric reasoning (proceedings of a conference held at the IBM UK Scientific Centre). Oxford University Press, 1989
19. Pereira FCN. Can drawing be liberated from the Von Neumann style? In: Caneghem M van, Warren DHD (eds) Logic programming and its applications. Ablex, 1986
20. Arya K. A functional approach to picture manipulation. Comput Graphics Forum 1984; 3: 35–46
21. Suffell C, Blount GN. Sketch form data input for engineering component definition. In: Woodwark JR (ed) Geometric reasoning (proceedings of a conference held at the IBM UK Scientific Centre). Oxford University Press, 1989
22. Fisher RB, Orr MJL. Geometric constraints from $2\frac{1}{2}$D sketch data and object models. In: Woodwark JR (ed) Geometric reasoning (proceedings of a conference held at the IBM UK Scientific Centre). Oxford University Press, 1989
23. Brooks RA. Symbolic reasoning among 3D models and 2D images. Artif Intell 1981; 17: 285–348
24. Jared GEM. Shape features in geometric modelling. In: Pickett MS, Boyse JW (eds) Solid modelling by computers. Plenum, 1984
25. Kyprianou LK. Shape classification in computer aided design. PhD thesis, University of Cambridge, 1980
26. Martin RR, Howells DI. Relational algebra, relational calculus and computational solid geometry. In: Earnshaw RA (ed) Theoretical foundations of computer graphics and CAD. Springer, Berlin Heidelberg New York, 1988
27. Gray PMD. Logic, algebra and databases. Ellis Horwood, 1984
28. Jared GEM. Recognising and using geometric features. In: Woodwark JR (ed) Geometric reasoning (proceedings of a conference held at the IBM UK Scientific Centre). Oxford University Press, 1989
29. Husbands P, Mill F, Warrington S. Part representation in process planning for complex

components. In: Woodwark JR (ed) Geometric reasoning (proceedings of a conference held at the IBM UK Scientific Centre). Oxford University Press, 1989

30. Woodwark JR. Some speculations on feature recognition. In: Kapur D, Mundy JL (eds) Geometric reasoning. MIT Press, Cambridge, MA, 1989

31. Woo TCH. Computer understanding of design. PhD thesis, University of Illinois, 1975

32. Murray JL, Miller MH. Knowledge-based systems in process planning and assembly design. In: Woodwark JR (ed) Geometric reasoning (proceedings of a conference held at the IBM UK Scientific Centre). Oxford University Press, 1989

33. Wordenweber B. Automatic mesh generation of two and three dimensional curvilinear manifolds. PhD thesis, University of Cambridge, 1981

34. Bonney M, Taylor N, Case K. Using CAD and expert systems for human workplace design. In: Woodwark JR (ed) Geometric reasoning (proceedings of a conference held at the IBM UK Scientific Centre). Oxford University Press, 1989

35. Fleming AD. A representation of geometrically toleranced parts. In: Woodwark JR (ed) Geometric reasoning (proceedings of a conference held at the IBM UK Scientific Centre). Oxford University Press, 1989

36. Durrant-Whyte HF. Concerning uncertain geometry in robotics. In: Kapur D, Mundy JL (eds) Geometric reasoning. MIT Press, Cambridge, MA, 1989

37. Canny J. Constructing roadmaps of semi-algebraic sets, part I: completeness. In: Kapur D, Mundy JL (ed) Geometric reasoning. MIT Press, Cambridge, MA, 1989

38. Milenkovic VJ. Verifiable implementations of geometric algorithms using finite precision arithmetic. In: Kapur D, Mundy JL (eds) Geometric reasoning. MIT Press, Cambridge, MA, 1989

39. Guibas LJ, Stolfi J. Ruler, compass, and computer: the design and analysis of geometric algorithms. In: Earnshaw RA (ed) Theoretical foundations of computer graphics and CAD. Springer, Berlin Heidelberg New York, 1988

40. Barry M, Cyrluk D, Kapur D, Mundy JL. A multi-level geometric reasoning system for vision. In: Kapur D, Mundy JL (eds) Geometric reasoning. MIT Press, Cambridge, MA, 1989

Expert System for Engineering Design Based on Object-Oriented Knowledge Representation Concept

S. Akagi

Introduction

The process of computer-aided engineering design is characterized by interactive decision making between a designer and a computer so that design objectives are satisfied under various design constraints. The design process involves the recognition, formulation and satisfaction of constraints. They are continually being added, deleted and modified throughout the design process. The constraints are often numerous, complex and contradictory [1]. A designer must manage these constraints so that design objectives are well satisfied. However, it is difficult to manage them in the case of designing a large and complex engineering system.

Expert systems promise a powerful tool for engineering design. They can provide designers with assistance during the conceptual stage of design as well as detail design. In addition, they can help close the gap between novice designers and experienced designers by providing qualitative and quantitative support.

In this chapter, an object-oriented knowledge-based system is introduced. The model for the design process is constructed using networks composed of knowledge elements, i.e. constraints, which are modularly represented as objects. This modelling results in determining of design variables in a flexible manner during the design process. The system also includes diagnostic functions in order to improve the original design model. The system was encoded in Common Lisp, which was combined with Fortran programs for numerical computations and graphics. These functions of the system are also useful for engineering calculations with FEM, geometric modelling as well as engineering design. Finally, examples of applications to a preliminary ship design and a marine power plant design will be presented [2–4].

Related Works

The design process is characterized by the managing of constraints so that the design objectives are satisfied as mentioned above. An object-oriented system is suitable for building up and managing constraints. Work on constraint-based expert system has been reviewed in detail by Serrano and Gossard [1].

One of the first attempts to manage constraints in design problems was the work by Sutherland [5]. His well-known SKETCHPAD system used interactive computer graphics and a constraint system. SKETCHPAD was not artificial intelligence (AI) oriented. It used a propagation scheme combined with a relaxation scheme. Sutherland's work dealt with equality constraints, had no logical capabilities, or explanation facilities [1].

Steele and Sussman's work [6] was AI oriented. They presented a language for the construction of almost-hierarchical constraint networks and a simple solution technique called local propagation. They used constraint propagation on linear algebraic constraints. Their system was very similar to an object-oriented system, but not strongly related to design-oriented problems.

Alan Borning's THINGLAB [7] was a simulation laboratory based on constraints. THINGLAB used an object-oriented approach. Object-oriented programming [8] itself was developed independently from design or constraint problems. Object-oriented representation is a knowledge representation technique as well as a coding technique. There are several ways of implementing this; for example, Smalltalk is a language which is designed specifically to be object-oriented [9, 10]. On the other hand, object-oriented systems can also be constructed using other languages such as Lisp or C, e.g. Flavor [11] and Loops [12] in Lisp or Objective-C in C language. The system introduced in this chapter is similar to the latter one.

After THINGLAB, expert systems developed for design have usually adopted the object-oriented concept as well as constraint-based treatment. The reason is that a design model using an object-oriented concept is very suitable for managing the constraint-based feature. Some of the developed expert systems are for mechanical design. Gossard and Serrano [1,13] developed MATHPAK for preliminary design of mechanical systems using the constraint-based concept. Brown and Chandrasekaran's Class II concept [14] is for routine design, for which an example of air cylinder design was demonstrated. Ward and Seering [15] developed a machine design system using constraint-oriented modelling. Chan and Paulsen's expert system for truss design [16] and Elias's expert system for aircraft design [17] are also constraint-oriented.

Developed System

Modelling for the Design Process

The design process is characterized by a sequence of several steps which are composed of the following phases: "building a model for the design", "simulation using the model", "evaluating the result of the simulation" and "correcting the original model". This sequence is repeated step by step until the design conditions are satisfied completely. The process is illustrated typically in Fig. 4.1. In the design model, especially for large systems, design variables are usually related to each other and form a complicated network. In the design procedure shown in the figure, it is necessary to determine the value of design variables so as to satisfy the constraints which are specified by the relations among design variables. In this system, the concept of object-oriented programming is introduced to manipulate the network. Namely, the individual design knowledge elements are modularized and represented as "objects" (Fig. 4.2). This is a typical knowledge representation used in the field of AI [8]. To determine the values of design variables, the well-known "message-

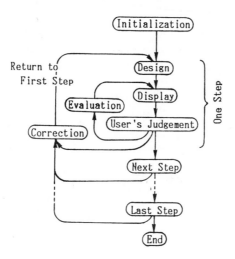

Fig. 4.1 Design process.

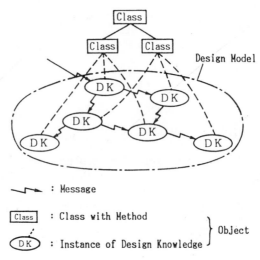

Fig. 4.2 Object-oriented model for design knowledge.

passing" procedure is introduced reflexively. For example, when the value of a design variable is to be determined, the message to request its determination is passed to the corresponding object in which the algorithm for calculating its value is described. If the values of other design variables are needed in the calculation, messages are passed also to the objects corresponding to them. On the other hand, when the value of some design variable must be corrected in this procedure, the values of the other related design variables are corrected automatically by the message-passing function tracing in the reverse direction of the above message passing. Furthermore, as shown in Fig. 4.2, the objects are classified into "class" and "instance". A class is the concept which is common to the design variables. In the system, the class is described in the "general" subsystem, shown in Fig. 4.3, where the basic design procedures are defined. On the other hand, an instance represents the knowledge element corresponding to each design variable included in a "specific" design subsystem. The knowledge element is usually acquired from a design handbook or the designer's expertise.

System Configuration

The configuration of the developed system is shown in Fig. 4.3. The system is separated into two subsystems: the general subsystem ① which is commonly applicable to various kinds of design problems, and the specific subsystem ② which corresponds to the specific design

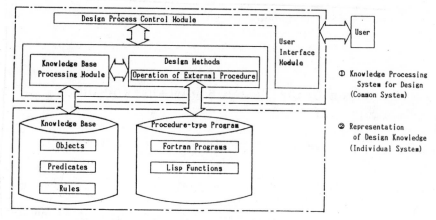

Fig. 4.3 Configuration of the expert system.

problem. The former manipulates the knowledge base during interaction between a user and the system. The latter contains a set of objects for the specific design knowledge, the predicates for controlling the design process, and the rules for the diagnostic evaluation of the design results generated in the design process. The functions of these types of knowledge representation enable the system to be generally applicable to various kinds of engineering design problems.

The general subsystem manages the knowledge base described in the specific subsystem through the interactive design procedure between a user and the system. The general subsystem consists of the control module for the design process, the processing module for the knowledge base and the user interface module.

The control module for the design process activates the predicates for controlling the design process which are described in the specific subsystem. The predicates written in a simple form, as will be shown later, are automatically transformed by this module into Lisp functions.

The processing module for the knowledge base activates and manipulates the design knowledge elements described as objects in the specific subsystem. Design variables are determined through the activation of the design knowledge elements in the sequence of design steps. Moreover, the system can be connected to external programs written in Fortran. Engineering design problems usually include extensive numerical computations and graphic representations which are encoded in Fortran, and hence the connection to external subroutine programs is essential for this kind of expert system. Data for programs encoded in Lisp and Fortran are automatically transferred through a data file.

The user interface module consists of programs for manipulating the multi-window system in a workstation which supports the man–machine interactive function of the system. The general subsystem is independently programmed from the specific (design) subsystem, and hence it can be adapted for various kinds of design.

Details of the function of the system will be introduced below.

Representation of the Design Knowledge and the System Function

Object-oriented knowledge representation and its manipulation in the design process

The design procedure is modelled as a relational network consisting of a set of design variables, i.e., instances as shown in Fig. 4.2. These design variables are represented as objects, which are manipulated by message passing as mentioned above.

The mechanism for manipulating instances is illustrated as follows. The instances described in Fig. 4.2 are assigned to the "objects" in the system, and their relations are represented using a relational network as illustrated in Fig. 4.2. Figure 4.4 represents an example of a procedure for a ship design (see below), where the value of instance "light weight (ship's self-weight)" is determined based on the principle of object-oriented manipulation as follows. Firstly, the message ① is passed to the instance "light weight" in order to activate the method "DECIDE"

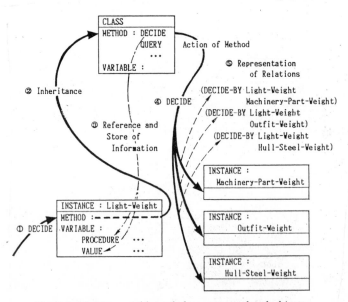

Fig. 4.4 Inheritance of knowledge among related objects.

which determines the value of that instance. This method is defined in its class ② and inherited by it. The message ④ is also passed automatically to the related instances, i.e. "machinery part weight", "outfit weight" and "hull steel weight", introduced by the design knowledge ③ which is described in the instance "light weight". Similar relations are also used in the automatic deletion of values as given by the following method "DELETE". This procedure can support a designer flexibly in the design process.

The various "methods" built in the system, including the above-mentioned "DECIDE" method, are summarized as follows:

- "DECIDE": to determine the value of the assigned design variable based on the procedure which is described in the instance (Fig. 4.4).
- "QUERY": to query the value of the design variable and to request to display it.
- "DELETE": to delete the value of the design variable and also to delete the values of the other design variables related to it.
- "SET-UP": to ask a user whether the value of a design variable is to be determined in the system or it is to be input.
- "INPUT": to require a user to input the information such as additional constraints.
- "DISPLAY": to display the value of the design variable which has been implemented or produced in the system.

Examples of the expressions for the elements of design knowledge, i.e. the instances, are shown in the following section for the case of ship design.

Rule representation for diagnosis

The design variables must be refined during the iterative design process by continuing trial and error procedures. The refinement of the design variables is achieved on evaluating the design results based on the diagnostic rules, which are represented by "if—, then—" statements [2, 3]. Examples of the rules for design diagnosis are shown in the following section.

Design process represented by predicates

The design process shown in Fig. 4.1 proceeds in several steps. These steps are organized from the aforementioned knowledge representations, i.e. objects and rules. In order to manipulate the design process, it is required to set up in advance which design variables are to be determined and at which step the obtained results are to be displayed in the course of the design process. This is controlled by means of predicates. The predicates used in the system have the following forms:

- (STEP ⟨N⟩ ⟨name of the 'object'⟩): to determine the value of the design variable – "object" – in the step N.
- (DISPLAY ⟨N⟩ ⟨name of the 'object'⟩): to display the value of the design variable – "object" – in the step N.
- (GRAPHIC-DISPLAY ⟨N⟩ ⟨description to call the Fortran program⟩): to display the graphic result by calling the Fortran program.

Examples of these predicates are also shown in the following section for the case of ship design.

Application to Ship Design

The design system (shell) explained above was applied to a preliminary ship design. A detailed description is given in this section.

Characteristics of Preliminary Ship Design and Its Model with AI Technique

Characteristics of preliminary ship design

A key characteristic of ship design is the complexity of the system being designed. Typically, ship design includes a large number of design variables, each of which is related to other design variables in different ways [18]. In the design process the procedures are not limited to mathematical computations, but include determination of various characteristics, e.g. geometrical form, spatial arrangement, etc. This complexity prevents a designer from evaluating all the design variables in detail. The design process is therefore broken down into a number of stages. The earliest stage corresponds to preliminary design. In preliminary design, the principal details of a ship (Fig. 4.5) are determined so as to satisfy the design requirements, e.g. size, speed, economy and safety. These procedures are characterized by iterative decision-making to determine the design variables. This decision-making is also characterized by the compromising of various design objectives. This is a type of informal optimization. In this procedure, a designer's free decision-making is allowed and this is more flexible than in conventional mathematical optimization. This process is more correctly called "satisfying" rather than "optimization" [19]. The procedure for satisfying is carried out by repeating trials to satisfy design objectives; this is called the design spiral [20]. The concept is that the repeating trials of the design procedure look like a spiral staircase.

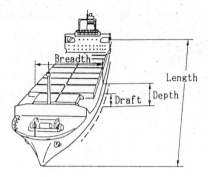

Fig. 4.5 Principal details of a ship.

The process characterized by the spiral is not actually one pass, but forms a network as illustrated in Fig. 4.6. The design variables composing the network relate to one another. In the figure the ship's service speed and deadweight (the weight of cargo to be loaded in a ship) are the

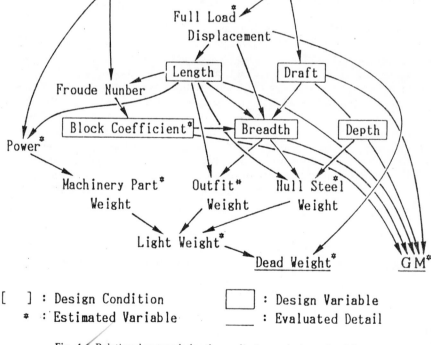

Fig. 4.6 Relational network for the preliminary design of a ship.

given design requirements. The ship's length, draught, depth, breadth and block coefficient (i.e. the parameter of fineness of a ship's hull) are the principal details (i.e. the design variables) to be determined (Fig. 4.5). These relate to the dependent (state) variables, e.g. displacement, Froude number (dimensionless speed/length ratio), hull steel weight, etc. The deadweight is calculated by relating it to the other design variables, and is evaluated by comparing it with the given design requirements. The design variables are then modified to satisfy the design requirements. In this investigation, a supporting technique using an object-oriented concept is introduced to the design process.

Model of ship design using object-oriented concept

The design process detailed above is characterized by sequential repetition as shown in Fig. 4.1. The procedure is performed using the network. To represent the network in a computer system, the concept of an "object-oriented" procedure, explained above is introduced. The design variables shown in Fig. 4.6 are modularized into the design knowledge elements and represented as "objects" (Fig. 4.2).

System Configuration and Function

The configuration of the developed system has a similar feature to that illustrated in Fig. 4.3. The specific subsystem ② (the knowledge base) in the figure describes the ship design for this case.

The subsystem contains a set of objects for describing ship design knowledge, predicates for controlling the design process, and rules for diagnostic evaluation of design results generated in the design process. The objects not only represent the design knowledge elements as mentioned above, but also include the ship's data and candidate designs for the ship's main engine. Fortran programs written as objects include the graphic representation of the ship's hull form, the calculation for the ship's resistance and propulsion, the loading tables for cargoes etc.

The predicates for controlling the design process indicate the order of the design steps which are described in advance by a user for the preliminary design of a ship.

The rules for diagnostic evaluation are represented by "if —, then —" statements which evaluate the results of the design achieved using the design knowledge. Details of the treatment of the subsystem will be given below.

Representation of Design Knowledge and its Manipulating Method

Object-oriented representation of design knowledge

The ship design procedure is modelled as a relational network consisting of a set of design variables, i.e. instances as shown in Fig. 4.5. These design variables are represented as objects manipulated by message passing, as mentioned above. Figure 4.7 illustrates an example of ship design knowledge. In the figure, the symbols L, B, ..., v_s correspond to the instances in Fig. 4.2. In the instances, the knowledge element to determine the value of a design variable is described using a mathematical formula or a symbolic representation. The knowledge is manipulated by a "method" inherited from the class. For example, the instance F_n (Froude number) includes the algebraic formula = $v_s/(9.806 \, L)^{1/2}$ (v_s = ship's speed, L = ship's length). The values of v_s and L are determined using the instances v_s and L, respectively. Their values are automatically determined by the programs described in the general subsystem ① shown in Fig. 4.3.

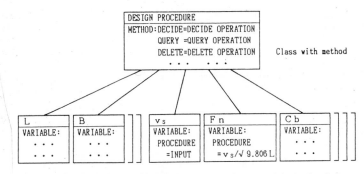

Fig. 4.7 Object-oriented knowledge.

Figure 4.8 shows examples of the programming description of objects respectively. Item (a) in the figure represents the Froude number, item (b) the approximate block coefficient, item (c) the design data of a type-ship and item (d) the prismatic coefficient respectively. These illustrate: (a) a mathematical expression; (b) a conditional expression; (c) the retrieving of design data and (d) the calling of external programs encoded in Fortran.

```
;
(INSTANCE (NAME FROUDE-NUMBER FN)
    (CLASS DNA)                                                    ··· ( a )
    (INSTANCE-VARIABLE
        (PROCEDURE ( R-VS // SQRT ( L * 9.806 ) )         )   )      )
;
(INSTANCE (NAME APPROX-BLOCK-COEFFICIENT A-CB)
    (CLASS SNA)                                                    ··· ( b )
    (INSTANCE-VARIABLE
        (PROCEDURE
            ( ( : DECIDE-BY-OTHER APPROX-BLOCK-COEFFICIENT )
              * --> ( A-DISP-F // ( 1.025 * L * DF * B ) )
            ( METHOD-OF-DPP = BY-TYPE-SHIP )
              --> ( CB OF TYPE-SHIP )
                    MEAN-CB                           )            )
        (INPUT-SENTENCE
            " HOW MUCH IS BLOCK COEFFICIENT ? " ) )               )
;
(INSTANCE (NAME TYPE-SHIP)
    (CLASS SNNA)                                                   ··· ( c )
    (INSTANCE-VARIABLE
        (PROCEDURE
            ( SEARCH ( ABS ( DEAD-WEIGHT OF *** - R-DW )
                     + ABS ( VS OF *** - R-VS-KNOT )
                              * R-DW // R-VS-KNOT ) )
            ( OBJECT-OF-SEARCH ( INSTANCE OF KIND-OF-SHIP ) ) ) ) )
;
(INSTANCE (NAME PRISMATIC-COEFFICIENT CP)
    (CLASS DNA)                                                    ··· ( d )
    (INSTANCE-VARIABLE (PROCEDURE ( $CP CP-CURVE ) ) )          )
;
```

Fig. 4.8 Examples of instances.

Rule representation for diagnosing the design process

The refinement of design variables is achieved by the evaluation of design results, which is executed with diagnostic rules represented by "if —, then —" statements. Figure 4.9 shows examples of rules for design diagnosis. Rule 21-2 means that if the calculated ship's deadweight is greater than 1.05 times that of the required value, then the calculated value is diagnosed to be "too large". The results of diagnosis are translated into simple English to advise the user during the design process.

Design process represented by predicates

Design proceeds by manipulating the network as previously mentioned. In order to control the design steps, it is necessary to set up in advance which design variables are to be determined and when the obtained results are to be displayed in the course of the design. This is done by using predicates. Examples of predicates are shown in Fig. 4.10. In the figure, the predicate "STEP 0 LENGTH" indicates determination of the ship's length at the step 0 in the design process. This predicate passes

```
;
(RULE21-2 :  IF ( &GREATERP DW ( 1.05 ÷ R-DW ) .)
             THEN ( &CREATE-PREDICATE
                    ( TOO-LARGE DW ( 1.05 ÷ R-DW ) ) )   )
;
(RULE32-1 :  IF ( &LESSP HEIGHT-OF-DOUBLE-BOTTOM ( B // 16.0 ) )
             THEN ( &CREATE-PREDICATE
                    ( TOO-SMALL HEIGHT-OF-DOUBLE-BOTTOM
                                           ( B // 16.0 ) ) ) )
;
(RULE34-2 :  IF ( &GREATERP TRIM-ON-FULL-LOAD   0.30 )
             THEN ( &CREATE-PREDICATE
                    ( TOO-LARGE TRIM-ON-FULL-LOAD   0.30 ) ) )
;
(RULE34-3 :  IF ( &LESSP   TRIM-ON-FULL-LOAD -0.05 .)
             THEN ( &CREATE-PREDICATE
                    ( TOO-SMALL TRIM-ON-FULL-LOAD -0.05 ) ) )
;
(RULE36-1 :  IF ( &LESSP    ( B // D ) 1.60 )
             THEN ( &CREATE-PREDICATE
                    ( TOO-SMALL ( B // D ) 1.60 ) ) )
;
(RULE36-2 :  IF ( &GREATERP ( B // D ) 1.82 )
             THEN ( &CREATE-PREDICATE
                    ( TOO-LARGE ( B // D ) 1.82 ) ) )
;
```

Fig. 4.9 Knowledge representations with rules for design diagnosis.

the message "DECIDE" to the object "length" in the system as explained in Fig. 4.4. The predicate "DISPLAY 0 KIND-OF-SHIP" in Fig. 4.10 displays the design result by passing the message "DISPLAY" to the object "kind of ship". Using the above predicates, the intended design process can be set up in advance of the design procedure. This gives the

```
;
(STEP 0 LENGTH)           (STEP 0 BREADTH) (STEP 0 DEPTH)
(STEP 0 APPROX-FULL-LOAD-DISPLACEMENT)
(STEP 0 FULL-LOAD-DRAFT) (STEP 0 APPROX-BLOCK-COEFFICIENT)
(STEP 0 MEAN-CB)
(DISPLAY 0 KIND-OF-SHIP) (DISPLAY 0 REQUIRED-DEAD-WEGHT)
(DISPLAY 0 REQUIRED-VS-KNOT)
(DISPLAY 0 APPROX-FULL-LOAD-DISPLACEMENT)
(DISPLAY 0 LENGTH)         (DISPLAY 0 BREADTH)
(DISPLAY 0 DEPTH)          (DISPLAY 0 FULL-LOAD-DRAFT)
(DISPLAY 0 APPROX-BLOCK-COEFFICIENT)
;
(STEP 1 APPROX-GM-ON-FULL-LOAD) (STEP 1 APPROX-GM-ON-BALLAST)
(STEP 1 APPROX-DEAD-WEIGHT)
(STEP 1 NEEDED-VG-ON-GREN-FULL-LOAD)
(STEP 1 APPROX-VG-WITH-TOP-SIDE-TANK)
(DISPLAY 1 KIND-OF-SHIP) (DISPLAY 1 REQUIRED-DEAD-WEIGHT)
(DISPLAY 1 REQUIRED-VS-KNOT)
(DISPLAY 1 APPROX-FULL-LOAD-DISPLACEMENT)
(DISPLAY 1 NEEDED-VG-ON-GREN-FULL-LOAD)
(DISPLAY 1 APPROX-VG-WITH-TOP-SIDE-TANK)
;
```

Fig. 4.10 Predicates for the knowledge of design process.

user flexibility in building up the intended design process and this would be effective for various kinds of design.

Supporting Method for Design Procedure

In order to determine the value of design variables, a trial and error technique is applied as mentioned in the previous sections. However, it is difficult to determine the variables effectively using only trial and error procedures. To support the procedure, several supporting techniques are built in the system. One is a "forward-tracking method", a kind of sensitivity analysis using a network representation, and another is a "backward-tracking method", a kind of optimization using sequential linear programming (SLP).

Supporting method with forward tracking: sensitivity analysis using a network

The relationship among design variables can be represented as a network composed of objects as shown in Fig. 4.6. To see how this method is

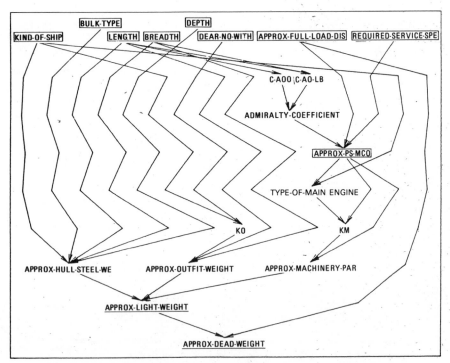

Fig. 4.11 Representation of the variables relating to "approx. deadweight".

used as a supporting method in determining the design variables in the network, consider the following example. Firstly, if a design variable is assigned to be modified, then the related variables are picked up automatically using the network manipulation [2]. For example, if the design variable "approx. deadweight" is assigned, the variables related to "approx. deadweight" are represented hierarchically as shown in Fig. 4.11. If a designer wants to change the value of "approx. deadweight" by modifying the value of the variable "length", the sensitivities in the value of "length" are examined to find its influence on the "approx. deadweight" and other related variables. Figure 4.12 shows the influence caused by a unit-change of "length" on the other variables as network representations. The designer can estimate the appropriate value of "length" so that the value of "approx. deadweight" enters an acceptable region by referring to the value of the sensitivity. The latter is determined by using the value of the gradient of the function which is described in the instance of the variable. The value can be calculated by a "symbolic differential" expression encoded in Lisp.

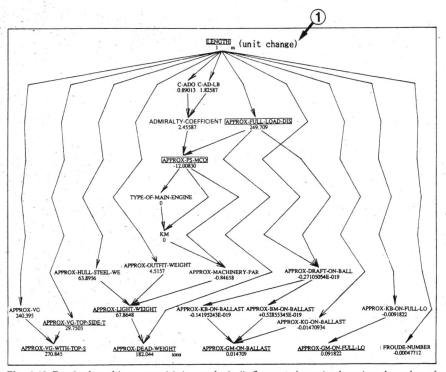

Fig. 4.12 Forward tracking – sensitivity analysis (influence of a unit–changing the value of "Length" ① of the ship on other variables).

Supporting method with backward-tracking optimization using SLP

In iterative design procedures, the above-mentioned forward-tracking method would be useful in most cases. In some cases, however, it would not be sufficient to support the modification of design variables in order to satisfy the design requirements. In this case, a backward- or inverse-tracking method becomes useful. An optimization calculation using SLP is applied for this purpose. Figure 4.13 shows an example of the application of the method. In the figure, the "breadth" is determined by using SLP to the value of 27.26 m which satisfies the required value of "approx. deadweight" of 38 000 tons. The method can be applied when all the design conditions are modelled in mathematical formulation. The linearization of the objective function and constraints for SLP are executed based on the symbolic differentiation of algebraic expressions described in objects.

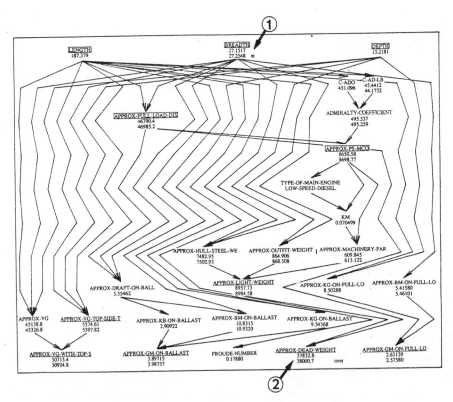

Fig. 4.13 Backward tracking – optimization using SLP (the value of the ship's "breadth" ① corresponding the approx. deadweight of 38 000 tons ②.

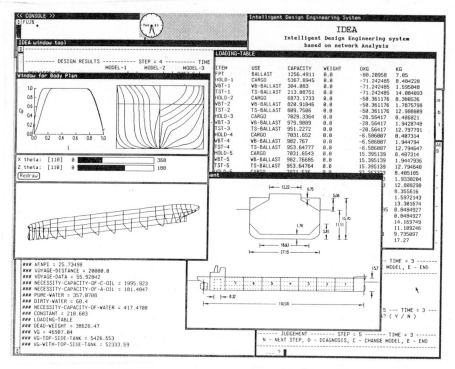

Fig. 4.14 An example of system execution.

Example of System Execution

An example of the execution of the system is shown in Fig. 4.14 for the case of preliminary design of a bulk carrier ship. The process of preliminary design of a ship is summarized in Fig. 4.15. In the process, an iterative procedure is adopted to determine the ship's principal details. After that, the design steps are carried out to select the main engine, determine the hull form and so on.

A sequential representation of the design is shown in Fig. 4.16. In the figure, the sign "⟩⟩⟩" designates the response from the system, the sign "...?" is the user's input, and the sign "###" designates the treatment done in the system, respectively. Explanations of the design example illustrated in Fig. 4.16 are given below:

① The design conditions are given as the ship's deadweight and speed.
② A model of the ship is constructed from the design conditions based on the expert knowledge in the system, and the ship's principal details are indicated.
③ More detailed design variables are estimated.

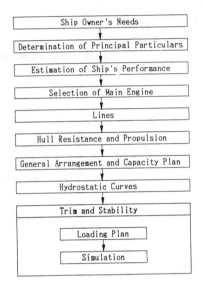

Fig. 4.15 Process of the preliminary design of a ship.

④ The diagnosis rules are applied to evaluate the results. The model is found not to satisfy the requirement exactly on two points; the shortages of deadweight and tank volume.

⑤ The displacement of the ship is now set up to satisfy one of the above conditions, i.e. the requirement on deadweight. The original displacement and its related variables are deleted by using the network procedure as mentioned above.

⑥ New principal details of the ship are indicated and compared with those of the original model. Shortage of the deadweight has been corrected.

⑦ Another problem, i.e. the tank volume, is improved by increasing the ship's depth.

⑧ Next, a model for the main engine is selected [2].

Following the above procedure, the hull form and the general arrangement of the ship are determined as displayed in Figs 4.17 and 4.18 respectively. After that, the final principal details of the ship are determined. Moreover, various kinds of ship performance, i.e. the ship's resistance, the required propulsion power and the stability under different loading conditions, are then calculated more exactly. Figure 4.19 is a graphical representation of the hydrostatic curve which is used for stability estimates. Figure 4.20 is a graphical representation of the stability in the fully loaded condition.

The system was encoded in Lisp, and implemented in the mainframe computer system NEC ACOS1000. After that, the system was ported to a Sun workstation.

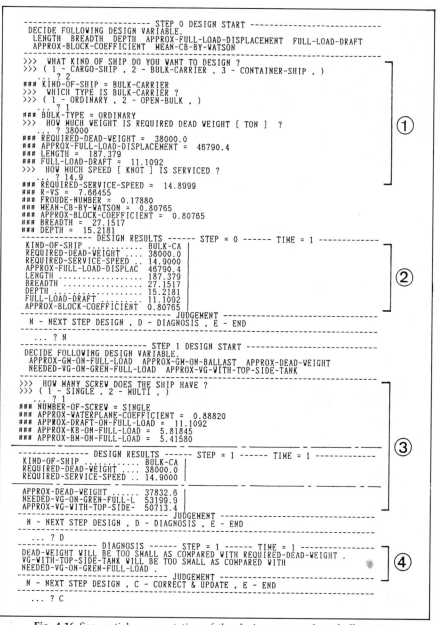

```
------------------------ STEP 0 DESIGN START ------------------------
DECIDE FOLLOWING DESIGN VARIABLE.
   LENGTH  BREADTH  DEPTH  APPROX-FULL-LOAD-DISPLACEMENT  FULL-LOAD-DRAFT
   APPROX-BLOCK-COEFFICIENT  MEAN-CB-BY-WATSON
------------------------------------------------------------------
>>>  WHAT KIND OF SHIP DO YOU WANT TO DESIGN ?
>>>  ( 1 - CARGO-SHIP , 2 - BULK-CARRIER , 3 - CONTAINER-SHIP , )
     ... ? 2
### KIND-OF-SHIP = BULK-CARRIER
>>>  WHICH TYPE IS BULK-CARRIER ?
>>>  ( 1 - ORDINARY , 2 - OPEN-BULK , )
     ... ? 1
### BULK-TYPE = ORDINARY
>>>  HOW MUCH WEIGHT IS REQUIRED DEAD WEIGHT [ TON ]  ?
     ... ? 38000
### REQUIRED-DEAD-WEIGHT = 38000.0
### APPROX-FULL-LOAD-DISPLACEMENT =  46790.4
### LENGTH =  187.379
### FULL-LOAD-DRAFT =  11.1092
>>>  HOW MUCH SPEED [ KNOT ] IS SERVICED ?
     ... ? 14.9
### REQUIRED-SERVICE-SPEED =  14.8999
### R-VS =  7.66455
### FROUDE-NUMBER =  0.17880
### MEAN-CB-BY-WATSON =  0.80765
### APPROX-BLOCK-COEFFICIENT =  0.80765
### BREADTH =  27.1517
### DEPTH =  15.2181
-------------- DESIGN RESULTS ------ STEP = 0 ------ TIME = 1 -------------
KIND-OF-SHIP ............... BULK-CA
REQUIRED-DEAD-WEIGHT .... 38000.0
REQUIRED-SERVICE-SPEED .. 14.9000
APPROX-FULL-LOAD-DISPLAC  46790.4
LENGTH ................  187.379
BREADTH ...............  27.1517
DEPTH .................  15.2181
FULL-LOAD-DRAFT .......  11.1092
APPROX-BLOCK-COEFFICIENT 0.80765
------------------------ JUDGEMENT ------------------------
 N - NEXT STEP DESIGN , D - DIAGNOSIS , E - END
------------------------------------------------------------------
     ... ? N
------------------------ STEP 1 DESIGN START ------------------------
DECIDE FOLLOWING DESIGN VARIABLE.
   APPROX-GM-ON-FULL-LOAD  APPROX-GM-ON-BALLAST  APPROX-DEAD-WEIGHT
   NEEDED-VG-ON-GREN-FULL-LOAD  APPROX-VG-WITH-TOP-SIDE-TANK
------------------------------------------------------------------
>>>  HOW MANY SCREW DOES THE SHIP HAVE ?
>>>  ( 1 - SINGLE , 2 - MULTI , )
     ... ? 1
### NUMBER-OF-SCREW = SINGLE
### APPROX-WATERPLANE-COEFFICIENT =  0.88820
### APPROX-DRAFT-ON-FULL-LOAD =  11.1092
### APPROX-KB-ON-FULL-LOAD =  5.81845
### APPROX-BM-ON-FULL-LOAD =  5.41580
-------------- DESIGN RESULTS ------ STEP = 1 ------ TIME = 1 -------------
KIND-OF-SHIP ............... BULK-CA
REQUIRED-DEAD-WEIGHT .... 38000.0
REQUIRED-SERVICE-SPEED .. 14.9000

APPROX-DEAD-WEIGHT ......  37832.6
NEEDED-VG-ON-GREN-FULL-L  53199.9
APPROX-VG-WITH-TOP-SIDE-  50713.4
------------------------ JUDGEMENT ------------------------
 N - NEXT STEP DESIGN , D - DIAGNOSIS , E - END
------------------------------------------------------------------
     ... ? D
-------------- DIAGNOSIS ------ STEP = 1 ------ TIME = 1 -------------
DEAD-WEIGHT WILL BE TOO SMALL AS COMPARED WITH REQUIRED-DEAD-WEIGHT .
VG-WITH-TOP-SIDE-TANK WILL BE TOO SMALL AS COMPARED WITH
NEEDED-VG-ON-GREN-FULL-LOAD .
------------------------ JUDGEMENT ------------------------
 N - NEXT STEP DESIGN , C - CORRECT & UPDATE , E - END
------------------------------------------------------------------
     ... ? C
```

Fig. 4.16 Sequential representation of the design process for a bulk carrier.

continued

```
---------------- SELECT DESIGN VARIABLES TO BE CORRECT ----------------
>>> PLEASE INPUT VARIABLE NAME WANTED TO CORRECT !
    ... ? A-DISP-F
```
⑤
```
---------------- DELETE & SET-UP DESIGN VARIABLES ----------------
>>> HOW MUCH WEIGHT [ TON ] IS FULL LOAD DISPLACEMENT ?
>>>    PRE-VALUE  NOW: 46790.4
    ... ? 47990.4
### APPROX-GM-ON-BALLAST IS DELETED .
### APPROX-KB-ON-BALLAST IS DELETED .
```
```
--------------- DESIGN RESULTS ------ STEP = 1 ------ TIME = 2 ------------
                         PAST  1              NOW
KIND-OF-SHIP ............           |   BULK-CA
REQUIRED-DEAD-WEIGHT ....           |   38000.0
REQUIRED-SERVICE-SPEED ..           |   14.9000
APPROX-FULL-LOAD-DISPLAC  46790.4  C+   47990.3
LENGTH .................. 187.379   +   192.185
BREADTH ................             |   27.1517
DEPTH ...................             |   15.2181
FULL-LOAD-DRAFT .........             |   11.1092
APPROX-BLOCK-COEFFICIENT             |   0.80765
APPROX-GM-ON-FULL-LOAD ..  2.63139   +   2.67485
APPROX-GM-ON-BALLAST ....  3.89716   +   3.96671
TYPE-OF-MAIN-ENGINE ..... LOW-SPE       LOW-SPE
APPROX-PS-MCO ........... 8650.56   -   8598.05
APPROX-LIGHT-WEIGHT ..... 8957.73   +   9285.61
APPROX-HULL-STEEL-WEIGHT  7482.95   +   7792.36
APPROX-OUTFIT-WEIGHT .... 864.906   +   887.088
APPROX-MACHINERY-PART-WE  609.864   -   606.163
APPROX-DEAD-WEIGHT ...... 37832.6   +   38704.7
NEEDED-VG-ON-GREN-FULL-L             |   53199.9
APPROX-VG-WITH-TOP-SIDE- 50713.4   +   52014.0
------------------------------------ JUDGEMENT ---------------------------
 N - NEXT STEP DESIGN , D - DIAGNOSIS , E - END
```
⑥
```
    ... ? D
---------------- DIAGNOSIS ------ STEP = 1 ------ TIME = 2 -------------
VG-WITH-TOP-SIDE-TANK WILL BE TOO SMALL AS COMPARED WITH
NEEDED-VG-ON-GREN-FULL-LOAD .
------------------------------------ JUDGEMENT ---------------------------
 N - NEXT STEP DESIGN , C - CORRECT & UPDATE , E - END
```
```
    ... ? C
---------------- SELECT DESIGN VARIABLES TO BE CORRECT ----------------
>>> PLEASE INPUT VARIABLE NAME WANTED TO CORRECT !
    ... ? DEPTH
```
```
---------------- DELETE & SET-UP DESIGN VARIABLES ----------------
>>> HOW LONG [ METER ] IS DEPTH ?
>>>    PRE-VALUE  NOW: 15.2181
    ... ? 15.7
### APPROX-GM-ON-FULL-LOAD IS DELETED .
### APPROX-KG-ON-FULL-LOAD IS DELETED .
```
⑦
```
-------------------------- STEP 2 DESIGN START --------------------------
DECIDE FOLLOWING DESIGN VARIABLE.
 MAIN-ENGINE
```
```
### ETA-RPM =  0.45999
### RPM =  89.9203
### CHECK-RPM-LIST-M*E IS DECIDED .
### CHECK-POWER-LIST-M*E IS DECIDED .
>>> DO YOU DESIGNATE MAKER ?
>>> ( 1 - SULZER-RTA , 2 - UE-LA , 3 - MAN , 4 - NO-DESIGNATED , )
    ... ? 4
### MAKER-OF-M*E = NO-DESIGNATED
>>> OK! I FOUND CANDIDATE OF MAIN-ENGINE. 9 ENGINES SATISFY THE REQUIREMENT.
```
⑧
```
>>> I RECOMMEND NEXT CADIDATES FROM POINT OF TOTAL-COST [ YEN ] .
    C-M*E-1   6UEC60LA  416212710
    C-M*E-6   5RTA62    427538870
    C-M*E-3   4RTA68    428360150
>>> INPUT NAME OF MAIN ENGINE .
    ... ? C-M*E-1
### MAIN-ENGINE = C-M*E-1
```

Fig. 4.16 (*continued*).

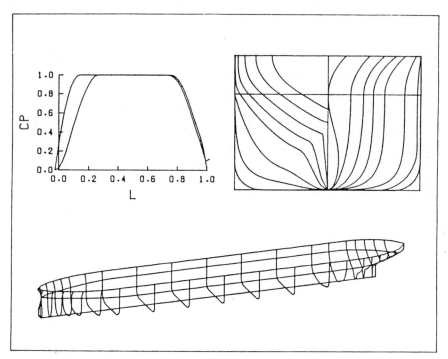

Fig. 4.17 Display of lines.

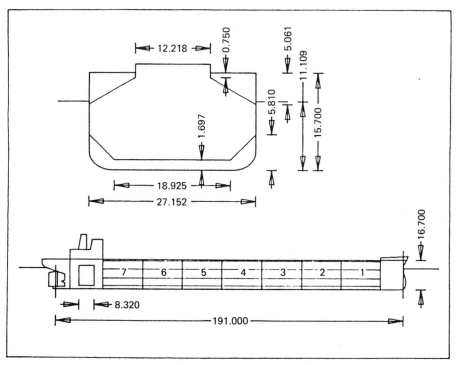

Fig. 4.18 Display of general arrangement.

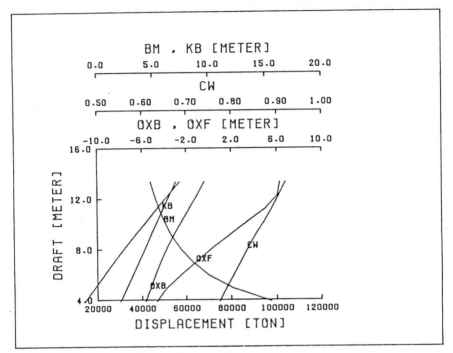

Fig. 4.19 Display of hydrostatic curves.

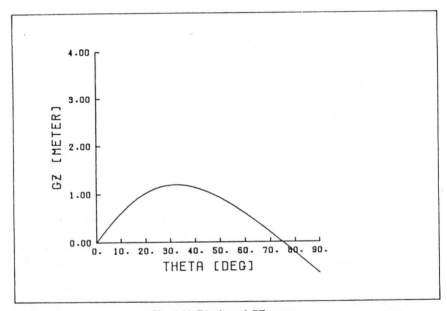

Fig. 4.20 Display of GZ curve.

Application to Power Plant Design

The design system (shell) explained previously was also applied to marine power plant design. A brief introduction is given in this section.

Characteristic of Power Plant Design and Its Task

In general, the process of preliminary design of a power plant consists of (1) selecting the candidates for machinery and equipment composing a plant and (2) optimizing their size and characteristics.

For this system, the object-oriented knowledge representation explained previously is adopted for supporting the design process flexibly and in a user-friendly manner. The developed system also provides hybrid functions combining numerical computations and graphics, as well as AI techniques.

The process for designing power plants is divided into several steps as illustrated in Fig. 4.21 [21]: namely,

1. To search for the candidate machines constituting the plant.
2. To determine the acceptable numbers and sizes of machines.
3. To select the optimal combination of the machinery considering the initial and operational cost.
4. To display the design result including the rough layout of the plant.

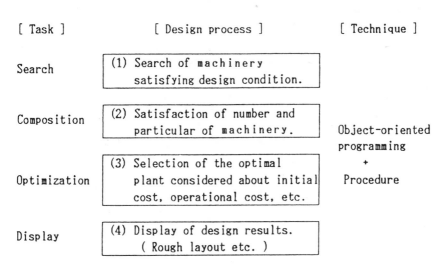

Fig. 4.21 Design process and task for power plant design.

In the above process, it is important to select candidates correctly and
to determine their optimal size and numbers reasonably.

Modelling of the Data and Design Knowledge

Modelling of the machinery data

Selection of candidate machines composing a power plant is usually
performed by systematically searching a database from the general to the
individual domains of machinery data. In order to select the candidates
reasonably and effectively, machinery data are to be represented
hierarchically in the form of a data object. This process is explained
using an example of the design of a marine power plant. Figure 4.22
represents an example for selecting the candidates for the main engine

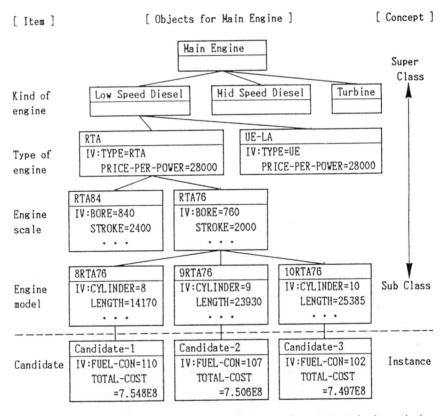

Fig. 4.22 Object-oriented representation for machinery data structure (main engine).

of a marine propulsion plant. In the figure, the structure of the machinery data is divided into the "superclass" and "subclass" hierarchy of data. This data structure is well suited for selecting the candidate machinery composing a power plant. The main engine is selected by matching items from the top to the bottom of the figure. It proceeds as follows: first, "the kind of an engine" is determined, and then "the type of an engine" is searched within the subclasses of "the kind of engines". After that, "scale of an engine" and "the actual model of an engine" are determined successively in the same way. The characteristics of machinery are defined and described in each level of the figure. The superclass holds the common characteristics as its attributes. The subclass inherits them from the superclass and holds the individual characteristics as in any object-oriented knowledge representation scheme.

Modelling of the power plant configuration

The model of the configuration of a power plant is formed by composing the data of various machines. An example of the data structure of a power plant is illustrated in Fig. 4.23. The composition of this model is

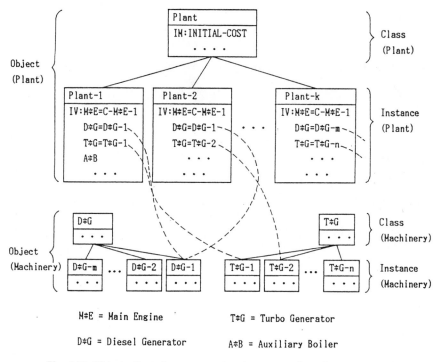

M≠E = Main Engine   T≠G = Turbo Generator

D≠G = Diesel Generator   A≠B = Auxiliary Boiler

Fig. 4.23 Object-oriented representation for power plant data structure.

as follows. The class "plant" indicates the framework of the plant configuration, and its instances, of which the attributes indicate the names of the machinery composing the plant, correspond to the candidate plants. Using this type of modelling, the machinery characteristics can be automatically inherited from the referred objects indicated in the instances of the machinery data as shown in Fig. 4.23.

Modelling of the design knowledge

The design knowlege applied to search the design candidates is divided into units, i.e. design knowledge elements. The design knowledge elements illustrated as "DK" in Fig. 4.2 compose a design model which forms a network. These knowledge elements are modelled and manipulated as "objects" in the system. The objects are encoded in the system using the general expert system for design as shown in Fig. 4.3. In the system, these objects are manipulated by "message-passing" functions as described in Fig. 4.4. Figure 4.24 illustrates an example of the manipulations to search for a candidate main engine composing a marine power plant. As shown in the figure, the objects of the design knowledge act on the respective data objects of machinery data, which corresponds to searching the levels of the hierarchy in Fig. 4.22 in order to determine the candidates for a main engine.

Configuration of the Expert System for Power Plant Design

System configuration

The configuration of the developed system is similar to that illustrated in Fig. 4.3. The system consists of two subsystems: the common subsystem ① (knowledge processing module) and the plant design subsystem ② (knowledge base). The latter consists of a set of objects including the design knowledge and the machining data, and predicates for controlling the design process of searching for the candidate machinery. The Fortran programs, which are used for graphics and optimization calculation, are also called automatically from the knowledge base ②.

Description of knowledge and system function

Figure 4.25 illustrates some examples of describing design knowledge for a marine power plant. The knowledge is represented in the form of instances, i.e. objects. Example (a) of Fig. 4.25 represents the object in which the formula to calculate the number of revolutions of the main

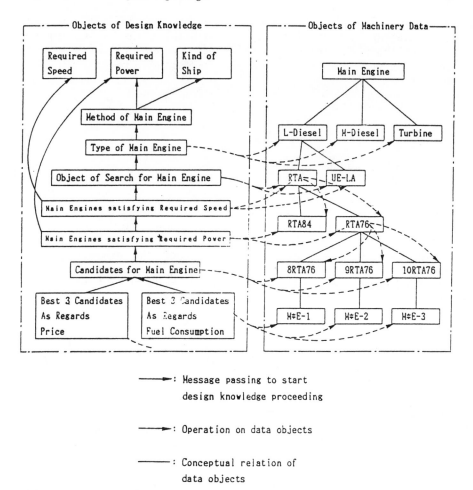

────►: Message passing to start
design knowledge proceeding

────►: Operation on data objects

──── : Conceptual relation of
data objects

Fig. 4.24 An example of manipulations to search for the candidates for a main engine.

engine is described. The calculation procedure takes the form $V_s * 60 *$ $0.514/(0.65 * d_F)$, a familiar Fortran-like expression, which means that the number of revolutions of the main engine is to be calculated using the ship's speed V_s and ship's draught d_F. In the system, the messages to request the values of V_s and d_F based on this description are automatically passed to the objects "V_s" and "d_F" respectively. Example (b) represents the "rule-type" knowledge by which the kind of main engine is determined; the rule states that "if the kind of a ship is a ferry, then a medium-speed diesel is to be selected . . . ". The description of this type of knowledge takes a familiar "If —, then —" form. Example (c) shows the procedure for searching the candidates for a main engine to satisfy

```
(INSTANCE (NAME RPM-RTA)
    (CLASS DNA)
    (IV (PROCEDURE ( R-VS-KNOT * 60.0 * 0.514 // ( 0.65 * DF ) ) ) )
)
:
                              ( a )

(INSTANCE (NAME TYPE-OF-MAIN-ENGINE T-ENGINE)
    (CLASS SNNA)
    (IV (PROCEDURE
        (
        ( KIND-OF-SHIP = FERRY )          --> 'MID-SPEED-DIESEL
        ( ( KIND-OF-SHIP = TANKER
          ; KIND-OF-SHIP = LNG )
            & ( A-PS-MCO > 55000 ) )  --> 'TURBINE
        ( A-PS-MCO < 55000 )          --> 'LOW-SPEED-DIESEL
            ) ) )
)
                              ( b )

:
(INSTANCE (NAME CHECK-RPM-LIST-M*E)
    (CLASS DNNL)
    (IV (PROCEDURE
        ( SATISFY? (SEND-MESSAGE *** 'GET-VALUE 'CM 'CHECK-RPM) )
        ( OBJECT-OF-SEARCH ( SUBCLASS OF OBJECT-OF-SEARCH-M*E ) ) ) )
)
                              ( c )

:
(INSTANCE (NAME BEST-3-FUEL-CONSUMPTION-LIST-M*E BEST-3-NENPI-LIST-M*E)
    (CLASS DNNL)
    (IV (PROCEDURE ( SELECT-BEST-N 3 ( FUEL-CONSUMPTION OF *** ) )
        ( OBJECT-OF-SEARCH ( SATISFY-LIST-M*E )          ) ) )
    (IM (DISPLAY   ( CALL DISPLAY-CAND        )
        ( ITEM FUEL-CONSUMPTION   )                              ) )
)
                              ( d )
```

Fig. 4.25 Examples of objects for design knowledge.

the required shaft revolutions. Finally, example (d) represents the procedure to determine the best three candidates for a main engine which have the lowest fuel consumption among the candidates satisfying the design conditions.

Figure 4.26 illustrates the description of the machinery data objects which are previously represented in Fig. 4.22. The characteristic data of a main engine are described in the form of an object as shown by example (a). The data for a diesel generator are described as example (b). The class of a power plant and its instances are described as examples (c) and (d) respectively. In these examples, typical descriptions of "instance variable (IV)" for machine data and of "instance method (IM)" for determining machine characteristics are shown.

Figure 4.27 gives examples of predicates which control the design process. The design process is modelled as a sequence of several design steps. The predicate in Fig. 4.27 means that the kind of the ship, the speed of the ship, the draught and the required propulsive power of the ship, are determined respectively in the design step 0, and the main engine is selected in the design step 1 and so forth. The design results are then displayed in the design steps 0 and 1 . . . respectively.

```
;
(CLASS (NAME RTA76)
   (SUPERCLASS RTA)
   (IV (BORE              760.0                                                         )
       (STROKE            2200.0                                                        )
       (SCALE             76                                                           )
       (SPEED             ((R1  98.0) (R2  98.0) (R3  71.0) (R4   71.0)) )
       (FUEL-CON100       ((R1 127.0) (R2 121.0) (R3 126.0) (R4 121.0)) )
       (FUEL-CON85        ((R1 125.0) (R2 121.0) (R3 124.0) (R4 120.0)) )
       (PME               ((R1 16.93) (R2  92.9) (R3 16.83) (R4 12.83)) )
       (HEIGHT            12560.0                                                       )
       (WIDTH             4100.0                                                        )
       (POWER-PER-CYLINDER
                          ((R1 3680.0) (R2 2020.0) (R3 2650.0) (R4 2020.0)) )
       (CYLINDER-RANGE ( 4 * 10 12 )                                           ) )
   (IM (LENGTH0        (LAMBDA (N)
                          (COND ((LESSP N 9)
                                  (PLUS 8370.0
                                          (TIMES 1450.0 (DIFFERENCE N 4))) )
                                (T (PLUS 16655.5
                                          (TIMES 1455.0 (DIFFERENCE N 4))) )) ) )
                                                                              ) )
)
```

(a)

```
(CLASS (NAME D*G-TYPE2)
   (SUPERCLASS DIESEL-GENERATOR)
   (CM (SEARCH-D*G        (CALL    SEARCH-D*G)                         ) )
   (IV (SCALE             2                                          )
       (OUTPUT            ((LOWER-LIMIT 500.0) (UPPER-LIMIT 1000.0)) )
       (FUEL-CONSUMPTION  ((LOWER-LIMIT  3.8) (UPPER-LIMIT  300.0)) )
       (COEFFICIENT       ((C0 5.769) (C1 0.2123) (C2 0.0))          ) )
   (IM (PROPERTY0         (LAMBDA (N)
                          (PLUS (TIMES 0.2123 N) 5.769))              ) )
)
;
```

(b)

```
(CLASS (NAME PLANT)
   (IV (INITIAL-COST      NIL
                          (ACCESS ((BEFORE IF-NIL-THEN-METHOD&PUT))) )
       (RUNNING-COST      NIL
                          (ACCESS ((BEFORE IF-NIL-THEN-METHOD&PUT))) )
       (TOTAL-COST        NIL
                          (ACCES  ((BEFORE IF-NIL-THEN-METHOD&PUT))) ) )
   (IM (INITIAL-COST      (CALL CALCULATE-INITIAL-COST)              )
       (RUNNING-COST      (CALL CALCULATE-RUNNING-COST)              )
       (TOTAL-COST        (CALL CALCULATE-PLANT-TOTAL-COST)          )
       (DISPLAY-PLANT     (CALL DISPLAY-ARRANGEMENT)                 ) )
)
```

(c)

```
(INSTANCE (NAME PLANT-106)
   (CLASS PLANT)
   (IV (EXHAUST-GAS-ECONOMIZER E*E-CANDIDATE1   )
       (SHAFT-GENERATOR         S*G-CANDIDATE1   )
       (TURBO-GENERATOR         T*G-CANDIDATE4   )
       (AUXILIARY-BOILER        A*B-CANDIDATE9   )
       (DIESEL-GENERATOR        D*G-CANDIDATE1   )
       (INITIAL-COST            +0.92934131E+009 )
       (NUMBER                  106              )
       (RUNNING-COST            +0.52210513E+008)
       (TOTAL-COST              +0.98155183E+009) ) )
```

(d)

Fig. 4.26 Examples of data objects for machines.

Several candidate machines composing the power plant are selected by using the design knowledge as mentioned above. After that, the optimal one is selected from the candidates considering the initial cost and the annual operating cost of the power plant. For this purpose, a design optimization procedure is adopted to minimize the annual operating cost under various constraints such as the power demands,

```
(PREDICATE INITIAL-DATA
    (STEP 0 KIND-OF-SHIP)
    (STEP 0 R-VS-KNOT)
    (STEP 0 DF)
    (STEP 0 A-PS-MCO)
    (STEP 1 MAIN-ENGINE)
        ⋮

    (DISPLAY 0 KIND-OF-SHIP)
    (DISPLAY 0 R-VS-KNOT)
    (DISPLAY 0 DF)
    (DISPLAY 0 A-PS-MCO)
    (DISPLAY 1 A-PS-MCO)

        ⋮
```

Fig. 4.27 Predicates for control of design process.

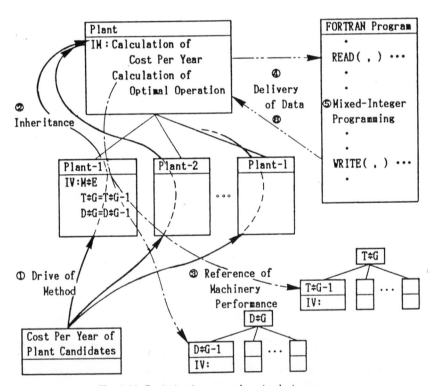

Fig. 4.28 Optimization procedure in design process.

the performance relations of the machinery composing the power plant and so forth. Figure 4.28 illustrates the procedure for the optimization. The calculation to minimize the cost is performed using objects in which the methods to calculate the cost are described. The messages for ① calculating the cost are passed to the respective objects corresponding to the candidate plants and machine data. In the optimization calculations, the constraints, e.g. performance relations of the machinery, usually include the on–off condition of the machinery. This type of optimization can be formulated as a mixed integer programming problem [1] by assigning binary variables to the on–off condition. The optimization calculation is performed by calling the Fortran program from the main system (programmed in Lisp), as illustrated in Fig. 4.28.

Example of System Execution

An example of the execution of the system is shown in Figs 4.29 and 4.30 for the case of a marine power plant design. The process of marine power plant design is separated into two parts; (1) the design of a main-engine system, and (2) the design of a heat and electric power generating (cogeneration) plant.

Figure 4.29 illustrates the design steps of (1) for selecting a main engine. After the design conditions are displayed, 9 candidates for the main engine are searched. The optimal main engine is then selected from the different three viewpoints for cost estimation. Firstly, three candidates are recommended in the order of lower fuel cost. Secondly, a different three candidates are recommended in the order of lower initial cost. Thirdly, three candidates are recommended in the order of lower total cost. Finally, the C-M*E-2 (6RTA84) type engine with the lowest total cost is selected by a designer, and its configuration and characteristics are displayed.

Figure 4.30 illustrates the design steps of (2) for designing the heat and electric power generating plant. After the design conditions are indicated, the kind of machinery forming the power plant is displayed. The candidate power plants are then composed. The 108 different candidates satisfying the design conditions are generated using the design knowledge to select the combinations of machinery. However, the numbers of candidates of plants are still too large to finalize the design. Then, the optimization calculation is applied, and a short list of 10 candidates is determined and arranged in order of lower initial cost and lower total cost. Finally, the 106th plant is selected as the best one by a designer, and its rough layout and the characteristics of machinery composing the plant are displayed.

The system was encoded in Lisp, and implemented on a mainframe NEC ACOS1000 computer system. After that, the system was ported to a Sun workstation.

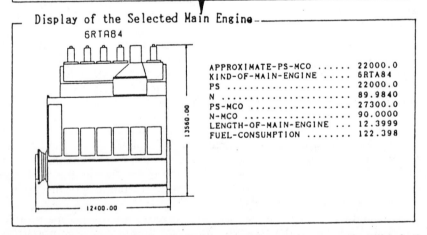

┌─ Design Condition ──────────────────────┐
│   KIND-OF-SHIP .... CONTAIN   │
│   REQUIRED-SERVICE-SPEED .. 22.0000   │
│   FULL-LOAD-DRAFT .... 11.5999   │
│   APPROXIMATE-PS-MCO ...... 22000.0   │
└──┘

┌─ Search of Candidates for Main Engine ──────────┐
│   >>> OK! I FOUND CANDIDATE OF MAIN-ENGINE.   │
│   9 ENGINES SATISFY THE REQUIREMENT.   │
│   *******CANDIDATE-OF-ENGINE*******   │
│   C-M*E-1 -- 5RTA84   │
│   C-M*E-2 -- 6RTA84   │
│   C-M*E-3 -- 7RTA84   │
│   C-M*E-4 -- 8RTA84   │
│   C-M*E-5 -- 7RTA76   │
│   C-M*E-6 -- 8RTA76   │
│   C-M*E-7 -- 9RTA76   │
│   C-M*E-8 -- 10RTA76   │
│   C-M*E-9 -- 9RTA68   │
│   ************************************   │
└──┘

┌─ Selection of the Optimal Main Engine ──────────────────────┐
│   >>> I RECOMMEND NEXT CADIDATES FROM POINT OF FUEL-CONSUMPTION│
│   C-M*E-4   8RTA84   120.4606   │
│   C-M*E-3   7RTA84   121.1673   [GRAM/PS/HOUR] .   │
│   C-M*E-8   10RTA76   121.4716   │
│   │
│   >>> I RECOMMEND NEXT CADIDATES FROM POINT OF PRICE [YEN] .│
│   C-M*E-1   5RTA84   +0.63700000E+009   │
│   C-M*E-5   7RTA76   +0.72128000E+009   │
│   C-M*E-9   9RTA68   +0.74340000E+009   │
│   │
│   >>> I RECOMMEND NEXT CADIDATES FROM POINT OF TOTAL-COST [YEN] .│
│   C-M*E-2   6RTA84   +0.77563286E+009   │
│   C-M*E-1   5RTA84   +0.77752227E+009   │
│   .C-M*E-5   7RTA76   +0.78491508E+009   │
│   │
│   >>> INPUT NAME OF MAIN ENGINE .   │
│   ... ? C-M*E-2   │
└──┘

┌─ Display of the Selected Main Engine ──────────────────────┐
│   6RTA84   │
│   │
│   APPROXIMATE-PS-MCO ...... 22000.0   │
│   KIND-OF-MAIN-ENGINE ..... 6RTA84   │
│   PS .... 22000.0   │
│   N .... 89.9840   │
│   PS-MCO .... 27300.0   │
│   N-MCO .... 90.0000   │
│   LENGTH-OF-MAIN-ENGINE ... 12.3999   │
│   FUEL-CONSUMPTION .... 122.398   │
└──┘

Fig. 4.29 Sequential representation for design process for a marine power plant (1) (selection of a main engine).

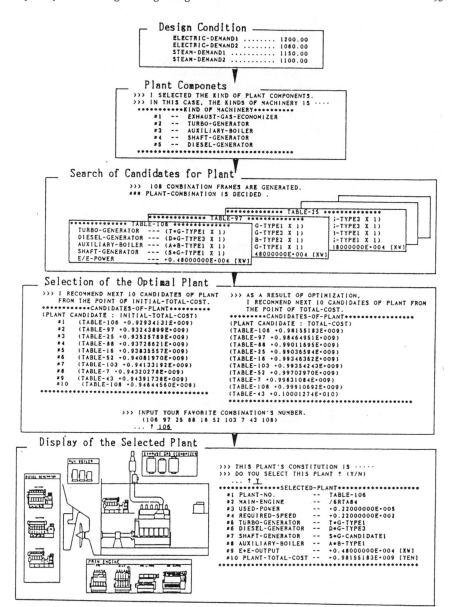

Fig. 4.30 Sequential representation for design process for a marine power plant (2) (design of heat and electric power generating plant).

Conclusion

Engineering design is a complex operation which includes various kinds of processing. The expert systems approach provides promising tools for expanding the effective application of computers to engineering design. In this chapter, an expert system using the object-oriented knowledge representation concept has been presented. This approach seems very suitable for the preliminary design stage for large engineering system design in which many constraints are handled to satisfy the design objectives.

Following an introduction of the general design system, which is effectively a shell, the applications to ship preliminary design and power plant design were presented. The former is characterized as a problem of determining design variables, i.e. the ship's principal details, interactively between a designer and a computer. The latter is characterized as a problem of searching for the candidate machines which compose a power plant based on an "intelligent" searching procedure. Through the above applications, the system's effectiveness has been clearly demonstrated.

References

1. Serrano D, Gossard DC. Constraint management in MCAE. In: Gero JS (ed.) Artificial intelligence in engineering: design. Elsevier Computational Mechanics Publications, 1988, pp 217–240
2. Akagi S, Fujita K. Building an expert system for engineering design based on the object-oriented knowledge representation concept. Proceedings 1988 ASME design automation conference, 1988, pp 287–295
3. Akagi S, Fujita K. Building an expert system for the preliminary design of ships. AI EDAM, 1987; 1: 191–205
4. Akagi S, Fujita K, Kubonishi H. Building an expert system for power plant design. Proceedings 1988 ASME design automation conference, 1988, pp 297–302
5. Sutherland I. SKETCHPAD – a man–machine graphical interface. PhD thesis, MIT, 1963
6. Sussman GJ, Steele GL. Constraints – a language for expressing almost-hierarchical descriptions. Artif Intell 1980; 14: 1–39
7. Borning AH. THINGLAB – a computer oriented simulation laboratory. Stanford University, 1979
8. Cox BJ. Object oriented programming: an evolutionary approach. Addison-Wesley, 1986
9. Goldberg AJ, Robson D. Smalltalk-80: the language and its implementation. Addison-Wesley, 1983
10. Goldberg AJ, Robson D. Smalltalk-80: the interactive programming environment. Addison-Wesley, 1984
11. Weinreb D, Moon D. Object, message passing, and flavors. LISP machine manual, Symbolics Inc., 1981
12. Bobrow DG and Stefik M. The LOOPS manual, working paper, memo, KB-VLSI-81-13. Xerox Corp., 1983
13. Gossard DC, Serrano D. MATHPAK: an interactive preliminary design package.

Proceedings 1985 international computers in engineering conference and exhibit. ASME, 1985

14. Brown DC, Chandrasekaran B. Expert system for a class of mechanical design activity. In: Gero J (ed.) Knowledge engineering in computer-aided design. IFIP, 1987
15. Ward A, Seering W. Representing component types for design. In: ASME, 1987 design automation conference, Boston, MA
16. Chan WT, Paulsen Jr BC. Exploratory design using constraints, AI EDAM 1987: 1: 59–71.
17. Elias AL. Knowledge engineering of the aircraft design process. In: Kowalik JS (ed.) Knowledge based problem solving. Prentice Hall, 1986, pp 213–256
18. MaCallum KJ. Understanding relationships in marine system design. Proceedings 1st international marine system design conference, 1982, pp 1–9
19. Simon HA. The science of the artificial, 2nd edn. MIT Press, 1981
20. Baxton IL. Engineering economics and ship design. BSRA report, 1971
21. Akagi S, Tanaka T, Kubonishi H. A hybrid-type expert system for the design of marine power plants using AI technique and design optimization method. Proceedings 1987 ASME design automation conference, pp 425–433

Section B

Component and Product Design

This section contains four chapters. The first chapter, by Dong and Soom, treats the problem of component dimensional tolerancing and suggests AI techniques for automating the tedious tasks of tolerance analysis and synthesis. The second chapter, by Kroll et al., is about product design for assembly and describes an intelligent knowledge-based system for this purpose. The third chapter, by ElMaraghy, also deals with intelligent product design. The emphasis of the chapter, however, is on integrating design and manufacture. The chapter presents a framework for supporting intelligent product design and manufacture. The framework comprises a feature-based modeller for design, high-level design languages and an assortment of expert process planning programs. The Section concludes with the chapter by Huang and Brandon which focuses on the knowledge-based design of machine tools after considering wider issues of managing knowledge bases for machine design.

Chapter 5

Some Applications of Artificial Intelligence Techniques to Automatic Tolerance Analysis and Synthesis

Z. Dong and A. Soom

Introduction

Since no mechanical part can be manufactured with exact dimensions and perfect shape, tolerances, together with dimensions, are used to specify acceptable variations of part geometry. To ensure design functionality, assemblability and manufacturability, tolerances must be specified and checked. Currently, dimensions and tolerances are usually placed on design drawings of CAD systems in accordance with ANSI [1], ISO [2] or other national standards. Tolerance analysis is used to check the stack-up of related tolerances. The process of tolerance accumulation is modelled, and the accumulated tolerance is verified. Tolerance synthesis (or tolerance design), on the other hand, is the process of allocating tolerance values derived from design requirements in terms of functionality or assemblability among related design tolerances. It is a process of distributing a few known tolerance values.

Traditionally, tolerance analysis and synthesis have been conducted manually using tolerance charts [3]. Dimensional tolerances are grouped into dimension chains (related dimension/tolerance groups), and tolerance calculations are performed. It is often a time-consuming and error-prone process. This method is generally limited to linear dimensional tolerances. Also, tolerance charts are generally too complicated to be used in conjunction with statistical tolerancing. A number of analytical tolerance calculation methods and computer-assisted tolerance analysis programs have been developed [4–11]. Two-dimensional (2D) and three-dimensional (3D) tolerance calculations can also be performed using such programs. However, these tolerance analysis programs may require extensive effort on the part of the designer to set up tolerancing problems, or to construct equations for tolerance calculations. There are no guidelines for incorporating geometrical tolerances into these approaches. The computational methods used are often complex and slow. Also, most of these methods are aimed at tolerance analysis. Tolerance synthesis techniques, which can provide initial tolerance designs based on manufacturing knowledge have not been well studied. Methods of automating tolerance analysis and synthesis have not yet been developed. Although tolerances are specified on most drawings, tolerance analysis and synthesis techniques are not widely used in design.

While there are still a number of conceptual and mathematical difficulties associated with tolerance representation and verification, it is not inappropriate to consider ways in which artificial intelligence (AI) techniques could be used to support mechanical tolerancing activity. The application of AI techniques to two aspects of tolerance analysis and synthesis are discussed in this paper. These are: (i) the automation of tolerance information retrieval from computer-aided design (CAD) systems followed by tolerance relation analysis, and (ii) optimal tolerance design, incorporating manufacturing knowledge, with an expert system.

The modelling and implementation of tolerance relation analysis using the search tree technique is considered. The concept of a basic dimension/tolerance graph or the three branch graph (TBG) representation is described. The automatic generation of TBGs from retrieved CAD data is discussed. Search techniques for identifying related dimensions and tolerances are implemented. A design problem is used to illustrate the methodology through a prototype automatic tolerance analysis and synthesis program.

In considering optimal tolerance design, we examine the question of how to model required manufacturing knowledge. This includes the production cost–precision relations and production statistical distributions, and methods of associating manufacturing data with tolerance design using an expert system approach.

Tolerance Analysis and Synthesis Automation via Basic Dimension/Tolerance Graph Formation and Search

Tolerance analysis or tolerance synthesis can be divided into two main steps. The first step is tolerance problem interpretation. This includes tolerance information retrieval from a CAD system or mechanical drawings, followed by tolerance relation analysis, in which all tolerances related to a specified design function or assembly requirement are identified. The second step is to perform appropriate stack-up or tolerance distribution calculations. Tolerance relation analysis is a major obstacle to automatic tolerance analysis and synthesis.

The search tree technique [12–14] is often used in AI to model exploring problems. The availability of a number of efficient search methods associated with this technique makes it attractive for tolerance relation analysis. The identification of related tolerances by this technique can be made quite efficient compared with more conventional numerical methods.

Tolerance Relation Analysis Modelling Using the Search Tree Technique

The search tree technique is characterized by an initial state, a goal state, and a search space diagrammed as a tree, with each node representing an intermediate state. Each state is connected to its predecessor and successor states by arcs. To model and solve a problem using the search tree technique one must represent the problem as a search space (usually a tree), and implement a search scheme appropriate to that modelling.

The representation of dimension and tolerance information via a basic dimension/tolerance graph

To model the tolerance relation analysis problem using the search tree technique, the dimension/tolerance hierarchy of mechanical parts and assemblies is represented as a basic dimension/tolerance graph, called the three branch graph (TBG). The TBG of a part is illustrated schematically in Fig. 5.1.

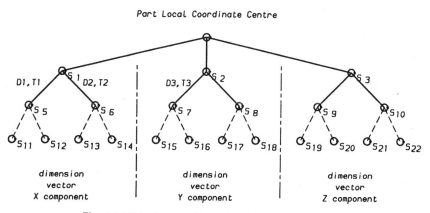

Fig. 5.1 TBG of a part dimension/tolerance hierarchy.

In the TBS representation, each dimension/tolerance of a designed part or assembly is represented as an arc. Surfaces (edges) associated with dimension and tolerance pairs are represented by nodes. Since a part or an assembly may be over-dimensioned, multiple-predecessors may exist for a node. A graph rather than a tree is used for the representation and subsequent search. Each graph in the TBG represents the dimension hierarchy of a part (or assembly) in one of the three orthogonal design coordinate directions (x, y, z). The root represents a part coordinate centre (local coordinate centre) or an assembly coordinate centre (global coordinate centre).

In order for the TBG approach to be applicable to general tolerancing work, the concept of basic tolerances is introduced. Basic tolerances are identical in form to dimensional tolerances, but specified along only one of the three orthogonal design coordinates. They are intended to cover all part geometry variations and assembly relations. These include dimensional tolerances along one of the three orthogonal design coordinates, functionally equivalent counterparts of nonlinear tolerances (e.g., angular dimensions and tolerances, dimensions and tolerances skewed with respect to design coordinates), geometric tolerances of location and orientation, and all specified assembly relations of physical contact and different levels of fit. Nonlinear dimensional tolerances (polar dimension/tolerance, angular dimension/tolerance) and geometric tolerances are transformed or otherwise converted to obtain basic tolerances.

The transformation of nonlinear dimensions and tolerances to basic dimensions and tolerances basically involves the calculation of polar dimensions and tolerances and angular dimensions and tolerances, projected onto x, y, z design coordinates. These transformations are not intended to provide an exact mapping of permissible variations. Original permissible tolerance zones are replaced by somewhat smaller tolerance zones to ensure a conservative approach. Some of the originally allowed variation is lost. We call these losses variation sacrifices. Since these variation sacrifices tend to be small for statistical tolerance analysis problems, typically less than 10%, this approach is acceptable in many situations.

The same principle is used in connection with geometric tolerance conversion. The variation originally specified as a circular tolerance zone by a geometric tolerance of location is converted into its exscribed square area with some variation sacrifice. The conversion of geometric tolerances of orientation is more complicated and case-dependent. However, the same principle is used.

Detailed discussion of these transformation and conversion methods is beyond the scope of this chapter.

Mechanical part and assembly TBG generation

Part and assembly TBGs can be automatically constructed after the design dimension and tolerance information has been retrieved from a CAD system and preprocessed. The construction of single part TBGs is straightforward. Assembly TBG construction includes the construction of all component part TBGs and the interconnection of these part TBGs. In the present approach, dimension and tolerance data are obtained from the textual information on the component part drawings, and the part assembly relations are inferred from the pictorial information (edge coordinates) on assembly drawings plus any basic tolerances that have been specified on the assembly drawing itself. No assembly description language is used and the system need not "understand" the assembly relations as long as the tolerance values used to specify these assembly relations are provided. Comprehensive assembly description work is replaced by an ordinary assembly drawing with some additional specifications required. Labelling of components in an assembly and the generation of a component list table is performed with the aid of an interactive program. Through this program, the positions of component parts in the assembly global coordinate are specified by probing the local coordinate centres of the components on the assembly drawing. The component list of the assembly is also generated in a database. Next, various assembly relations among components of an assembly are modelled via the mating relations of their surfaces. These can include coincidence, clearance, transition and interference as listed in Table 1.

Table 1. Modelling of assembly relations using basic tolerance

Physical condition	Classification	Tolerances
Contact	Coincidence	$T_u = T_l = 0$
Gap fit	Clearance	$T_u > T_l > 0$
Loose fit	Transition	$T_u < 0; T_l > 0$
Tight fit	Interference	$T_u < T_l < 0$

In the assembly drawing, the physical contact of two parts is specified by drawing the contacting surfaces of mating parts to be coincident. Other assembly relations, such as gap fit, loose fit and tight fit are either explicitly specified by a dimension line, with a tolerance representing the fit, or implicitly specified by the component dimensions and tolerances specified on the part drawings. Since it is the assembly result that is represented on the assembly drawing, the assembly sequence is implicitly determined. Various assembly sequences with possible different results may be represented on assembly drawings by using slightly different specifications.

The assembly TBG can be formed by the following procedure:

Step 1. Retrieve dimension and tolerance information from the assembly drawing and decompose nonlinear dimensions/tolerances.

Step 2. Generate the initial assembly TBG with the information obtained in step 1.

Step 3. Retrieve the assembly component list.

Step 4. For each component of the assembly:

Retrieve the part dimensions and tolerance information from the part drawing and decompose nonlinear dimensions/tolerances

Transfer the witness line or arrow line drawing coordinates of dimensions and tolerances into relative coordinates using the part local coordinate centre specified in the assembly drawing as the reference

Rescale all drawing coordinates of dimensions and tolerances to full scale

Change geometric tolerances into basic tolerances

Construct the part TBG with linear dimensions and basic tolerances from dimension/tolerance decomposition and geometric tolerance transformation (create a TBG node for each witness line drawing coordinate in the x, y, z directions)

Locate coordinates of a part local coordinate centre in the assembly global coordinate system

Transfer the coordinates of part TBG nodes into the assembly global coordinates via coordinate transformation

Rescale the coordinates of part TBG nodes to the assembly drawing scale

Download the part TBG to the existing assembly TBG according to its node coordinates in the assembly drawing to form an updated assembly TBG.

Step 5. Connect pairs of nodes on the current assembly TBG using a basic tolerance with zero tolerance value, if they share same coordinates (coincidence surfaces).

Step 6. Connect each node of the basic tolerances specified on the assembly drawing to represent the assembly relations to the nodes of transferred and rescaled part TBGs if they have same coordinates.

A comprehensive assembly TBG has now been formed.

Design dimension/tolerance relation analysis using TBG search

The introduction of the TBG representation permits related dimensions to be identified and grouped using search techniques. An interrogated clearance in tolerance analysis or a specified clearance between two surfaces in tolerance synthesis is treated as a basic dimension and is represented as two nodes on the graph. One surface is the initial node, and another surface is the goal node. Performing a node-oriented, path-remembering, depth-first search through an assembly TBG yields all related dimensions. Tolerance calculations for analysis or synthesis with these grouped dimensions then becomes relatively straightforward.

Another application of the TBG representation is design dimension redundancy verification. By specifying two surfaces in a design as the initial and the goal nodes and performing all possible depth-first searches in the TBG, one can identify a proper dimensional specification by the existence of only one path. Multiple paths imply over-dimensioning and absence of a path represents under-dimensioning. This method can eliminate the confusion that could be caused by the absence of necessary dimensions or the presence of conflicting and redundant dimensions.

Implementation of Basic Dimension/Tolerance Graph Formation and Search

Implementation of basic dimension/tolerance graph formation and search includes tolerance information retrieval from CAD systems, TBG representation in a computer database, and TBG search algorithm implementation.

Tolerance information retrieval from CAD systems

The first step in tolerance analysis and synthesis automation is to automate the retrieval of all dimension and tolerance information from CAD systems. This information retrieval work can be conducted from 2D CAD packages or 3D CAD modelling systems, from low level system dependent CAD databases, or high level general CAD databases, such as IGES, PDES, or from parametric (feature-based) CAD systems. In our implementation, IGES files from general CAD systems are used, since most of today's CAD systems are equipped with IGES interfaces. The IGES file structure makes dimension and tolerance information retrieval feasible, if not always easy.

In IGES, dimension and tolerance information is represented as annotation entities, such as linear, ordinate, point, angular, diameter, radius dimension entities; witness line, leader entity; or a general note entity. Each dimension and tolerance cluster includes the nominal dimension, tolerances, coordinates of the leader line and witness lines [15].

Two interface programs have been developed. The first one is a preprocessing IGES interface, through which all dimension and tolerance data on a CAD drawing are retrieved along with their witness lines or leader line coordinates. These coordinates are used to determine relations among dimensions and tolerances as discussed above. The second is a postprocessing IGES interface. Tolerance calculation results are used to compare and modify the original IGES file, and then produce a new CAD drawing showing the calculated tolerances. Tolerance analysis and synthesis questions are also passed through CAD drawings and IGES files. For example, in tolerance analysis, an interrogated clearance is specified as a regular dimension line with the dimension value as a question mark"?". This "dimension" is then treated as a resultant dimension which results from the stack up of a group of related dimensions. In the TBG, the two end nodes are not connected by an "arc", but exist as "nodes" associated with other dimensions/tolerances. The search will start and end at these two nodes. After tolerance calculations, the calculated tolerance is used to replace the question mark in the IGES file. Regeneration of the design drawing from this file feeds the tolerance analysis result back to the designer.

Two other interactive programs have been developed to assist the designer. One specifies geometric tolerances (generating the feature control frame) and the other specifies component name, location, and generating component list table in assembly design.

Graph representation and search implementations

The graph representation and search are implemented in a Lisp programming environment. Assembly, part, dimension and tolerance are stored as atoms with their property lists. The property list for an *assembly* A_i is:

Components:
$((P_1,$ X_Coor, Y_Coor, Z_Coor,$)$, $(P_2,$ X_Coor, Y_Coor, Z_Coor,$)$, ... $)$

Dimension_In_X_Direction:	$(D_{x1}, D_{x2}, \ldots)$
Dimension_In_Y_Direction:	$(D_{y1}, D_{y2}, \ldots)$
Dimension_In_Z_Direction:	$(D_{z1}, D_{z2}, \ldots)$
Node_In_X_Direction:	$(N_{x1}, N_{x2}, \ldots)$
Node_In_Y_Direction:	$(N_{y1}, N_{y2}, \ldots)$
Node_In_Z_Direction:	$(N_{z1}, N_{z2}, \ldots)$

A *part* P_j has the same property list as an assembly, except for the components property.
The property list for a *dimension* D_k is:

Magnitude:	(Nominal_Dimension)
Tolerance:	(Lower_Tol, Upper_Tol)
End_Nodes:	(Node1, Node2)
End_Nodes_Coor:	(Node1_Coor, Node2_Coor)

For a *node* N_l:

      Neighbor: $(N_p, N_q, \ldots)$

      Belongs_To: $(D_i, D_j, \ldots)$

An Example

In order to illustrate the method, a sample tolerance analysis problem is presented. A grinder head design is defined by an assembly drawing, and several component part drawings, including the housing, spindle, front cap, rear cap, etc. The assembly relations, component list table and overall dimensions are specified on the assembly drawing as shown in Fig. 5.2.

The design details with dimensions and tolerances of each component are specified in part drawings as shown in Figs 5.3–5.6.

The component labels on the assembly drawing are generated by an interactive program, rather than simply drawn. Through this program, the locations of the component part local coordinate centres in the assembly coordinate system are also specified (not shown). A component list of the assembly is generated in the database using drawing names.

The tolerance analysis problem can be stated as: Given the designs of the housing, spindle, front and rear caps, and knowing the dimensions and tolerances of purchased front and rear bearings, (both $20^{+0.01}_{-0.01}$ mm), will the rubber washer (component No. 7) properly seal the grinder head?

To solve this problem, a tolerance analysis of the clearance between the housing and the front cap is needed. A dimension line is placed across the two surfaces with the question mark in place of a dimension value as shown in Fig. 5.2. Entering the automatic tolerance analysis and synthesis environment, dimension and tolerance data are retrieved from the CAD databases through the IGES interface and the dimension/tolerance information pre-processor. The assembly TBG is formed following the procedures discussed above under TBG generation. The assembly TBG, generated automatically in the x direction only is shown in Fig. 5.7 (parts which are not associated with this problem are not shown).

Through the TBG search, only nine dimensions and tolerances are found to be related to the "gap" in question. The tolerance stack up calculation is then performed using these nine related dimensions/tolerances. Using a worst-case approach, in this instance, the result shows that the possible variation of the "gap" is $2^{-0.1}_{-0.52}$ (or 1.48–1.9 mm). Since the rubber washer is 2 mm thick, with a 0.1–0.6 mm depression required to seal properly, the tolerance design is verified to be good. Otherwise, modifications of related design dimensions/tolerances would be required. The calculated value $2^{-0.1}_{-0.52}$ replaces the "?" in grinder head assembly drawing, providing feedback to the designer.

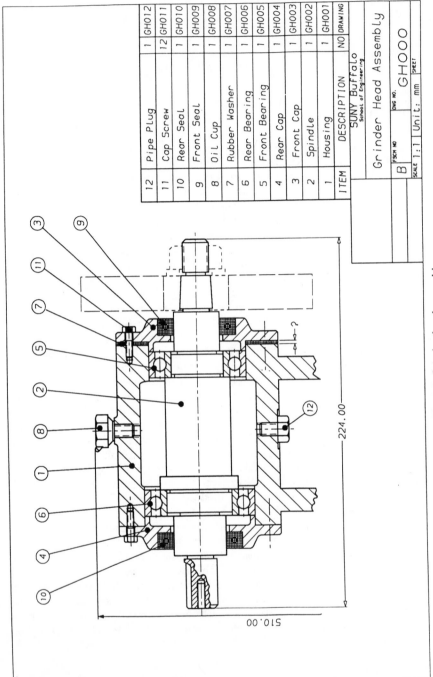

12	Pipe Plug	1	GH012
11	Cap Screw	12	GH011
10	Rear Seal	1	GH010
9	Front Seal	1	GH009
8	Oil Cup	1	GH008
7	Rubber Washer	1	GH007
6	Rear Bearing	1	GH006
5	Front Bearing	1	GH005
4	Rear Cap	1	GH004
3	Front Cap	1	GH003
2	Spindle	1	GH002
1	Housing	1	GH001
ITEM	DESCRIPTION	NO	DRAWING

SUNY Buffalo
School of Engineering

Grinder Head Assembly

B   FSCM NO   DWG NO.   GHOOO
SCALE 1:1   Unit: mm   SHEET

224.00

510.00

Fig. 5.2 Grinder head assembly.

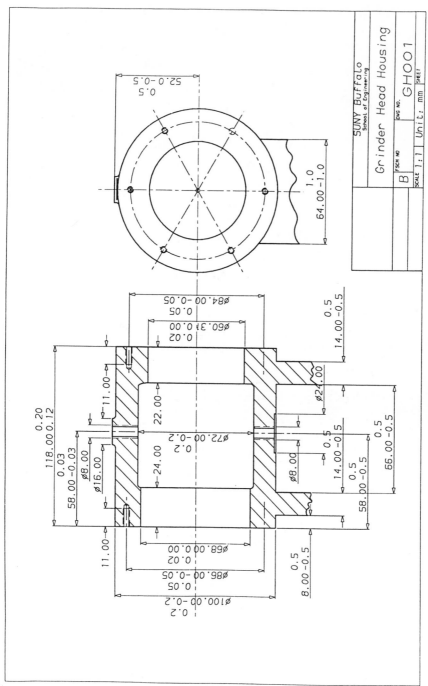

Fig. 5.3 Grinder head housing.

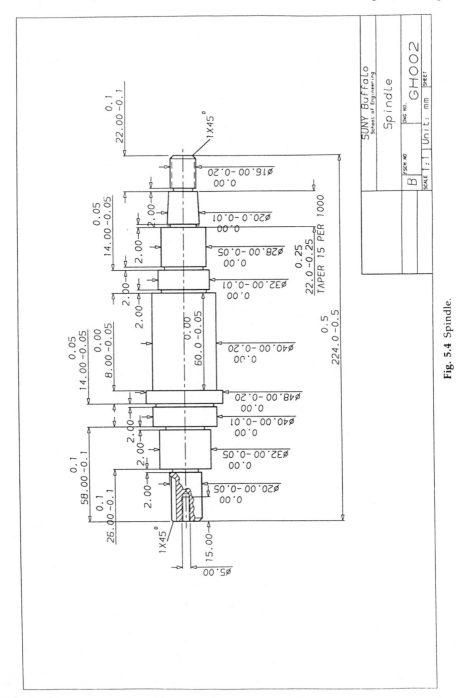

Fig. 5.4 Spindle.

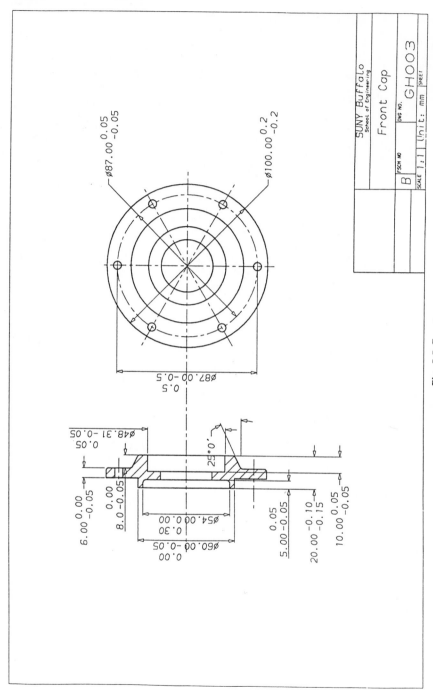

Fig. 5.5 Front cap.

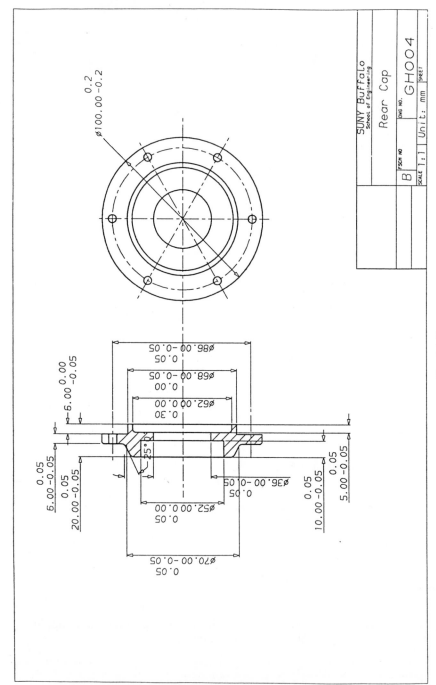

Fig. 5.6 Rear cap.

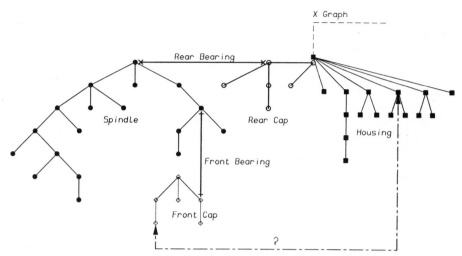

Fig. 5.7 Assembly TGT (of associated parts) in X branch.

Optimal Tolerance Design Incorporating Manufacturing Knowledge via an Expert System

While tolerances that have been specified can be compared with design requirements and modified on a trial and error basis, the initial distribution of tolerances among dimensions and geometric feature specifications is more difficult. As much as this task is a part of design activity, it is strongly related to manufacturing. Manufacturing knowledge should be incorporated into the tolerance allocation process. Traditional tolerance synthesis is based on some criterion, such as "equal precision" or "proportional scaling" [21], or may simply rely on the designer's judgement and experience. A good design may be obtained, while an "optimal" design, based on quantitative measures, cannot be expected. A number of elements must be combined for automated optimal tolerance design:

- *Manufacturing knowledge acquisition*: collection and classification of empirical production data.
- *Mathematical formulation*: mathematical modelling of production data.
- *Mathematical model and design tolerance association*: matching a particular design tolerance to the appropriate production data.
- *Solution method*: structuring the optimization problem and applying appropriate mathematical programming techniques.

Past work on optimal tolerance design has mostly focused on the mathematical formulations and associated optimization methods [16–20].

They have been well-summarized by Chase and Greenwood [21] and Wu et al. [22]. Methods of determining the parameters of these mathematical models and the means of associating these mathematical models with tolerances in design have not received very much attention.

In this section, we discuss methods to classify and utilize limited production data, to determine parameters of production cost–precision relations and production distributions, and automatically to associate dimensions and tolerances with these relations and distributions.

We propose to use a rule-based expert system [24, 25] with a numerical routine interface as the vehicle to incorporate and apply this manufacturing knowledge to tolerance design. This seems appropriate since the production data association process is of the deductive and reasoning type, whereas tolerance calculation is, of course, numerical. Therefore, an expert system with a numerical routine interface provides useful capabilities not offered by a purely numerical approach. Another advantage of expert systems is related to the fact that manufacturing knowledge and production data comprise dynamic information. Changes in production environment may result not only in new cost–precision data, but also new rules or logic to perform tolerance design. For example, with a piece of new equipment, a different part material or a different production site, more preconditions may have to be satisfied to associate production data with a particular design tolerance. With the separation of the inference program and knowledge base, an expert system provides the required flexibility to handle these situations conveniently.

Optimal Tolerance Design with Production Cost–Precision Models

Production cost–precision model

It is recognized that the smaller the allowed tolerance in production, the higher the manufacturing cost. Several mathematical formulations have been proposed to describe cost–precision relationships [22]. In our approach, an exponential function $g_i(\delta_i)$, is used to model the empirical production data. The formulation is

$$g_i(\delta_i) = A_i e^{-k_i(\delta_i - \delta_{0i})} + g_{0i}, \quad \delta_{ia} < \delta_i < \delta_{ib} \tag{1}$$

This model provides flexibility of curve shape and position control. A_i, k_i, δ_{0i}, and g_{0i} are four parameters controlling the shape and position of the cost–precision curve. They can be determined by curve fitting to experimental data. δ_{ia} and δ_{ib} define the valid region of the exponential curve. An earlier exponential model is due to Speckhart [23]. The present model differs from this earlier one by including the curve position control parameters and the range of validity.

Formulation of an optimal tolerance design problem

Given the design requirement defined as a clearance, δ_R (resultant dimension/tolerance) and the n related tolerances $\delta_1, \ldots, \delta_n$ (component dimensions/tolerances), an optimal tolerance design problem can be formulated. Among these component dimensions and tolerances, some tolerances $\delta_{p+1}, \ldots, \delta_n$ may be prespecified to meet other design requirements, while the remaining tolerances $\delta_1, \ldots, \delta_p$ are to be determined. A method of assigning values to these undetermined tolerances to obtain least manufacturing costs can be formulated as:

$$\min_{\text{w.r.t.} \delta_i} \quad G(\delta_1, \ldots, \delta_p) \tag{2}$$

Subject to

$$\delta_1 + \delta_2 + \ldots + \delta_n = \delta_R \tag{3}$$

and

$$\delta_{ia} < \delta_i < \delta_{ib}, i = 1, \ldots, p, p \leq n \tag{4}$$

where

$$G(\delta_1, \ldots, \delta_p) = \sum_{i=1}^{p} g_i(\delta_i) = \sum_{i=1}^{p} A_i e^{-k_i(\delta_i - \delta_{i0})} + \sum_{i=1}^{p} g_{0i} \tag{5}$$

$G(\delta_1, \ldots, \delta_p)$ is the total cost of producing the mechanical features associated with the adjustable design tolerances.

A multi-variate, nonlinear constrained optimization problem is formed. The final term is a constant and may be dropped off from the optimization. By solving the optimization problem, the optimal design tolerances can be determined.

Incorporating Manufacturing Knowledge into Tolerance Design

As was discussed above, the manufacturing knowlege required in optimal tolerance design may be divided into two categories: (i) production cost–precision relations, and (ii) statistical production distributions. It is necessary to develop both the mathematical formulations of production data and the rules to associate these models with design tolerances.

To simplify this association, all mechanical features are grouped into four basic features. There are plane features, external rotational surface features, hole features, and location features. In fact, most mechanical features can be either directly represented as or decomposed into these four basic features. The main exceptions are noncylindrical curved surfaces. The connection between a design tolerance and the production process to be used to produce that basic feature is established. The production data of that process can then be associated with the tolerance.

Production cost–precision relations of the four basic features are shown in Fig. 5.8. These relations are based largely on empirical data compiled by Trucks [26]. Cost–precision relations of production operations including rough turning, semi-finish turning, finish turning, cylindrical grinding, centreless grinding, cylindrical honing, drilling, boring, surface grinding and internal grinding are plotted. These data are not always complete and some are as much as 30 years old. However, they do provide quantitive measures of relative production costs of different processes for present purposes.

The cost–precision curves shown in Fig. 5.8 are based on relative production cost. For example, the production cost of rough turning with $\delta = \pm 0.03$ is set to be equal to one. The production costs of all empirical manufacturing data are converted into relative production cost with respect to this reference. Each basic feature has a continuous production cost–precision curve which may encompass several production operations. The first curve in Fig. 5.8 is the external rotational surface feature cost–precision curve. It covers the production operations of rough turning, semi-finish turning, finish turning, cylindrical grinding and cylindrical honing. The second curve is the hole feature cost–precision curve. Only data for internal grinding are available. Therefore, production data for the

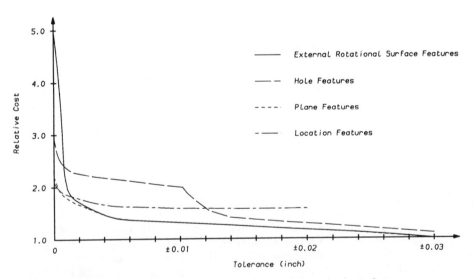

Fig. 5.8 Empirical production cost–precision relations for basic features.

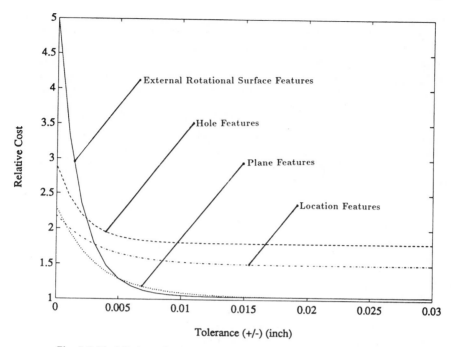

Fig. 5.9 Modelled production cost–precision relations for basic features.

external turning have also been used here for internal turning. For the plane surface feature generation, only face grinding data are available. Data for external turning have also been used to cover the range of face milling and face turning. The last curve represents the location feature. It is directly obtained from the plot of "cost relationship of true position tolerances in drilling and boring". Valid ranges of these models are also obtained from Trucks [26]. There is a definite need for updated and more complete cost–precision data to support tolerance design.

The curves generated from the corresponding exponential equations are plotted in Fig. 5.9. They provide a good match to the empirical curves of Fig. 5.8. The parameters of these curves are listed in Table 2.

Production cost–precision and distribution model parameters for basic mechanical features

For the basic features discussed above, parameters describing the production cost–precision relations and statistical production distributions are listed in two tables. The lengths are given in inches.

Exponential curve parameters A, k, δ_0, g_0, δ_a, δ_b of the production cost–precision relations described in equation (1) are listed in Table 2. The parameters of four fitted curves in Fig. 5.9 provide data for normal

Table 2. Exponential curve parameters A, k, δ_0, g_0, δ_a, δ_b

Dimension range	Feature type							
	T_1 Ext. rotational		T_2 Hole		T_3 Plane		T_4 Location	
$D < 1.5$	3.96	−550	1.8	−500	1.73	−320	0.68	−300
	0.00	0.79	−0.001	1.55	−0.001	0.79	0.0	1.25
	0.00015	0.008	0.00015	0.008	0.00015	0.008	0.00015	0.008
$1.5 \le D < 20$	3.96	−550	1.8	−500	1.73	−320	0.68	−300
	0.00	1.04	−0.001	1.8	−0.001	1.04	0.0	1.5
	0.0003	0.02	0.0003	0.02	0.0003	0.02	0.0003	0.02
$D \ge 20$	3.96	−550	1.8	−500	1.73	−320	0.68	−300
	0.00	1.29	−0.001	2.05	−0.001	1.29	0.0	1.75
	0.0007	0.03	0.0007	0.03	0.0007	0.03	0.0007	0.03

applications. Since the production cost of mechanical features would ordinarily increase with the increase of workpiece dimension (size), parameters are adjusted for small features and large features.

For statistical tolerance calculations, an advanced dimension distribution model is required to represent manufacturing knowledge. The beta distribution has been adopted here because of its ability to approximate a large variety of distribution shapes. It has been used previously in modelling statistical tolerances [4]. Instead of tolerance modelling per se, the major concern here is to determine the appropriate values of the beta distribution parameters and to associate the distribution parameters with a design tolerance.

Parameters of the beta statistical distribution model α and β are determined by production data in the manufacturing knowledge base. For example, a uniform distribution is approximated by, $\alpha = \beta = 1.0$. For normal distributions, one can use: (i) the best fit over the range $[0, 1]$, yielding $\alpha = \beta = 4.65$, or (ii) the best fit at the ends, equivalent to $\delta = \pm 3.1\sigma$, fit range: $[0, 0.15]$ and $[0.85, 1.0]$, yielding $\alpha = \beta = 4.10$. We normally use the latter fit.

When measured statistics are unavailable, a distribution between uniform distribution and normal distribution is recommended. In this case, $\alpha = \beta = 2.8$. For well-controlled processes, $\alpha = \beta = 3.5$ to 4 can be used. For uncertain processes, $\alpha = \beta = 1.5$ to 2 can be used. Since distributions tend to become somewhat flatter when the workpiece size increases, smaller α and β values, representing flatter distributions, are used to adjust for these changes in distribution. On the other hand, mechanical features with small tolerances tend to be produced with high precision tools, providing more concentrated distributions. Larger α and β values are then used. Due to the maximum material criterion often used to specify design features, or

Table 3. α and β values of the beta distribution-based distribution model

Dimension range	T_1 Ext. rotational		T_2 Hole		T_3 Plane		T_4 Location	
Small tolerance: $0.0005 \leq \delta < 0.003$								
$D < 1.5$	3.90	3.60	2.10	2.30	3.90	3.60	2.10	2.10
$1.5 \leq D < 20$	3.85	3.55	2.05	2.25	3.85	3.55	2.05	2.05
$D \geq 20$	3.80	3.50	2.00	2.20	3.80	3.50	2.00	2.00
Medium tolerance: $0.003 \leq \delta < 0.01$								
$D < 1.5$	3.85	3.55	2.05	2.25	3.85	3.55	2.05	2.05
$1.5 \leq D < 20$	3.80	3.50	2.00	2.20	3.80	3.50	2.00	2.00
$D \geq 20$	3.75	3.45	1.95	2.15	3.75	3.45	1.95	1.95
Large tolerance: $0.01 \leq \delta < 0.03$								
$D < 1.5$	3.80	3.50	2.00	2.20	3.80	3.50	2.00	2.00
$1.5 \leq D < 20$	3.75	3.45	1.95	2.15	3.75	3.45	1.95	1.95
$D \geq 20$	3.70	3.40	1.90	2.10	3.70	3.40	1.90	1.90

due to tool wear during production, skewed distributions associated with mean shift may occur. External mechanical features, such as external rotational surfaces, blocks, etc. tend to be oversize, resulting in right-skew distributions, i.e. $\alpha > \beta$ and internal mechanical features, such as holes, slots, etc. tend to be undersize, resulting in left-skew distributions, i.e. $\alpha < \beta$. Normal distribution can also be used directly with $\delta = \pm 3\sigma$. Both beta distribution and normal distribution calculations are implemented through library programs. The recommended distribution parameters α and β of the beta distribution are listed in Table 3.

Design tolerances and basic feature association

The association of basic features with design tolerances can be performed in connection with the dimension and tolerance retrieval through IGES. Diameter dimension/tolerance data clusters are associated with external rotational surface features and hole features. Further distinctions can be made by requiring that a text "h" (for hole features) be inserted on the drawings at the end of diameter dimension and tolerance texts. Geometric tolerances of location are associated with the location features. Dimensions and tolerances specified by other kinds of data clusters are usually associated with plane features. As parametric design systems become more widely available, more complicated associations between design tolerances and manufacturing knowlege can be made.

Master production rules used in manufacturing knowledge association

After dimensions and tolerances have been retrieved from CAD databases, and related dimensions and tolerances have been found through the TBG search, each tolerance is assigned to a production cost–precision model and a production distribution model (if statistical calculations are required). Production rules based on Tables 2 and 3 can take the following forms:

Type I rules: If a tolerance is in the related dimension/tolerance group, and its value is to be determined, assign the production cost–precision model parameters (A, k, δ_0, g_0, δ_a, δ_b) according to its basic feature type.

Type II rules: If a statistical tolerance calculation is required, the tolerance is in the related dimension/tolerance group, and its value is to be determined, assign the distribution parameters α and β according to the basic feature type, tolerance range and nominal dimension.

Type III rules: If there is any change of the tolerances in related dimension/tolerance group after the tolerance calculation, decide whether the iterative tolerance calculation is necessary.

Rules in the first two groups perform the design tolerance and production data association. Rules in the last group handle possible changes of deduction conditions.

Overview of the Prototype Automatic Tolerance Analysis and Synthesis System

The tolerance analysis and synthesis automation and the optimal tolerance design methods discussed in this chapter have been implemented in a prototype automatic tolerance analysis and synthesis system. The structure of this system is illustrated in Fig. 5.10.

At the top centre is the expert system core. It can perform inference based on the current state of the tolerance design, stored in the dynamic manufacturing database, and the production rules describing the manufacturing environment, stored in the static manufacturing database.

The CAD database of a design will be transferred and stored in the standard IGES format. Then, as shown on the left-hand side of the figure, tolerance information is retrieved and processed as was described above. For an assembly, the TBG is created from its component TBGs and associated assembly relations. Design tolerances which have been specified or estimated will be stored in the dynamic manufacturing database for further calculation.

As shown on the top right-hand side of the figure, the generic manufacturing data required for tolerance analysis and tolerance synthesis

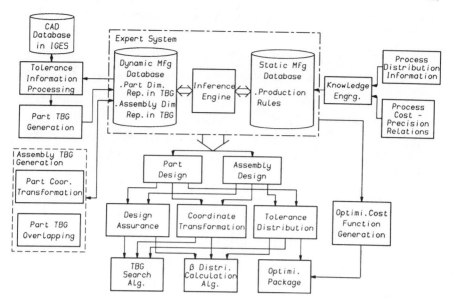

Fig. 5.10 Prototype automatic tolerance analysis and synthesis system.

is acquired, classified, and stored in the static manufacturing database in the form of production rules of the expert system.

When one of the tolerance design problems is selected from the centre of the figure, a TBG search is first performed to identify the related dimensions and tolerances. The necessary inference is then performed to assign each tolerance appropriate production model parameters. Next, tolerance calculations applying either the worst-case approach or the statistical approach to the analysis or synthesis problem are performed. The calculated tolerances are stored in the dynamic manufacturing database and fed back to the designer.

Acknowledgement. One of the authors, A. Soom, would like to acknowledge the National Science Foundation, Grant No. DMC-8603511, which partly supported this work.

References

1. American National Standard Institute. Dimensioning and tolerancing. ANSI Y14.5M, The American Society of Mechanical Engineers, New York, 1982
2. ISO-129 Dimensioning, ISO-1101 Geometrical tolerancing
3. Wade OR. Tolerance control in design and manufacturing. Industrial Press, New York, 1967

4. Bjorke O. Computer-aided tolerance. Tapir Publishers, 1978
5. Hillyard RC and Braid IC. Analysis of dimensions and tolerances in computer-aided mechanical design. Comput-Aided Des 1978; 10: 161–166
6. Hoffmann P. Analysis of tolerance and process inaccuracies in discrete part manufacturing. Comput-Aided Des 1982; 14: 83–88
7. Parkinson DB. Tolerancing of component dimensions in CAD. Comput-Aided Des 1984; 16: 25–32
8. Dong Z, Soom A. Automatic tolerance analysis from a CAD database. Paper presented at the 12th design automation conference in Columbus, Ohio, 1986, ASME 86-DET-36
9. Turner JU, Wozny MJ. A mathematical theory of tolerances. In: Wozny M, McLaughlin H, Encarnacao J, (eds) Proceedings IFIP G 5.2 working conference on geometric modelling for CAD applications, North-Holland, 1987
10. Greenwood WH, Chase KW. A new tolerance analysis method for designers and manufacturers, J Eng Indust 1987; May: 109–113
11. An introduction to VSA^{TM} variation simulation analysis software. Applied Computer Solutions, Inc., Clair Shores, MI, 1987
12. Nilsson NJ. Principles of artificial intelligence. Springer-Verlag, 1982
13. Winston PH. Artificial intelligence. Addison-Wesley, 1984
14. Charniak E, McDermott D. Introduction to artificial intelligence. Addison-Wesley, 1985
15. Smith B, Wellington J. Initial graphics exchange specification (IGES). Version 3.0, NBSIR 86-3359, US Department of Commerce, April, 1986
16. Spotts MF. Allocation of tolerances to minimize cost of assembly. ASME J Eng Indust 1973; August: 762–764
17. Bandler J. Optimization of design tolerances using nonlinear programming. J Optim Theory Applic, 1974; 14: 99–114
18. Michael W, Siddall JN. The optimization problem with optimal tolerance assignment and full acceptances. ASME J Mech Des, 1981; 103: 842–848
19. Michael W, Siddall JN. The optimal tolerance assignment with less than full acceptances. ASME J Mech Des 1982; 104: 855–860
20. Parkinson DB. Assessment and optimization of dimensional tolerances. Comput-Aided Des 1985; 17: 191–199
21. Chase KW, Greenwood WH. Design issues in mechanical tolerance analysis. Manuf Rev 1988; 1: 50–59
22. Wu Z, Elmaraghy WH, Elmaraghy HA. Evaluation of cost-tolerance algorithms for design tolerance analysis and synthesis. Manuf Rev 1988; 1: 168–179
23. Speckhart FH. Calculation of tolerance based on a minimum cost approach. ASME J Eng Indust 1972; 94: 447–453
24. Forsyth R. Expert systems principles and case studies. Chapman and Hall, 1984
25. Waterman DA. A guide to expert systems. Addison-Wesley, 1986
26. Trucks HE. Designing for economical production. Smith HB (ed), Society of Manufacturing Engineers, Dearborn, MI, 1976

## Chapter 6

Intelligent Analysis and Synthesis Tools for Assembly-Oriented Design

E. Kroll, E. Lenz and J.R. Wolberg

Introduction

Designing products for easier assembly is recognized as an important part of "simultaneous engineering": the process of making products meet functional, manufacturing, quality and cost targets at the early stages of design. Considering the assembly aspects of existing and new products may lead, for example, to the use of simpler robots with fewer tools and grippers, and less costly fixtures, or indeed, to abandoning robots in favour of alternative process equipment.

Although the knowledge of the area is fairly established, current design-for-assembly methods exhibit basic weaknesses when viewed as design aids, because they do not include adequate models of the product and the assembly process. Automatically advising the designer about how to improve the product must be based upon comprehensive knowledge of two areas: understanding the functional role and mechanical principles associated with each part, and having a satisfactory model of the relevant assembly process with regard to operations, tools, etc. But acquisition of such information by a computer program poses some difficulties. First, the amount of knowledge is enormous. Second, it consists of formal as well as heuristic aspects. Third, both the domain knowledge and the particular circumstances of the design process may be constantly changing: designing a product to be assembled by a simple SCARA-type robot, for example, could not be the same as for an assembly cell with a pair of six degree-of-freedom robots. Moreover, such a program should also simulate decision-making stages of human design processes which are certainly not well understood.

In this chapter, a knowledge-based consultant system is described which improves upon the current methods [1]. By efficiently utilizing extensive domain knowledge, the system can "understand" high-level issues concerning the structure and character of the product and the goals of the specific design process, and provide design guidance accordingly. The emphasis in this work is on creating suitable models for the product and its assembly requirements, and on integrating them in an environment of analysis and synthesis tools, thus modelling a complete design process. As appropriate to a knowledge-based solution of any problem, the scope of the system is limited. If one tries to develop a completely general system, the knowledge base becomes unwieldy. The current system has therefore been limited to the class of axisymmetric, layered-structured products (e.g., electric motors). Emphasis is also put on robotic assembly, although most of the knowledge and the methodology presented are relevant to other assembly processes as well. Being a prototypical implementation, programming is in Turbo-Prolog on an IBM-XT. Clearly, a full implementation could utilize a more powerful software–hardware combination.

Related Work

Previous work from several areas is pertinent to the current research. These include contemporary design-for-assembly methods, ways of representing parts and assemblies, and systems for automatic assembly planning. Various investigations into knowledge-based design method-ologies are also closely related, but will not be surveyed here.

The methods which have been suggested and implemented in the past for guiding product design-for-assembly represent two different approaches to the problem. The first is a qualitative approach which presents the designer with general rules and guidelines accompanied by illustrated examples [2–6]. This method is often considered too general to be practically applied during design. Merely presenting the guidelines to the engineer, whether on paper or on a computer monitor, falls short of providing a useful methodology. Moreover, general rules are by nature more prone to misinterpretation.

The second approach is quantitative, assigning time periods, cost and numerical codes to various part characteristics and assembly operations [7–11]. This method requires very specific information such as the expected production rate, the cost of assembly hardware, and symmetry properties of components. Its two main drawbacks are the very implicit way of identifying design improvements, making the method more suitable for analysis than for actual design work, and, even more fundamental, its inability to treat products at a higher level than the individual parts. As a result, configurative design can only take place by elimination or integration of components. *Despite their weaknesses, the experience accumulated by using both approaches, together with numerous case studies (see [12] and [13] for recent examples) are the principal source of assembly-oriented design knowledge to be incorporated in any related work.*

More recently, the knowledge-based approach has also been applied to the design-for-assembly problem. The work by Swift [14] is oriented towards advising the designer of difficulties in the automatic handling of components, suggesting remedies, and estimating the cost of the required handling equipment. Jakiela and Papalambros [15] are more concerned with the integration of a knowledge-based design-for-assembly consultant system in a conventional CAD environment. There, as soon as the designer adds a feature to the part, the system numerically estimates the corresponding ease of assembly, and is capable of suggesting improvements in an optimized manner. However, most of the knowledge acquired by these systems is based on the quantitative methods, rendering them oriented more towards analysing and improving the detailed design of individual components, and less for configurative-level product design. Another work [16] suggests improving the qualitative approach by using an expert system for the introduction of relevant design-for-assembly rules to designers.

Advanced techniques for describing parts and products have been separately studied. Many researchers concede that a feature-based representation of geometry is most appropriate for engineering applications [17,18]. However, it seems that each application requires its own set of features which are meaningful to the particular activities involved [19]. A controversy exists between those preferring extraction of the required features from conventional computer-aided design (CAD) data [20], and those who are in favour of designing with features in the first place [21].

High-level data structures to represent assemblies rather than individual parts have not been well established. Traditionally, one would state the location and orientation of each component in the assembly by a homogeneous transformation matrix, an awkward and error prone process. The idea of specifying high-level relationships among components has been described by Wesley et al. [22]. Lee and Gossard [23] have suggested the use of mating conditions such as "against" and "fits" to express the relationships inside an assembly. Rocheleau and Lee [24] have shown how assemblies could be represented in a hierarchical tree structure, and each component's location and orientation computed, after mating conditions have been determined.

The common approaches to the problem of automatic generation of assembly plans fall into two categories: those using "heavy" tools and very little "understanding", and those requiring comprehensive knowledge of the product structure and characteristics, but reliant on the user to provide it. A typical system of the first kind starts with a conventional CAD representation of parts and assemblies, and simulates a disassembly process [25–27]. The program attempts to "move" parts away by calculating new bodies, having the swept shape of the original parts in the direction of disassembly, and continuously checking for interference between them and all other, still assembled parts. Reversing the order of parts "removal" from the assembly then yields an assembly sequence. But the required disassembly trials can be quite exhaustive, and the whole process lacks "understanding" of real-life machine parts. For example, the program might try to "pull" a bolt by its head, not knowing that it has to be rotated while counter-torqued at the nut.

A way to overcome the drawbacks of the "unsmart" methods has been presented in another work [28]. There, contact relationships between parts are entered by the user of a computer program to create a "liaisons" diagram. An elaborate question-answering process is then used to determine precedences between part installation moves. It is at this stage that the real understanding of the product structure is required, something that is left for the user to exhibit by answering the following two questions for each liaison. (1) What liaisons must be completed prior to accomplishing the current liaison? (2) What liaisons must be left until after the current liaison? Answering these questions is certainly nontrivial even for experienced engineers. Finally, sequences of establishing all the liaisons are algorithmically generated subject to the constraining precedences. In this fashion, all possible assembly sequences are deduced.

A more theoretical basis for correlating a product's description with its sequence of assembly has been suggested by Haynes and Morris [29]. Assemblies are first defined based upon the concept of degrees of freedom of components (actually, an expanded set of mating conditions). Then, assembly operations are regarded as reductions of these degrees of freedom, enabling the generation of a task-level robot program for assembling the product.

High-Level Product Modelling

Although common practice for constructing the main body of knowledge in an expert system is to use rule-based programming, two other issues concerning knowledge representation remain to be solved: (1) What should be the starting point for design? (2) What data structures will be used to describe the product throughout the process?

Ideally, the description of a product is stored in a CAD database. But contemporary CAD systems usually do that at a much lower level than what might be considered convenient for design tasks and human communication. For example, engineers rarely think in terms of solid volumetric primitives, but rather they may specify features having engineering meaning like "equally spaced holes on a pitch circle", "V-belt pulley", etc. Consequently, we decided to start with natural feature-based engineering descriptions for the parts. Besides defining the exact geometry of each part, a topological characterization of the product is also required. Four "standard" mating conditions serve to define relationships among parts and their features: "above" (to describe planar contact), "loosely-fits-in" and "tightly-fits-in" (for the link between cylindrical surfaces), and "threaded-in".

To represent the complete assembled product, a structured objects scheme has been devised, with each part constituting a "frame" with "slots" for several attributes. Some of the slots are filled with the aid of a preprepared "library" of features which contains definitions of "basic bodies" (e.g., hollow cylinder, cup, hex nut) and features like eccentric threaded hole, cylindrical cavity, etc. When choosing a certain feature from the menu, the user is prompted for the required parameters of that feature, with default values shown as well. Some use is also made at this stage of inheritance of properties from stored prototypes of standard components to their instances. For example, the frame-like data structure for the bearing in Fig. 6.1 might be:

```
FRAME TYPE:  part
   NUMBER:  7
     NAME:  6304 ball bearing
 QUANTITY:  1
     TYPE:  standard rolling bearing
 MATERIAL:  irrelevant
 GEOMETRY:  basic body: rolling bearing(52,20,15)
               features: "outer ring", ring(52,40,15,0)
                         "rolling elements", ring(40,30,15,0)
                         "inner ring", ring(30,20,15,0)
 TOPOLOGY:  above("inner ring", shaft("shoulder"))
            tightly-fits-in("outer ring", housing("bearing
            mount"))
            assembled location: world(0,0,−37.5)
```

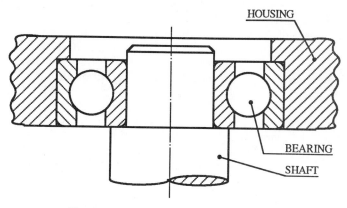

Fig. 6.1 A ball-bearing, housing and shaft.

The geometric description consists of the basic body from the library, with suitable parameters (outer diameter, inner diameter, width), and features with corresponding values. Note the definition of the rolling elements feature as a "ring", which is actually the shape of the envelope, but is satisfactory for our purposes. The topology slot defines the mating conditions of each feature with other part features. What seems to be missing, like the outer ring being "below" the housing, is actually entered as an "above" condition when the housing is described. Each mating condition name appears with two arguments: the name of the feature (e.g., "inner ring") which belongs to the current part, and a compound term having another part's name ("shaft") with its relevant feature ("shoulder").

Such representation ensures relatively compact data structures while still maintaining a lot of expressive power. However, having a thorough description of the product does not mean such information is needed at all times. The human design process is characterized by having varying levels of product abstraction at different stages. Starting with general concepts of the product's structure, the description is gradually refined by including more and more details. As described in the next section, the present system employs various representations.

Design Process Modelling via System Architecture

The major consideration while developing the architecture of the consultant system has been a desire to model the design process adequately. From simulating design-for-assembly processes of several example products, a model of the required phases has been established and implemented via the system architecture. Inherent characteristics of human design processes, such as the varying levels of product abstraction and the iterative nature, have also been accommodated.

It turns out that even after a product description has been created, some analysis must still precede the actual synthesis work. The single most important factor determining the ease of assembly of products seems to be their assembly plan, i.e., the sequence of putting together the individual components, and the required part installation operations. In addition, the parts and subassembly composition of the product is also closely related. Thus, the first analysis stage is concerned with generating sequences of assembling the product to serve as a basis for improved structures. It is clear that the system must be capable of "understanding" the role and function of each part in the product if its advice is to be "intelligent". So, prior to considering design changes, functional analysis of the product must also be completed.

   The system software consists of nine programs or modules, two library
files, two working memory files, a connection to another computer, and
a main program to control the whole design process. This file structure
is schematically shown in Fig. 6.2. Normally, an existing or a new design
description is interactively entered, as explained in the previous section,
using the Initial Design module. Next, the product is analysed from the
assembly sequence viewpoint with the aid of the rule-based Assembly
Sequence Analyser module in two stages: first, a data structure representing an "exploded" layout of the product is created from its
assembled description now located in the working memory file. Then,
the exploded view is used for the automatic generation of assembly
sequences. The operation of this module is described in more detail in
the next section.

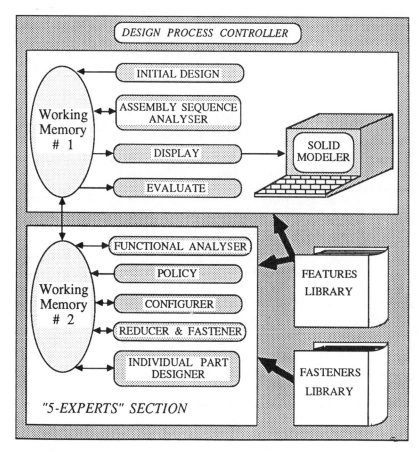

Fig. 6.2 System architecture to model the design-for-assembly process.

As one can hardly imagine an engineering design process carried out in words alone, provision is made through the Display module to draw the exploded view of the product on a graphics terminal. Each part's feature-based description is converted there to constructive solid geometry (CSG) format. (Note that turning an abstract description into a more detailed one is relatively easy.) Upon completion of proper scaling, the whole CSG file is sent via a communication link to another computer running a conventional solid modeller.

At this stage the user makes the decision whether to proceed to the actual design-for-assembly session. If so, the Evaluate module is bypassed, the current working memory is saved to serve as the basis for improved designs, and the system activates its "5-experts" section arranged in a hierarchical "blackboard" architecture. This means that several programs, simulating "experts" or "knowledge sources" in subdomains of the problem, share a common working memory, or "blackboard", where the design evolves. These "experts" have a distinct hierarchical ranking among them, with each module treating the product at a lower abstraction level (or higher detail level) than its predecessor.

Before design can start, functional analysis (as opposed to the structural analysis performed earlier) should be completed. This is accomplished by the Functional Analyser module, which employs extensive knowledge about various machine elements and their function, in order to distinguish between parts that perform supporting functions such as fasteners and spacers, and functionally important parts (e.g., housings, bearings, shafts). The second "expert", the Policy module, which is the first of the real synthesis programs, interrogates the user as to the general goals of the design process. This stage simulates company policy-level decision making, and produces goals to be satisfied later. Such policy goals might be: "try to design the product to have no subassemblies" or "do not change part X as it is an interfacing component to something else". Besides these general goals, the program also asks the user whether the product could be made a "throw-away" type of product, or if it should be easily dismantled for maintenance and repair. This has a significant influence when choosing fastening methods.

The third "expert" module, the Configurer, plays perhaps the most important role in the design process: making the initial decisions regarding the structure of the overall product. As with human design processes, the earlier the decision, the more significant is the effect that it will have on the results. Based on the functional model of the product created earlier, the program attempts to plan new configurations as instructed by the policy-level goals. This involves a complex procedure of checking the structure of the product to pinpoint conflicts between it and the intended simpler assembly sequence, presenting these findings to the user and suggesting remedies, kinematically sorting the functional model of the product (e.g., a stationary housing and a rotating shaft should be connected via a bearing), hypothesizing about "stacking" and "fastening" requirements, checking the latter against heuristic constraining rules, and finally, establishing goals to be satisfied by consecutive modules.

Down a level in the hierarchy, the Reducer & Fastener "expert" is responsible for two essential functions in any design-for-assembly process: reducing the number of parts in the product, and choosing suitable fastening methods. Although no truly superfluous components should exist in products, some parts (e.g., spacers) might be eliminated if their function can be incorporated as features of other parts. The program now uses information accumulated during previous stages, together with its own knowledge base, to make sensible recommendations such as integrating bearing-support spacers into shafts as steps, or into other parts, but not into the bearings themselves. The second task, choosing the fasteners, is somewhat more difficult. Here, previous goals define both the fastening needs and the product character (e.g., easily disassembled, throw-away). The program consults a library file of fastening methods with their characteristics, picks the potential ones, sorts them according to their ease of assembly, and asks the user to indicate a preference. When the choice is made, that fastener is incorporated into the product model, and goals are accordingly established for the next module.

The last "expert", the Individual Part Designer, embodies all previously accepted recommendations in each part design. Although some rules can accomplish this automatically, the program is highly interactive, consulting the user at each step. To some extent, this allows the human designer's creativity to participate in the process, after the intention of the program becomes clear. Besides satisfying the already established goals, general improvements in the handling, orientation, and assembly characteristics of each part can take place in this module. These improvements include turning asymmetric parts into symmetric ones, or exaggerating their asymmetric features, adding chamfers and tapers, "closing" the ends of springs, thus preventing parts from tangling and nesting during feeding, etc.

The product evolution by the last three "expert" modules includes a backtracking capability. When an "expert" module cannot satisfy goals dictated to it from further up in the hierarchy, the cause of failure is returned via the main control program to the higher "expert" for reconsideration of these decisions. If all goes well, the design process controller activates again the Assembly Sequence Analyser and Display modules. Then the last program, the Evaluate module, compares the original product design with the improved one, and produces a summary of the assembly needs such as the required operations, fixtures, tools and grippers, and the sequence with which the product is to be assembled.

Assembly Sequence Analysis

To predict the effect that any design decisions will have on the final product, a means of establishing the assembly requirements of evolving designs and identifying their "weak" points is needed. However, the problem of directly planning the assembly sequences is either very complicated computationally, or requires an "expert" user. The approach presented here is different, and involves two phases: an "exploded" layout of the assembled product is first found from its topological and geometric description by a graph-based technique. Then, "simple" sequences for putting together the individual components are planned by utilizing the theoretical as well as heuristic knowledge of mechanical components and assembly processes. An example demonstrates the application of the method to an actual product.

Generation of Exploded Views

Exploded views are a very common graphical aid in comprehending the structure of products, and are frequently used by engineers. Although experience shows that manual planning of such layouts is not particularly difficult, devising a computerized procedure for that purpose is nontrivial. Finding the order of the parts in a uni-axial, vertical exploded view, is analogous to asking: Out of all the parts in a product, which should be *above* which? Such a statement logically leads to using the "above" mating conditions of the product's topology. Demanding that each part in an assembly will appear at least once as one of the arguments of an "above" condition is quite sensible, because during assembly, a part usually has to be located against some planar surface. Conveniently, that portion of the initial information which expresses the "above" mating conditions, and the solution to the problem, i.e., the exploded-view layout, are represented as directed graphs. An Above graph is defined as the directed graph whose vertices are all the components in an assembly, and its edges (P_i, P_j) imply that part P_i is located directly above part P_j, with planar contact between them. Referring to actual assemblies which consist of real machine parts, and in view of the semantic meaning conveyed by an Above graph, it can be shown that normally, Above graphs have no loops, and are connected.

Examining the nature of the desired solution (i.e., a data structure describing the single-spaced "stack" of components of the exploded view), it is clear that the parts of the product should actually be sorted according to their location along the centreline of "explosion". The problem can now be stated as follows: The Above graph of a product with n parts will have n vertices, and at least $n-1$ edges. Correctly transform this graph to an n-edge degenerate directed graph having only one edge emerging from each vertex except the last, and only one edge entering each vertex except the first. This last graph will represent the desired exploded-view layout. The structure of products as expressed by their Above graphs often renders two kinds of problematic situations: multiple edges emerging from a single vertex, and multiple edges entering a single vertex. With the first kind we seek a geometry-based ranking among the "offspring", and with the second, among the "parents".

The Above graph transformation procedure starts by "positioning" the assembly drawing with the product's centreline vertical. The "explosion" planning will spread the parts along this height- or Z-axis using rules of transformation, which themselves rely on geometry-dependent functions. For example, $height-overlapping(Part_1,Part_2)$ checks whether two parts overlap along the Z-axis, and $translatable(Part_1,Part_2)$ checks whether assembling either one of the two parts interferes with the assembly of the other, or if any assembly sequence is possible. This is based upon recognizing the existence and size of a "central-through-bore" feature of one part (e.g., the inside diameter of a rolling bearing, or the net inside diameter of "screws-on-a-pitch-circle") and comparing it to the maximum diameter of the other part in a plane perpendicular to the assembly centerline.

The rules of transformation distinguish between "translatable" parts (i.e., the order of "explosion" is unimportant), where a heuristic sorting is called upon, and "nontranslatable" cases, where the correct solution is unambiguous. A heuristic rule of the first kind, applied to the situation of Fig. 6.3a (the corresponding Above subgraph is also shown) is:

IF   two parts (1 and 2 in the figure) are both "parents" of a third part (3),

AND these two parts (1 and 2) are "translatable",

AND they have the same Zmin (i.e., their lower Z coordinates are identical),

AND they have different Zmax,

THEN sort them according to their Zmax.

The actual sorting referred to here involves deleting one edge of the Above graph, and adding a new edge between the "parents", as shown in Fig. 6.3b.

Having "nontranslatable" parts, but with one of them having a (too small) central through bore, sorting is performed by comparing the height location of that bore with the maximum-diameter feature of the other part (i.e., the feature "responsible" for that part's maximum lateral dimension). Four rules apply to parts with no central bores: the first describes a situation where there is no "height overlapping" between the parts. In this case, parts are simply sorted according to their relative height location. The remaining three rules handle "height overlapping" parts as shown in Fig. 6.4. If one part extends beyond a height limit

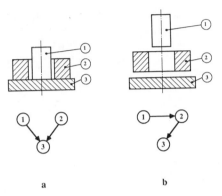

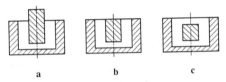

Fig. 6.3 Heuristic sorting of the "translatable" parts in **a** yields the exploded view of **b**.

Fig. 6.4 The three cases of "height-overlapping" parts with no central through-bores.

(Zmin or Zmax) of the other, case (a), then the solution is trivial. In case (b), one height limit is equal for the two parts, and the sorting is performed according to the other limiting height. The most difficult case is (c), where one part is confined within the height limits of the other. The only possible method to apply here is to identify the relative location of the "blocking feature" (e.g., the "bottom" of the cup-shaped part). This is accomplished by using knowledge about the structure of certain basic bodies and features in the parts' definitions.

These relatively simple sorting rules are readily applicable to actual products. Figure 6.5 is the assembly drawing of an electric alternator which consists of fourteen parts, arbitrarily assigned with letters as part names. Figure 6.6a describes the Above graph of the alternator in the arbitrarily set "upright" position. With vertex M chosen as the graph traversal starting point, there is no problem getting from M to N (the edge (M,N) is distinct), but the edges (N,L) and (N,A) are not distinct, representing a "multiple offspring" problem as marked in the figure. It is found that parts L and A are "nontranslatable" because the central bore of each of them is too small to allow the passage of the other, so a new edge (L,A) is introduced by the program, and the existing (N,A) edge is deleted. (Clearly, there is no physical contact between parts L and A. In this case, the meaning of the new edge is that L is above A in the exploded view.) After this partial transformation, the graph looks

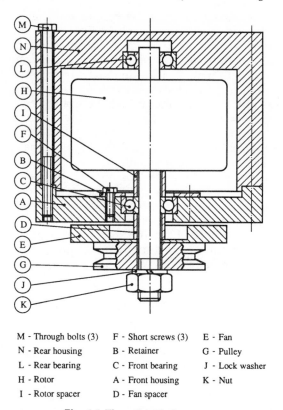

M - Through bolts (3)   F - Short screws (3)   E - Fan
N - Rear housing   B - Retainer   G - Pulley
L - Rear bearing   C - Front bearing   J - Lock washer
H - Rotor   A - Front housing   K - Nut
I - Rotor spacer   D - Fan spacer

Fig. 6.5 The original alternator.

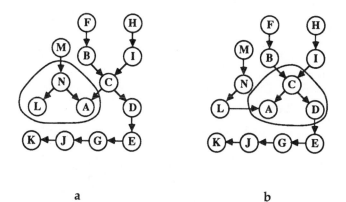

a   **b**

Fig. 6.6 The initial Above graph **a**, and the graph after the first transformation step **b**.

as in Fig. 6.6b. This process continues with eight additional transformation steps until the desired degenerate graph is obtained. The reconstruction of the exploded view thus obtained is shown in Fig. 6.7. As with any algorithm of sorting by pairwise comparisons, the current procedure is polynomially bounded, making it quite efficient computationally.

The basic procedure can be extended to include some special cases. For example, an initially disconnected Above graph (a possibility when temporary support is provided during assembly and removed later) is first converted to an equivalent connected graph by examining the geometry of the parts involved and introducing imaginary Above edges. Statically indeterminate structures might also pose some difficulties, resulting in multiple edges between two vertices, closed paths or multiple paths (more than one path between two vertices) in the Above graph. The procedure employs several means, based on "understanding" of part geometries, to resolve these problems. Finally, the special cases of a uni-axial assembly where one or more parts are still assembled (and "exploded") perpendicular to the centreline (e.g., a Woodruff key on a shaft) are also accommodated.

Interference of Selected Assembly Sequences

A large number of dimensionally and geometrically feasible assembly sequences exist even with small part-count products. Most of these sequences, however, would be practically useless for the product designer or manufacturing engineer as they involve redundant assembly operations or ignore some constraints. The present system requires assembly sequences only to serve as a basis for designing improved product configurations, and therefore, unnecessary complications can be avoided and parts are not grouped to create subassemblies unless mandatory.

The problem to be solved at this stage is formulated as follows. Given a complete geometric and topological description of a product, and a list of part names corresponding to its "exploded" layout, construct useful assembly sequences for that product. For each subassembly in these sequences, recursively generate its own assembly sequences, until the level of individual parts only has been reached. Although in practical planning of assembly sequences, subassemblies might sometimes be desirable, the theoretically optimal product structure is one which allows the mounting of each individual component on top of the other, eliminating excessive fixtures, handling operations, and robotic arms. But the structure of real products is often not that simple, with some components being impossible to assemble individually. In this case, indirect assembly of such parts is necessary by first incorporating them into independent subassemblies. The problem of constructing subassemblies out of individual parts can be quite complex theoretically. However, most such groupings of parts cannot constitute actual subas-semblies because of practical constraints. Extracting the necessary information from the product description to "understand" these constraints is therefore at the heart of the present methodology.

While scanning the input description of a product, the program is currently capable of identifying three types of potential subassemblies: a group of parts held together by dedicated fasteners such as bolts and nuts; two or more parts connected by special techniques such as press-fitting, adhesive bonding, soldering and so on (e.g., a housing or shaft with a rolling bearing press-fitted onto it); and any two groups of the previous types which share at least one part and thus can constitute a single potential subassembly.

However, not all potential subassemblies defined so far can indeed be used, for some constraints regarding theoretical limitations as well as heuristic aspects should be satisfied first. Consider, for example, a nut which is to be tightened onto a bolt. Access to both the head of the bolt and to the nut is checked for by the program, using several rules about part geometry and topology. A second type of practical constraint is having a suitable part to be mounted first onto a fixture at either one of the potential subassembly's ends. Thus, screws on a pitch circle can be the base component only if their mating part is nuts. Other parts that fall into the same category are, for example, retaining rings and rolling bearings, which are obviously inconvenient in serving as base components.

Some of the constraints are heuristic by nature. For example, assembling a rolling bearing simultaneously on a shaft and in a housing (the last two parts being already assembled) is theoretically possible, but difficult to accomplish in practice. Moreover, if a bearing has already been fitted into a housing and the shaft is to be assembled next, then the housing and bearing subassembly is suitable for assembling in a certain direction only – the one which keeps the bearing supported against the housing shoulder. Trying to assemble the same housing and bearing subassembly in the opposite direction has heuristically been determined as forbidden by the program, to prevent the accidental dislocation of the bearing with respect to the housing. The same reasoning is also applicable to a shaft and bearing subassembly.

The internal representation of the exploded view of Fig. 6.7 is the following list of part names (from top to bottom): [M,N,L,H,I,F,B,C,A,D, E,G,J,K]. The program first attempts to find an assembly sequence that starts with placing the three through bolts (part M) with their heads down onto a fixture, just to find out that this would render the assembly of part A impossible. Turning now to the other "end", the nut (part K) is found to be suitable for assembly onto a fixture, and parts J,G,E and D can easily be assembled then in that order, for each has an "above" mating condition with its predecessor. But the continuation of such a "simple" (i.e., no subassemblies) sequence is prevented by part A, not being "above" (in the mating condition sense) any of the parts already "assembled". Therefore the program attempts to assemble part A indirectly, as a component of a subassembly.

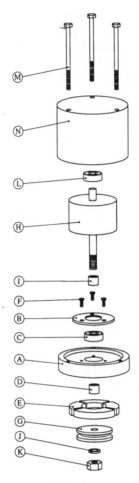

Fig. 6.7 The exploded view of the alternator as found by the system.

Potential subassemblies which contain part A at one of their "ends" are: [M,N,L,H,I,F,B,C,A], [C,A] (a bearing press-fitted into a housing), and [F,B,C,A]. The first is proven to be impossible: assembling it requires the tightening of the rotor shaft (part H) against the nut (part K) held by the fixture when no access to the now enclosed rotor (by the housings N and A) exists. The second subassembly is not suitable for the proposed assembly direction, but the last potential subassembly can be used, so the program "adds" parts F,B,C and A as a subassembly.

Proceeding with the assembly process simulation, there is no problem in assembling part I and then H. The bearing (part L) does not have an "above" mating condition with any of the previous parts, and so another subassembly is required. The "bearing-mount" type of subassembly, consisting of the bearing and the housing (part N) where the former is press-fitted and supported against a shoulder in the latter, is suitable for the proposed assembly direction, so the following sequence for assembling the product is tentatively established: mount the nut, part K, first onto the fixture; add to it parts J,G,E and D in this order; assemble on them a subassembly which consists of parts A,C,B and F; add parts I and H in this order; then a subassembly of parts L and N; and finally tighten the bolts, part M.

Next, sequences for putting together each of the two subassemblies are to be established. The short screws, part F, are obviously unsuitable to go first onto a fixture, but part A is. So a "simple" sequence for the first subassembly is: mount part A on a fixture, add the bearing (part C), then part B, and finally tighten the screws (part F). Heuristically forbidding the other bearing (part L) from serving as a base part, the only possibility is to mount the housing, part N, first, and force-fit the bearing into it. This would of course require turning the last subassembly over before joining the rest of the assembly. This concludes the recursive process of planning the assembly sequence, as the level of individual parts only has been reached.

Structural Synthesis Example

Working through the example of the alternator of Fig. 6.5 is now continued to demonstrate structural changes made to this product in order to facilitate assembly by a single robot. The basis for improvements is the stored high-level product description, together with the results of the previous assembly sequence analysis.

In the design-for-assembly mode, the Functional Analyser module was called upon, to identify parts G, E, A, C, H, L and N as "functional parts". This involved the creation of findings such as: "The 'rotor' (H) is a shaft mounted between two bearings, parts C and L", and "The 'front bearing' provides bidirectional location of the 'rotor', and the 'rear bearing' is a nonlocating bearing". It also involved the establishment of goals such as: "The two housings should be fastened together", and "The 'pulley' and 'fan' should be axially and circumferentially fastened to the 'rotor' shaft". Then the user indicated to the Policy module the desire to assemble the product with the simplest robot possible, and maintain the easily-disassembled character of the product. The former led to establishing the goal: "Try a 'simple' assembly sequence", which means avoiding subassemblies.

The next program, the Configurer, had trouble trying a "simple" assembly sequence with the current functional model of the product, so its first suggestion to the user was to change to a cross-location arrangement of the "rotor" in the bearings (i.e., each bearing locates the shaft in another direction). The user, judging that such an arrangement is acceptable for a relatively short shaft with low thermal expansion, responded affirmatively. The knowledge used at this stage of reasoning is similar in essence to the one serving the assembly sequences analysis phase, except it is now used for synthesis. For example, when the "cross-location" recommendation was accepted, the program attempted to configure such a structure for an "upright" sequence of assembly, with the front bearing locating the "rotor" shaft "downward" and the rear bearing "upward". This failed because the "front housing" had then to be assembled before the front bearing, but the former part could not have the required support when mounted on its predecessors. (An intricate arrangement of mounting the "front housing" on the same fixture as the "pulley" and "fan", and bringing it to its exact final location when the "rotor" is tightened, is actually possible, but is currently beyond the system's capabilities.) So, the other possibility, i.e., the front bearing locating "upward" and the rear bearing locating "downward", was considered next. This would solve the problem of assembling the "front housing" (it would be mounted after the front bearing), but would have required the rear bearing to be assembled after both the rear housing and the shaft had already been mounted, causing a heuristic failure. An "upside-down" sequence was therefore attempted, again while checking both "cross-location" possibilities. In that case, one configuration was found to be impossible, because the front bearing had to be mounted on the shaft and in the housing simultaneously, but the other configuration seemed to be feasible. With that, a "simple", "upside-down" assembly sequence was hypothesized by the Configurer to be possible if a step were to be incorporated in the "rotor" shaft to support the "rear bearing", along with the two spacers (parts D and I) added to the product model (The "rotor spacer" was to provide the required support for the front bearing, and the "fan spacer" to allow the mounting of the "fan".)

In attempting to reduce the number of parts in the product, only the two spacers proved to be candidates for elimination, from which the user found it difficult to integrate the "fan spacer" (D) in another part, but accepted the recommendation made by the Reducer & Fastener module to eliminate the "rotor spacer" (I) and introduce a step in the shaft in its place. Then, the system suggested suitable fasteners for an easily disassembled product, from which the user chose "screws from the 'front housing' side" for the housings, and "splines with a retaining ring" for the "pulley" and "fan". A drawing of the improved design which permits a "simple" stacked assembly, starting with the "rear housing", is shown in Fig. 6.8. The next obvious stage of treating the handling and orientation of each individual component is omitted from the current example because of the similarity to conventional design-for-assembly techniques.

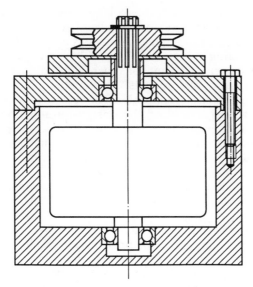

Fig. 6.8 A suggested improved alternator design.

Comparison of the original design with the new one shows that the part-count was reduced from 18 to 12, the three fixtures that were needed, one of them quite complicated (with an hexagonal cavity for the "nut" and a spring-loaded plunger for the "rotor" shaft), were replaced by one simple fixture, and the relatively difficult "flip-over" operation was eliminated, with all remaining assembly operations executable by a four degree-of-freedom robot. From the part manufacturing viewpoint, the "rotor" is now more complicated to produce, having additional steps, splines, and a retaining ring groove. The "pulley" and "fan" also require internal splines, but it is believed that the additional fabrication cost is more than balanced through other savings resulting from the redesign.

Conclusion

Implementing a knowledge-based approach to the solution of a configurative design problem has proven to require a high-level method of representing the design subject in terms similar to those in which the domain knowledge is formulated, and which allow useful dialogue with a human designer. Complete modelling of the design process, implemented via the system architecture, has been shown to be essential for providing advice where suggestions to the engineer are tailored to suit specific circumstances and product character. Contrary to the traditional view of expert systems as a large collection of IF-THEN rules, it seems that for a computerized design aid to exhibit more than just "shallow" knowledge, a set of relevant analysis and synthesis tools should be integrated in a single environment.

Although the current system has been limited to the layered and axisymmetrical family of products, the methodology introduced is believed to be generally applicable. The essence of the knowledge pertinent to all levels of solving the design-for-assembly problem has been classified, and processes for utilizing this knowledge have been determined. Within the current framework, additional domain knowledge can be added under the existing classification to enhance the individual knowledge bases of the various system modules. Moreover, similar systems can be developed to fit into the needs of particular corporations, encompassing specific lines of products and accumulated design experience and practices.

Clearly, ultimate designs can only be established after detailed cost estimates have been completed. The present system, although considering product functionality and assemblability, does not employ means of quantifying evolving designs with regard to their manufacturing and assembly expenditure. Therefore, the system can best serve as a "front end" to more elaborate design consultants which also include various cost estimation modules.

Future research is still needed in establishing a more formal theoretical basis for correlating the configuration of products with their assembly requirements, and in providing complete tools for creating feature-based product descriptions. Advanced knowledge of assembly techniques, and ways of representing them have yet to be developed. For example, inertia-based tightening of the "nut" onto the "rotor" of the case study described in this chapter would probably have yielded additional assembly sequences. Modern fastening and fabrication methods such as plastic snap connections and insert moulding would also considerably enhance the field when incorporated and modelled in design-for-assembly consultant systems.

References

1. Kroll, E, Lenz E, Wolberg JR. A knowledge-based solution to the design-for-assembly problem. Manuf Rev 1988; 1: 104–108
2. Andreasen MM, Kähler S, Lund T. Design for Assembly, IFS Publications and Springer-Verlag, 1983
3. Laszcz JF. Product design for robotic and automatic assembly. In: Proceedings of robots 8 conference, Detroit, MI 1984; pp 6.1–6.22
4. Schraft RD and Bässler R. Considerations for assembly oriented product design. In: Proceedings 5th international conference on assembly automation, Paris, 1984.
5. Browne J, O'Gorman P, Furgac I, Felsing W, Deutschlaender A. Product design for

small parts assembly. In: Rathmill K (ed) Robotic assembly, IFS Publications and Springer-Verlag, 1985

6. Scarr AJ. Product design for automated manufacture and assembly. Ann CIRP 1986; **35/1**: 1–5
7. Boothroyd G, Dewhurst P. Design for assembly – a designer's handbook. Department of Mechanical Engineering, University of Massachusetts, Amherst, 1983
8. Miyakawa S, Ohashi T. The Hitachi assemblability evaluation method (AEM). In: Proceedings 1st international conference on product design for assembly, Newport, Rhode Island, 1986
9. Behuniak J, Graves RJ, Poli C. Design/production integration – systematic approach. In: Proceedings 8th international conference on assembly automation, Copenhagen, Denmark, 1987; pp 111–122
10. Poli C, Fenoglio F. Designing parts for automatic assembly. Mach Des 1987; 59: 140–145
11. Boothroyd G. Making it simple: design for assembly. Mech Eng 1988; 110 2: 28–31
12. Leu MC, Weinstein MS. A case study of robotic assembly for a printer compensation device. J Manuf Syst 1988; 7: 163–170
13. Schepacz C, Brand A, Lacour M. Design for automation: a case study. In: Proceedings 3rd international conference on product design for manufacture and assembly, Newport, Rhode Island, 1988
14. Swift KG. Knowledge-based design for manufacture, Kogan Page, 1987
15. Jakiela MJ, Papalambros PY. A design for assembly optimal suggestion expert system. In: Proceedings 7th international conference on assembly automation, Zurich, Switzerland, 1986; pp 341–350
16. Hernani JT, Scarr AJ. An expert system approach to the choice of design rules for automated assembly, Proceedings 8th international conference on assembly automation, Copenhagen, Denmark, 1987; pp 129–140
17. Pratt MJ. Solid modeling and the interface between design and manufacture. IEEE Comput Graphics Applic 1984; 4: 52–59
18. Walske S. Solid models link design and manufacturing. Mach Des 1988; 60: 52–55
19. Dixon JR. Designing with features: building manufacturing knowledge into more intelligent CAD systems. Proceedings manufacturing international '88, Atlanta, GA, 1988; pp 51–57
20. Henderson MR. Extraction of feature information from three dimensional CAD data. PhD thesis, Purdue University, 1984
21. Klein A. A solid groove: feature-based programming of parts. Mech Eng 1988; **110**: 37–39
22. Wesley MA, Lozano-Pérez T, Lieberman LI, Lavin MA, Grossman DD. A geometric modeling system for automated mechanical assembly. IBM J Res Dev 1980; 24: 64–74
23. Lee K, Gossard DC. A hierarchical data structure for representing assemblies: part 1. Comput-Aided Des 1985; 17: 15–19
24. Rocheleau DN, Lee K. System for interactive assembly modelling. Comput-Aided Des 1987; 19: 65–72
25. Ko H, Lee K. Automatic assembling procedure generation from mating conditions. Comput-Aided Des 1987; 19: 3–10
26. Sudhakar M, Faruqi MA. Octree models for robotic assembly planning and manufacturing process planning. Proceedings 3rd international conference on computer-aided production engineering (CAPE), Ann Arbor, MI, 1988; pp 556–571
27. Shpitalni M, Elber G, Lenz E. Automatic assembly of three-dimensional structures via connectivity graphs. Ann CIRP 1989; 38: 25–28
28. De Fazio, TL, Whitney DE. Simplified generation of all mechanical assembly sequences. IEEE J Robotics Automat 1987; RA-3: 640–658
29. Haynes LS, Morris GH. A formal approach to specifying assembly operations. Int J Mach Tools Manuf 1988; 28: 281–298

Chapter 7

Intelligent Product Design and Manufacture

H.A. ElMaraghy

Introduction

This chapter presents the concepts and philosophy underlying an intelligent product design and manufacturing (IPDM) system developed at the Centre for Flexible Manufacturing Research and Development at McMaster University. The overall objective of the IPDM research program is to: (1) investigate the basic issues involved in integrating the design and manufacturing task planning activities; and (2) demonstrate the feasibility and potential of applying artificial intelligence, expert systems and feature-based modelling concepts to these activities. The project addresses the fundamental problem(s) of establishing an effective link between computer-aided design (CAD) and manufacture (CAM).

The IPDM project is a result of many years of research efforts by several members of our group. The motivation for this research is manifold. Firstly, the current computer-aided design systems are merely geometric processors which manipulate entities such as points, lines and surfaces. They do not capture a great deal of the technical information essential for subsequent manufacturing task planning. Secondly, manufacturing process planning and the knowledge required for it are not well understood. However, effective solutions for automating this process in a generative fashion would have a significant impact on manufacturing productivity. Last, but not least, true integration between design and manufacturing task planning remains an elusive dream. Available process planners are simply interfaced to existing CAD systems with obvious inefficiencies, redundancies and duplication of efforts.

The IPDM Philosophy

The design process is initiated in response to a perceived need for products or devices to perform certain functions. The designer then generates several concepts and ideas to fulfil this need and proceeds to synthesize and detail the design accordingly. During this highly creative phase, the designer relies on both intuition and accumulated knowledge and expertise. The selected design features are chosen primarily to achieve the desired functions, by themselves or in relation to other features within the part or product. Once major features are decided upon, others follow for the purpose of linking the major features together or to facilitate their manufacture.

We wish to differentiate here between functional features and form features. Form features refer to recognizable shapes which cannot be further decomposed, otherwise they will reduce to meaningless geometric entities such as lines, points and surfaces. Form features, while affecting manufacturing processes, may or may not by themselves, have a functional purpose. Several research efforts to define and use features were recently reviewed [1]. An effort to standardize form features and develop a part data exchange system (PDES) similar to IGES is spearheaded by the US National Institute of Standards and Technology [2]. An equivalent European Standard, STEP, is also evolving. PDES, when complete, will be a standard for feature-based data representation and exchange, not a modeller. It is only concerned with the geometric representation of form features.

We believe that higher level functional features (such as keyseat, fillet, chamfer, splined shaft, threads, etc.) are more natural to use by designers than geometric abstractions and often meaningless form features. These *macro functional features* may be further decomposed into simpler or *micro form features* for the purpose of geometric representation or manufacturing task planning. However, such decomposition should be transparent to the designer. Some nonfunctional features may be deduced and created automatically, using expert knowledge, to join functional features or facilitate their manufacture. For example, undercuts may be needed to make external thread cutting feasible in some situations and rounded corners would be needed to facilitate assembly of components against right-angled shoulders.

Additional knowledge required to convert design specifications to manufacturing steps (i.e. process planning) is predominantly domain specific. However, process planning is also dependent on inherent attributes of functional features. A shaft's function, size and method of support suggest a certain fit class and consequently, specific dimensional and/or geometric tolerance requirements. Such attributes (whether deduced, inherited or specified) must be included in functional features definition, as well as known heuristic or basic knowledge about expected behaviour, relationship with other features or manufacturing alternatives.

Hypotheses

Based on the above discussion, the philosophy and hypotheses used in the IPDM project may be summarized as follows:

1. Design features are there for a reason. They are chosen for their functionality. Functional features embody the designer's intent and knowledge and express the designer's reasoning regarding relationships with other features and functional characteristics.

2. Geometric abstractions are not useful during the design creation and synthesis phases. The best design aids are those which allow the designer to function naturally, concentrating on achieving the part (or product) design purpose without resorting to abstractions (geometric or otherwise) which are foreign to the creative design and modelling activity.

3. Capturing only geometric design information at the modelling stage ignores part of the reasoning and logic associated with the choice of features. It causes loss of valuable information and calls for subsequent reinterpretation by planners.

4. Design features directly affect manufacturing methods and detailed plans. Features form, size, tolerance, material and surface finish attributes supplemented by manufacturing knowledge, resources and constraints determine the final manufacturing plan.

5. Manufacturing and process planning is an intelligent behaviour concerned with devising means of achieving desired goals. It combines problem solving, reasoning and conflict resolution methodologies and lends itself to the application of artificial intelligence and expert systems techniques.

Modelling with Functional Features in IPDM

The IPDM system consists of a modelling and design environment and a number of application specific generative planners and their associated knowledge bases as shown in Fig. 7.1. The design environment contains an interactive user interface, a feature-based part definition and modelling modules and interactive graphical display programs. Expert design and synthesis modules and supporting design analysis routines are currently being researched and developed. For example, an expert tolerancing advisor [3] and an automatic expert quadrilateral elements mesh generator [4] have already been developed. These modules are meant to provide

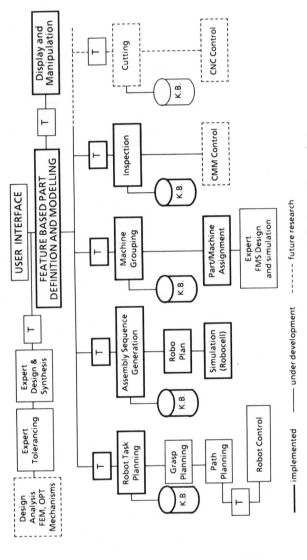

Fig. 7.1 IPDM – intelligent product design and manufacture.

the user with expert design advice on-line. The remainder of this section deals with the part/product definition and modelling.

The IPDM system is implemented on SUN 3/260C and SUN 4 computers. Various modules are written in C, Prolog and Suntools according to their requirements. More specific development tools (such as LEX and YACC) are used when appropriate.

User Interface

An interactive user interface provides the designer with a natural and user friendly design environment. It allows the user to enter design information graphically or using special design description languages. It provides a link to the display program for verifying the modelled objects. The user interface prompts the designer for missing information and checks for conflicting or incorrect data. Deduced design attributes or relations are provided automatically which reduces the amount of required input by the user, reduces errors and provides expert support for the designer. Any modification or corrections are done interactively.

The structure of the user interface has two separate sections: one is feature-dependent and the other is feature-independent [5]. The feature-dependent section contains the basic feature definition structure as well as utilities for listing, displaying and saving feature descriptions. The feature-independent section contains support functions such as reading existing features and many other help functions. Component level parameters are feature-independent and include name, units, material, surface finish and heat treatment specifications. This structure facilitates adding new features and modifying existing ones. A multiple window and graphical menu environment enhances the effectiveness of the user interface.

Functional Features

A library of standard features exists and has been modelled after an earlier prototype [5] with significant improvements. All features used in the IPDM modeller are functional features, examples of which are shown in Fig. 7.2. Designers would find it quite natural to model using these types of features.

Library features may be selected and combined to model a part. Standard components (such as bearings, gears, keys, etc.), may also be selected from tables and combined with modelled parts to form a product. This is done in an object-oriented frame data structure which also allows definition of custom features if an equivalent is not available in the data base. In the IPDM project, the following definitions are used:

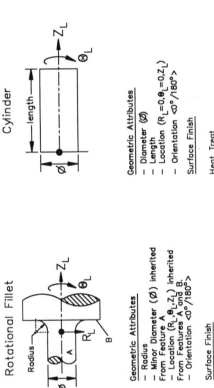

Fig. 7.2 Sample functional features used in IPDM.

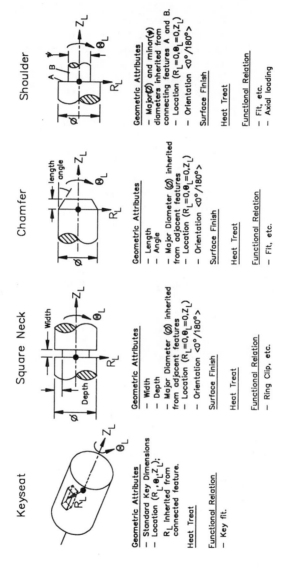

Fig. 7.2 (continued).

IPDM elements	Engineering design interpretation
● Product	A combination of components assembled together. Something that does not require any more processing as far as a manufacturer is concerned.
● Component	A unit combination of features. The component is defined completely in terms of its functional and geometric relationship to other components and its own constitutive features. Components may have standard elements, e.g. gears, bearings, etc., which do not have to be broken down into features.
● Feature	The basic building blocks of a component. Features are: Parametrically defined Located relative to other features Unitless. Dimensions, tolerances, surface finishes, heat treatment, colour; they can be assigned or inherited Features are divided into two categories: *Microfeatures* Form features Basic features not specific to any application Wide range of geometric complexity; from well defined geometries (e.g. cylinder, cube, cone, etc.) to complex geometries (e.g. sweeps, twists, sculptured surfaces, etc.) *Macrofeatures* Functional features (e.g. keyseat, shoulder, fillet, etc.) Compilations of microfeatures to suite application domain

Using the above concepts, a collection of features may be combined to form the model of the transmission shaft shown in Fig. 7.3. It should be emphasized here that feature models contain not only geometric data but also other functional, relational or generic manufacturing knowledge. Only a few feature attributes are shown in Fig. 7.3 to avoid clutter.

Design Languages

Two design languages to integrate design and manufacture have been developed. The first is called feature-based design description language (FDDL) [6]. It is designed to be used with a feature-based modeller. FDDL allows the designer to specify parts or product names and attributes such as material, surface finish and relationships with other features in

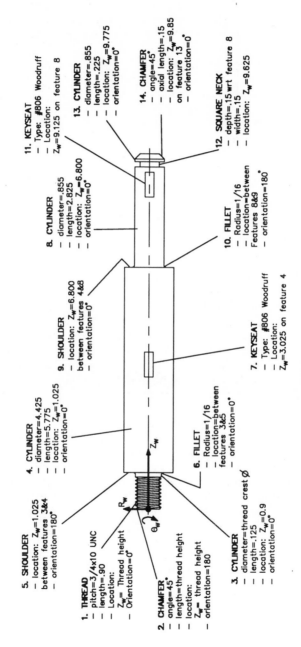

11. KEYSEAT
- Type: #806 Woodruff
- Location:
  Z_w=9.125 on feature 8

13. CYLINDER
- diameter=.855
- length=.225
- location: Z_w=9.775
- orientation=0°

14. CHAMFER
- angle=45°
- axial length=.15
- location: Z_w=9.85
  on feature 13
- orientation=0°

12. SQUARE NECK
- depth=.15 wrt feature 8
- width=.15
- location: Z_w=9.625

8. CYLINDER
- diameter=.855
- length=2.825
- location: Z_w=6.800
- orientation=0°

9. SHOULDER
- location: Z_w=6.800
  between features 4&8
- orientation=0°

10. FILLET
- Radius=1/16
- location=between
  Features 8&9
- orientation=180°

7. KEYSEAT
- Type: #806 Woodruff
- Location:
  Z_w=3.025 on feature 4

4. CYLINDER
- diameter=4.425
- length=5.775
- location: Z_w=1.025
- orientation=0°

5. SHOULDER
- location: Z_w=1.025
  between features 3&4
- orientation=180°

1. THREAD
- pitch=3/4x10 UNC
- length=.90
- Location:
  Z_w= Thread height
- Orientation=0°

2. CHAMFER
- angle=45°
- length=thread height
- location:
  Z_w= thread height
- orientation=180°

3. CYLINDER
- diameter=thread crest ∅
- length=.125
- location: Z_w=0.9
- orientation=0°

6. FILLET
- Radius=1/16
- location=between
  features 3&5
- orientation=0°

Fig. 7.3 Feature-based model of a transmission shaft [5].

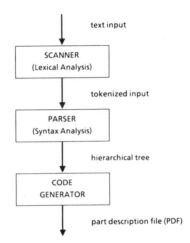

Fig. 7.4 Overall structure of design languages [7].

a textual manner. The FDDL vocabulary and syntax is consistent with commonly used engineering terminology and standards. It is easy to learn and use by designers. An improved design language, called part description language (PDL), has been designed and implemented on the SUN computer [7]. PDL is a data definition language which facilitates the description of parts and products in terms of attributes, relationships and feature decomposition. PDL input is in the form of a text stream, which may be stored in a text file or an output pipe from a concurrently executing process such as the graphical user interface. The PDL produces data records to be stored in a part definition file (PDF). This file has a frame structure to simplify description and data retrieval and allows attribute inheritance among products, components and features.

Both languages perform four basic operations: lexical analysis (scanning), syntax analysis (verification) and code generation for subsequent applications as shown in Fig. 7.4. The LEX and YACC development tools greatly simplified the design of the scanning and parsing phases. The remaining components of the language systems are written in C.

The part definition file structure used by PDL is extensible to allow the addition of features and attributes without affecting existing applications. Component data are stored as an arbitrary length record consisting of a number of variable length lists, each terminated by a definite string as shown below.

        PDL COMPONENT DATA FRAME
            attribute list:
                name
                class

```
            type
            function
            material
            heat treatment
            . . .
            ⟨terminator⟩
        relation list:
        relation name, component name
            . . .
            ⟨terminator⟩
        feature list:
            name
            parameter list:
                parameter name,value
                . . .
                ⟨terminator⟩
            location
            tolerance list:
                tolerance name, tolerance parameters
                . . .
                ⟨terminator⟩
            relation list:
                relation name, feature name
                . . .
                ⟨terminator⟩
            surface finish
            . . .
            ⟨terminator⟩
    . . .
    ⟨terminator⟩
```

The hierarchical structure of PDL facilitates the development of new user-defined features or composites of existing library features. Unlike FDDL, PDL allows arithmetic expressions consisting of integer, real and scientific-notational constants, binary and unary operators, exponentiation, subexpressions, functions and variable names. An if-then-else command structure has been added to the PDL languages in order to implement feature constraints and inheritance rules and checks.

Geometry Display

Modelled parts, products and selected features are displayed during the design session. Currently, a solid geometric modeller, PADL, from Cornell University, is being used for that purpose. Other surface/solid modellers are now being evaluated as a means of displaying more complex and sculptured shapes. The display shows pertinent data for features and components and supports rendering, shading and viewing facilities.

Knowledge-Based Process Planning

Manufacturing process planning is an essential link between design and manufacturing and is crucial to realizing their integration. It converts design data to work instructions. Variant process planning procedures can only handle parts which belong to predefined part families and retrieve and modify master plans as needed. Therefore, their application domain is narrow by definition. Generative methods generate, rather than retrieve, process plans using basic knowledge and heuristic rules.

Reasonable progress has been made in automating some aspects of the detailed, low level (macros) process planning activities particularly in metal removal operations. Research in automatic or generative process planning in other areas such as assembly and inspection has been slow, especially at the high level of task planning, action ordering and reasoning. Knowledge-based process planning methods are complex but can potentially offer much needed flexibility. Review articles [8–12] compare variant, generative, and knowledge-based planners and discuss their relative merits. The main hurdle in designing generative task planners which are truly integrated with the design function is the sheer complexity and variety of components to be produced and the diversity of manufacturing processes involved.

The IPDM modelling system, design languages and user interface all lead to the creation of a part/product database which consists of: (a) geometric form data, (b) functional and physical relationships between features, (c) technical attributes such as material, surface finish, tolerance, etc. and (d) generic knowledge regarding relations, functions or manufacturing processes and alternatives.

The IPDM modeller and associated languages are designed to be general, not application-specific. Yet not every process planning application will require the complete set of attributes and knowledge stored with each feature. Therefore, domain specific code generators (i.e. translators shown in Fig. 7.1 as T) are used to extract the required data selectively for each application using parsing, lexical analysis and syntactic pattern recognition techniques. This is then combined with additional relevant expert rules, knowledge and heuristics to construct automatically and generatively the appropriate process plan. Examples of the implemented planners are discussed in the next section.

Planning Coordinate-Measuring Machine Inspection Tasks

A knowledge-based generative inspection process planner for coordinate-measuring machines (CMMs) has been developed [13]. The planning system takes into consideration the CMM characteristics and capabilities, geometry and function of parts to be inspected, as well as size and geometric tolerancing principles. Syntactic pattern recognition methods [14] are used to identify various inspection features. Pattern primitives and grammar, suitable for representing geometric and tolerancing inspection features were selected. Clustering, grammatic inference and syntactic analyses were used to recognize the part features. Knowledge rules pertinent to inspection with coordinate-measuring machines and tolerancing theory were used to detail the inspection plan. The expert inspection planner determines the inspection datum and inspection priority where: (a) there is more than one datum or, (b) there are multiple features per datum. It groups inspection tasks to optimize probe usage, selects appropriate probes and orientation, and determines features accessibility for inspection based on part orientation and modelling data. A schematic of the inspection process-planning system is shown in Fig. 7.5.

Assembly Sequence Planning

The generation of the correct and feasible assembly sequence for any product requires a great deal of geometric and functional reasoning regarding component features and their interaction. Maintainability, ease of disassembly and life cycle considerations are embedded in the choice of design features. The choice of assembly sequence can significantly influence the productivity and efficiency of the assembly process as well

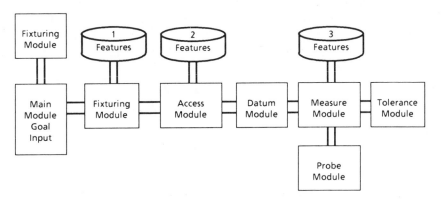

Fig. 7.5 Knowledge-based inspection planning system [13].

as subsequent product support after sale. Assembly sequence directly affects the choice of assembly equipment and the need for jigs and fixtures. It generally determines the complexity and reliability of the assembly process. Assembly planners, such as manufacturing engineers and technicians, rely on: (a) past experience, (b) knowledge of the operational functionality of the product being assembled, (c) knowledge of capabilities of existing equipment and (d) knowledge of the nature of parts mating modes and relevance of their geometric features. The reasoning process which goes on during assembly planning is very complex and not well understood, yet it lends itself to the application of artificial intelligence (AI) techniques.

The need for automated assembly-planning capabilities is well recognized in flexible and less structured manufacturing systems with small to medium size batches and frequent changeovers. With increased emphasis on global economy, careful examination of alternative feasible assembly plans and selecting the best can potentially pay off handsomely.

From a pure combinatorial viewpoint there may exist a very large number of sequences in which a product can be assembled. However, taking the previously discussed factors into consideration and, using a simultaneous engineering approach, there are usually very few assembly sequences (perhaps only one) which are considered best or most logical. Recent research into assembly sequence generation focused on generating all possible assembly sequences (for a given product) by representing links (liaisons) between various components using a graph constructed by asking the user questions about those links. A set of logical expressions were used to encode the directed graph of assembly states [15]. The number of questions asked to construct this graph is very large and can be overwhelming for any realistic product with a moderate number of parts. A simplified method for generating all mechanical assembly sequences based on the same concept was developed [16]. It requires answering two questions for every connection between two parts, which is a much smaller set of questions compared with the first method. Both these approaches depend on answers provided by human users. A computer program generates the questions and constructs the assembly precedence diagram [16]. A recent research effort developed an algorithm for automatically generating the precedence diagram and all assembly sequences without asking questions [17]. A proof of the correctness and completeness of the generated sequences has also been presented. This method employs a relational model of assemblies including attachments which connect pairs of parts and generates disassembly sequences (reversed assembly sequences).

Our approach has been to avoid generating all assembly sequences which is a time and resource intensive task. A great number of these sequences are either infeasible, useless or undesirable due to geometric, technical or other reasons and are eventually discarded. We carry out problem analysis by reasoning about geometries, operational functionality of the product and components, generic assembly knowledge and relations between parts to reduce drastically the number of assembly sequences produced. We are trying to understand, represent and use the human reasoning process and knowledge regarding mating conditions, relationship between features, and functional requirements. If currently available knowledge in our system is insufficient to create the necessary assembly sequence generatively; it may be supplemented by additional information obtained from the human user.

With this objective in mind, we have developed two *semi-generative* assembly sequence generators and used them to plan various assemblies, from abstract blocks to real products such as flash lights and direct current electric motors.

The first prototype uses the database generated by the IPDM feature-based modeller as well as a relations diagram generated by the user with interactive graphics on a SUN computer. Geometric reasoning is utilized to provide missing links and verify/question links in a relation diagram which are redundant or present conflicts [18]. The system automatically identifies base components in subassemblies as part of developing the overall plan and an ordered sequence of assembly tasks.

The second assembly sequence generator is driven by the design description language, FDDL [6]. The various components are described using language syntax which defines features, their relation with other features and the nature of such relation (e.g. type of fit, screw fastening, fill, etc.). After lexical analysis and syntax checking, assembly knowledge rules and geometric reasoning are used to identify base components, subassemblies and the final assembly sequence. In this prototype the knowledge base is not exhaustive or complete. Therefore, additional interaction, verification or information may at present be sought from the user in a questions and answers session. Figure 7.6 shows a DC motor, Fig. 7.7 shows a portion of a DC motor final assembly description using FDDL, and Fig. 7.8 includes the generated assembly sequence [6].

Both assembly sequence generators are designed ultimately to be generative in nature. It is important to note that the *semi-generative* description used here is due to the present incomplete knowledge base rather than the technique of process plan generation.

Generative Robot Task Planning

The assembly sequence generated by either one of the above planners can be used for manual, automatic or robotic assembly. For robotic assembly, a knowledge-based task planner ROBOPLAN [19] was implemented to convert the ordered assembly sequence (produced by any of the above assembly sequence generators) into robot motion plans as shown in Fig. 7.9. A description of objects in the robot workspace (robot world) is provided through the IPDM feature modeller. This initial state may represent the beginning of an assembly or a partially assembled

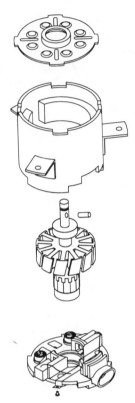

Fig. 7.6 Final DC motor assembly consisting of four subassemblies: two end caps, main body and armature.

product. ROBOPLAN uses first-order predicate logic to generate the robot motion plan required to achieve the goal state (final product assembly) and the subgoals specified by the assembly sequence. ROBOPLAN is concerned with motion planning, and such details regarding grasp planning, path planning etc. are contained in supplementary modules currently under development. The plan produced by ROBOPLAN may be verified using a simple graphic display program GRAPHPLAN [19] using blocks to represent three-dimensional parts and components. A more elaborate simulation of the assembly operation can be performed using the ROBOCELL simulator [20]. It accepts the object-level robot assembly plan and produces a colour dynamic display of the whole operation using a robot model selected from a library of five robots (see Fig. 7.10). Drivers and code generators for industrial robots such as PUMA and ADEPT which translate the assembly steps into robot specific control languages (e.g. VAL II) have been developed.

```
1 housing
[cylinder_side,50,weld,2,cylinder_side,50,weld,3,cylinder_side,50,weld,4,
bore,40,cure,5,bore,40,cure,6,bore,40,pufit,7,cylinder_front,50,contact,8]
eee.
2 bracket [bore,50,weld,1] eee.
3 bracket [bore,50,weld,1] eee.
4 bracket [bore,50,weld,1] eee.
5 magnet [cylinder_side,40,cure,1] eee.
6 magnet [cylinder_side,40,cure,1] eee.
7 end_cap
[cylinder_side,40,pufit,1,bore,30,pfit,10,squre_hole,8,mfit,12,squre_hole,5,
mfit,13,squre_hole,5,mfit,14,cylinder_side,4,pufit,15,cylinder_side,4,pufit,
16] eee.
8 o_com_endcap [bore,30,pufit,11,cylinder_front,50,contact_free,1] eee.
9 armature [cylinder_side,10,pfit,10,cylinder_side,10,pfit,11] eee.
10 bearing_b [bore,10,pfit,9,cylinder_side,30,pfit,7] eee.
11 bearing_b [bore,10,pfit,9,cylinder_side,30,pfit,8] ee.
12 power_connector [box,8,mfit,7] eee.
13 brush [box,5,mfit,7] eee.
14 brush [box,5,mfit,7] eee.
15 spring [bore,4,pufit,7] eee.
16 spring [bore,4,pufit,7] eee.
stop input eee.
```

Fig. 7.7 FDDL target language description of the DC motor to be used by assembly sequence planner.

The assembly sequence generation and robot task planning and simulation clearly demonstrate the integration between design by features, knowledge bases, inference rules and the generative process planning for a typical manufacturing process, i.e. assembly.

Generative Machine Clustering and Part Assignment

This is an application of the IPDM philosophy and system in production planning rather than process planning. An expert part/cell assignment system has been developed as a decision tool for the generative and dynamic assignment of parts to manufacturing cells directly from the feature-based models produced by the IPDM modeller [21]. The system consists of an analyser, synthesizer and recognizer as shown in Fig. 7.11. Pattern recognition, finite-state automata and domain-specific expert rules

Part Name	ID	Feature	Relation	Mate Part	ID
magnet	5	cylinder_side	cure	housing	1
magnet	6	cylinder_side	cure	housing	1
bracket	2	bore	weld	housing	1
bracket	3	bore	weld	housing	1
bracket	4	bore	weld	housing	1
bearing_b	10	bore	pfit	armature	9
bearing_b	11	bore	pfit	armature	9
spring	15	bore	pufit	end_cap	7
spring	16	bore	pufit	end_cap	7
brush	13	box	mfit	end_cap	7
brush	14	box	mfit	end_cap	7
power_connector	12	box	mfit	end_cap	7
bearing_b	10	cylinder_side	pfit	end_cap	7
housing	1	bore	pufit	end_cap	7
o_com_endcap	8	bore	pufit	bearing_b	11

The assembly sequence planning is finished.

Fig. 7.8 Generated assembly sequence of DC motor shown in Fig. 7.6 and described in Fig. 7.7.

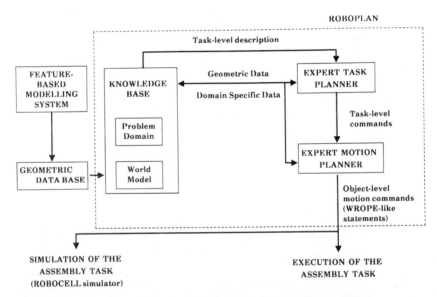

Fig. 7.9 Architecture of the ROBOPLAN task planner [19].

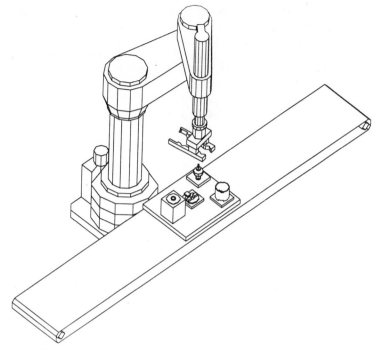

Fig. 7.10 Sample graphical output produced by ROBOCELL [20].

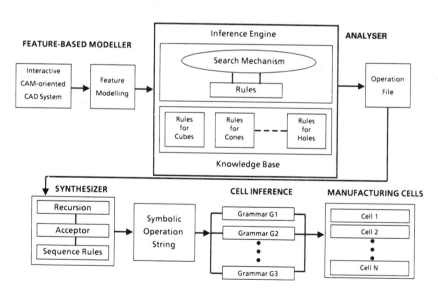

Fig. 7.11 Block diagram of expert parts/manufacturing cells assignment system [21].

are used in this system. Symbolic strings containing the design and manufacturing attributes of components are analysed, then cell inference and parts assignment are carried out. The system finds the best match between product features and machine production capabilities, takes into consideration the relative importance of various machines and examines all feasible alternatives for producing the part. Case studies for machining were used to illustrate the effectiveness of this approach [21]. This system can be used as a decision tool to assist production planning in batch cellular manufacturing environments.

Discussion

In spite of the existence of numerous sophisticated CAD systems, most designers feel that the design environment provided by these systems is foreign to their thinking process. Many experience a cultural shock when first asked to think cubes, cylinders and spheres in order to model brackets and axles using a constructive solid modeller. In addition, present CAD systems produce geometric databases and fail to capture the designer's knowledge and intent regarding functions, relationships and choice of features and attributes. Consequently, process plan generation presently requires extensive interpretation of design models and additional data. Such planners are merely interfaced with existing CAD systems. Manufacturing process planning is an essential link between design and manufacturing and is crucial to realizing their integration. To date, only very few computer-aided production-planning systems have been used by industry with limited impact on manufacturing. This is due to the complexity of the planning domain, lack of scientifically rigorous foundation for planning methods and absence of true integration with design.

Conclusions

Design is the intelligent activity of creating concepts and synthesizing details in order to create products which satisfy perceived needs. Therefore, the choice of *design features* is deeply rooted in their functionality, by themselves or in relation to other features. Meanwhile, design features drive the selection of manufacturing methods and process parameters.

This chapter has presented an intelligent product design and manufacture (IPDM) system framework and reviewed its feature-based modelling environment and part description languages. This environment is integrated with several generative process planners e.g. expert assembly, expert inspection, and expert robots task planners as well as a production planning tool for parts/cell assignment. Actual development of these modules has taken place and was reported in literature.

The IPDM framework lends itself to several other applications such as metal removal and intelligent design and synthesis. The potential for truly integrated design and manufacture has been clearly demonstrated in the IPDM project. Knowledge rules and heuristics about specific domains reduce the search space and produce workable plans by eliminating exhaustive search. Grouping features into classes representing objects and their inherited properties also speeds up and simplifies modelling and planning. The use of design languages and structured, frame-based object representation schemes proved very effective in achieving the goal of integrating design and manufacture.

The IPDM system is not, at present, a universal comprehensive modeller or planner. This was not the objective. It has been decided from the outset that features selection and inclusion will be driven by the implemented applications. As the scope of these applications increases, so will the number and type of features included in the IPDM system library.

Acknowledgements. The IPDM system research extends over many years during which several graduate students and research associates made many valuable contributions. The author would like to extend special thanks to Peihua Gu, Larry Hamid, Larry Knoll, Todd Pfaff, Jean-Michel Rondeau, Luc Laperriere and Zhang Wu. This research was funded primarily by the Canadian Natural Sciences and Engineering Research Council (NSERC) via several operating, strategic and CRD grants to H.A. ElMaraghy.

References

1. Shah JJ, Rogers MT. Expert form feature modelling shell. Comput Aided Des 1988; 20: 515–524
2. PDES form features information model; version 4 (draft), National Institute of Standards and Technology report, August 1988
3. ElMaraghy HA, Gu PH. Expert tolerancing consultant for geometric modelling. Proceedings ASME manufacturing international '88, Atlanta, Georgia, Product and process design Symposium, pp 17–22
4. Liu YC, ElMaraghy HA, Zhang KF. An expert systems for forming quadrilateral finite elements. Eng Comput, in press
5. ElMaraghy HA, Knoll L. IPDM – Intelligent product design and manufacture. Proceedings MAPLE 1990, Ottawa, Canada, pp 67–79
6. Gu PH, ElMaraghy HA, Hamid L. FDDL: A feature-based design description language. Proceedings ASME 1st design methodology conference, Montreal, Canada, 1989, pp 53–63

7. Pfaff T, ElMaraghy HA. Part definition language – PDL, FMRD report no. 5, Centre for Flexible Manufacturing Research and Development, McMaster University, Canada, 1988
8. Weil R, Spur G, Eversheim W. Survey of computer aided process planning systems. Ann CIRP 1982; 31: 539
9. Eversheim W. Survey of computer-aided process planning systems. CIRP Tech. Reports, The CIRP Annals, pp 407–611, 1985
10. Phillips R, Zhou X, Mouluswaran C. An artificial intelligence approach to integrating CAD and CAM through generative process planning. Proceedings ASME international computers in engineering conference, pp 459–463, 1984
11. Ham I, Lu S. Computer aided process planning: the present and future. Ann CIRP 1988; 37: pp 591–601
12. Proceedings 19th CIRP international seminar on manufacturing systems (major theme: CAPP). The Pennsylvania State University, 1987
13. ElMaraghy HA, Gu PH. Expert system for inspection planning. Ann CIRP 1987; 36: 85–89
14. Fu KS. Syntactical pattern recognition, theory and applications. Prentice Hall, Englewood Cliffs, NJ, 1982
15. Bourjault A. Contribution à une approche méthodologique de l'assemblage automatisé: elaboration automatique des séquences opératories, Thèse d'État, Université de Besançon Franche-Comté, France, 1984
16. Whitney DE et al. Computer aided design of flexible assembly systems, report no. CDSL-R-2033, Charles Draper Laboratory, MIT, Cambridge, MA, 1988
17. Homem de Mello LS, Sanderson AC. Automatic generation of mechanical assembly sequences. CMU-RI-TR-88-19 report, The Robotics Institute, Carnegie-Mellon University, Pittsburgh, PA, 1988
18. Laperriere L, ElMaraghy HA. An assembly sequence generator. Proceedings ASME flexible assembly conference, Montreal, Canada, 1989, pp 15–22
19. Rondeau JM, ElMaraghy HA. ROBOPLAN – intelligent task planner for robotic assembly. Proceedings ASME flexible assembly conference, Montreal, Canada, 1989, pp 23–30
20. ElMaraghy HA, Hamid L, ElMaraghy WH. ROBOCELL: a computer-aided robots modelling and workstation layout system. Int J Adv Manuf Technol 1987; 2: 43–59
21. ElMaraghy HA, Gu P. Knowledge-based system for assignment of parts to machine cells. Int J Adv Manuf Technol 1988; 3: 33–44

Specification and Management of the Knowledge Base for Design of Machine Tools and Their Integration into Manufacturing Facilities

G.Q. Huang and J.A. Brandon

Introduction

Engineering design problem solving, in general, is ill structured, i.e. high diversity, complexity and uncertainty are involved. Different strategies and techniques are employed in various design domains, and even within the same domain diversely by different designers. Design heuristics are difficult to organize in a form suitable to be coded into computer systems. The question of what underlies empirical and experiential expertise and skills in specific domains leads to a need for research on developing a general model of the design process. There seems to be little possibility however, in general, of achieving this goal without first experimenting with specific domain problems. A complete model of a manufacturing system must contain information describing both plant and product, reflecting the design process in systematic terms (both organizational and technological).

Machine tools have a twofold role both as products and elements in mechanical manufacturing. Thus research on the modelling of machine tools can undoubtedly contribute to the modelling of manufacturing systems. To introduce the general integration of manufacturing facilities into total systems, there will be a brief discussion on general aspects of modelling manufacturing systems. Then a methodology of developing a knowledge-based integrated representation model of manufacturing facilities is presented, and the related work is briefly reviewed. In the second section, aspects of specification and management of the knowledge base for design problem solving with special interest in mechanical manufacturing are discussed. A metaphor for machine tool design is outlined in the third section.

Domain Identification and System Modelling from Multiple Views

Manufacturing systems admit a wide variety of domain representations. A straightforward view is its "bird's-eye" layout. Alternatively, a more popular view for theoretical analysis is obtained by decomposing a manufacturing system into "generic functions" [1]. Whatever view is taken, a subdomain "manufacturing facilities", which determines manufacturing capabilities, can be identified. A further subdomain "machining facilities" may be defined within it, as shown in Fig. 8.1. Machining facilities comprise equipment performing machining functions to change product characteristics (especially geometry) by material removal.

For both human and computer problem solving in a complex environment such as a manufacturing system, representation models are necessary to accommodate various techniques to deal with the system complexity. Smith and Wang [2] develop the methodologies of Checkland [3] and IDEFO [4]. They propose a composite methodology capable of meeting requirements of qualitative modelling and of integration with quantitative analyses. A more fundamental approach is due to Kim [5] who proposes "mathematical foundations for manufacturing", aiming to develop a methodology for synthetic design. He discusses various models and techniques suitable for multiple-view modelling of manufacturing systems. The type of model exploited in a specific manufacturing system depends on how well it can be defined and what is the primary concern. In practice, choices are sometimes made on the basis of user's personal

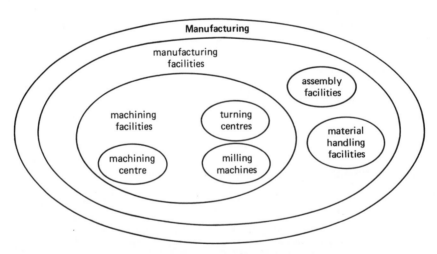

Fig. 8.1 Generic functions.

preferences rather than the sophistication of the methods. A brief summary is made by Seliger et al. [6].

The methodology that is followed here is developed from the work of Kim [5], Smith and Wang [1, 2], and the work about knowledge-based systems for selection and design of manufacturing facilities, which is reviewed towards the end of this section. Although the analysis can be generalized to complete manufacturing systems, this work concentrates on applying such a methodology to the modelling of a low-level subdomain (machining facilities such as machine tools).

There is no simple universal model satisfying all the requirements for all aspects of manufacturing systems. Instead, this can be compensated for by developing a basic abstraction model which can be extended into multiple application models by adding appropriate details. Such a model should be mathematically rigorous so that other mathematical processes can be applied later on. Taking the design of a mechanical product as an example, there is a reflexive relation between its geometry and functions. Once the geometry is determined, the functions are constrained and the manufacturing features are also roughly fixed by the technological environment. Therefore, the product model can be geometry-centred, function-centred or feature-centred. Function-oriented modelling implies explicit statements of functionality of a system with its structure unspecified or specified implicitly. It is often used in conceptual design and planning of new systems. Functional performances act as criteria for design solution evaluation. On the other hand, structure-oriented modelling (geometry-centred model, for instance) is increasingly exploited as the design is specified. A more general model to accommodate both functions and structures is such that if the functions of an unspecified structural entity in the model are known, then it is identified by its functional attributes; if the structure of a functional entity in the model is known, then it is identified by this structural attribute. Thus there is a duality relationship between structures and functions.

However, neither function- nor structure-oriented modelling are well suited for detailed study of operation of manufacturing systems. Activity-oriented modelling is required to communicate and control the rules and sequences of manufacturing activities. An activity is the realization of a specified set of functions by a set of structures within a set of defined environments in a certain order.

Knowledge-Based Selection and Design of Manufacturing Facilities

The consensus of the brief overview given above, on research about manufacturing, derives from many branches of science and has revealed that the definition of fundamental models of manufacturing is achievable but significant difficulties appear in implementation. The application of knowledge-based systems technology will reduce the above difficulties, and assure model consistency and integrity. Modelling a specific manufacturing system is usually company dependent. Thus company-specific expertise and skills in the design, operation and development of the system must be included in the model by separating the well understood or formal elements from the poorly understood or informal elements, in order to analyse its performance and behaviour. The formal elements are embedded in the system's knowledge base, while the informal elements consisting of specific human expertise are captured in an interactive way or in application programs. It should be noted that there are some general principles and disciplines under which manufacturing systems should be designed and operated by the appropriate arrangement of both formal and informal elements. Selection and design of manufacturing facilities is a process of matching manufacturing capabilities and product requirements under constraints of available resources.

Kilmartin and Leonard [7] have built a system for assistance in selecting advanced machine tools based on key machined components. Kusiak [8] discusses production equipment requirements problems. The ROBOSPEC system, developed by McGlennon et al. [9], helps in selecting appropriate robots for advanced manufacturing systems. Pham and Yeo [10] develop a prototype knowledge-based system for selecting grippers for robots. In addition, knowledge-based systems approaches have been applied in mechanical manufacturing from product design to the design of global manufacturing systems. Product design, process planning and operation design, tooling management, and fixture design are some of the application fields which have received attention. Dominic-I is an expert system developed by Dixon and Simmons [11] for designing mechanical products in an interactive manner by using an iterative redesign procedure. AIFIX was one of the first attempts made by Ferreira et al. [12] to use an expert system approach for designing fixtures. FADES is a knowledge-based system, developed by Fisher et al. [13] and used for the design of facilities. AIR-CYL is an expert system developed by Brown [14] to demonstrate a systematic theory of knowledge-based design problem solving in the field of mechanical product design. Identifying three classes of design problem solving, Brown maintains that expert systems technology is suitable for routine design of class 3 (qv.). The complexity of class 3 routine design problems can be high due to the large number of components. But it is relatively fixed and the ambiguity is low. Therefore, a well structured organization, conceptually isomorphic to the scheme of complexity decomposition, can be established for the problem solver which consists of a hierarchical collection of design agents including specialists, plans, tasks, steps and constraints. Mittal et al. [15] built the PRIDE system for designing paper-handling mechanisms. Conceptual design at an early stage is simply viewed as knowledge-guided search in a large space of possible solutions. Suitable solutions must be modified based on analyses. The builders of PRIDE believe that expert system technology is suitable for both of the above tasks required by design problem solving. They also consider the nature of collaboration (parallelism and series) between designers in a team, although few findings are implemented formally in PRIDE.

By implication, the knowledge-based systems approach can be employed in the design of manufacturing facilities such as machine tools. Here machine tool design has been taken as an example to demonstrate the applicability of knowledge-based systems methodology to manufacturing, because:

1. As a specific kind of mechanical product, machine tools themselves are designed, manufactured and assembled in the same manner as other kinds of product.
2. Being elements of manufacturing systems, machine tools also have a functional role.

The integrity of a machine representation model is extremely important not only to the modelling of itself, but also to the modelling of the manufacturing system in which it serves functions or is produced.

Specification and Management of the Knowledge Base for Machine Design

Design, in the context of intelligence and integrity, is primarily concerned with: the functions that the product must perform in service, the structure of the product, establishing the geometrical shapes of the parts which will make up the product, the materials from which they will be made, and aspects of the manufacturing and assembly processes mandatory to ensure the integrity of the design. The scope of knowledge-based design covers a number of aspects. This includes the use of computer-based systems to capture human expertise and to develop a concept for a product into a fully engineered design (either fully or partially automated).

The degree to which design can be automated depends on the state of a design problem and design knowledge. Design can be fully computerized with little or no human intervention if the design problem is well structured and the design knowledge is completely known and captured in the system in a way which facilitates effective management. Since design problem solving is ill structured, uncertain and complicated, human assistance is often needed when problem definition extends beyond the system capability. Most of the knowledge-based design systems begin their process by interactively interviewing the user. The purpose of the interview is to formulate the specific problem and tune the particular knowledge to suit it. Design decisions can be made simultaneously during the interviewing process or subsequently after it.

The system can leave the most uncertain decisions for the user to make.

A knowledge-based design system makes some decisions by applying stored knowledge to the information, to produce design proposals. It is the human operator who decides on the value of the proposal and ultimately takes the decisions. Therefore, the system must explain its reasoning to the designer as required, i.e. why a specific piece of information is needed, what it is currently concerned with, or how a decision is reached etc.

Domain Specification and Inference Strategies

The scope of domain definition and chosen inference mechanisms are of considerable importance for the successful implementation of intelligent knowledge-based system (IKBS) strategies. If the domain is too limited then solutions are likely to be trivial or if the inference mechanisms are too strongly constrained then creativity will be stifled. If the domain is defined too broadly or weak inference processes are used then design will be undirected and aimless.

As has been briefly discussed in the introductory section, the domain is identified as manufacturing facilities, and the concern is about their design and selection. The definition of these concepts is a dynamic process, the domain boundary itself shifting as new knowledge is incorporated.

Functions and structures and their combinations in a machine tool may be incorporated into design knowledge in terms of domain entities (objects), and treated as graph elements of a graphical abstraction model. The states of each entity are defined by a set of functional and/or structural attributes. It is not necessary for the entities to be available as marketable products in their own right, but abstracted as generic objects, a major form of design knowledge. A series of build-or-buy decision-making activities take place during the design process. Traditionally, domain objects at the primitive level are usually expressed integrally within the system for the use of designers, whereas the relations between them have to be constructed and manipulated by the designers themselves. A knowledge-based design system should be able to deal with the relational knowledge by means of inference strategies.

Design Classes

Design problems fall into certain classes according to their complexity, ambiguity, uncertainty, etc. By Brown's classification [14], a design problem is of class 3 if both its general form and detailed structures are known, and consequently the design knowledge can be captured by computer systems. It is of class 2 if its general form is known or partially known but it is not clear how it can be organized. Class 1 design is completely creative and inventive because the general form of an artefact and the knowledge about the solution is completely unknown. Across the spectrum of design problems, there are two main types of design methodology, parametric (variational, analytic) design method [16] and generic (conceptual, synthetic) design method [17]. Generally speaking, a parametric design method is suitable for class 3 routine design problems, and a generic design method for class 2. Both parametric and generic design methods can be accomplished by knowledge-based systems technology, with several possible variations depending on the domain in which the problems are defined.

Although creative design, in its complete sense, is out of reach, creativity can to a certain extent be explored in routine parametric design and conceptual design. Creativity in knowledge-based design systems is based on the theory of combinatorics [17]. A wide variety of complicated systems can be generated from a small and finite set of elements termed as system primitives. System primitives must be clearly defined, described and represented in the database and knowledge base. However, it may take an unacceptable time to manage system primitives by purely mathematical methods since their complexity increases exponentially with the number of primitives. It is also possible to generate meaningless results. Therefore, the knowledge guiding the search should also be identified in the domain and represented in the system.

To increase the efficiency, domain primitive objects can be clustered into groups. Group technology, as an engineering philosophy [18, 19], has penetrated into almost all aspects of manufacturing planning and design, from gripper design and selection for robots, to assembly modelling, to equipment design and selection, to facility layout arrangement, to process planning, even to the planning and design of entirely new systems. Although it is not termed "group technology", a similar methodology has also been exploited in computer-assisted problem solving for identifying generic tasks and activities [20]. Taking object-oriented programming as an example, an abstract generic object represents a set of concrete objects which share similar properties, and therefore a set of appropriate operations and methods can be defined for this class of objects.

Management of Uncertainty and Complexity

Uncertainty and complexity are twin problems involved in design. Even human designers have difficulties in dealing with the problems of complexity and uncertainty, and therefore they should not leave all these problems to computer systems, which may amplify the problems themselves. As a decision maker, a knowledge base management system should have the ability of uncertainty management, perhaps with the help of human designers. Strategies and techniques used in diagnostic inference are hard to realize in a form directly suitable for design problem solving. Complexity and uncertainty of design problem solving can derive from many origins, for example: from different design disciplines, from different domain perspectives, or from problem hierarchical features. Different formulations of the same problem may lead to different degrees of uncertainty and complexity. In such complex cases, the system should support various schemes for formulating the problem from multiple problem views, and have the ability of choosing a proper (acceptable or optimal) one in order to minimize the uncertainty and complexity.

The emphasis shifts as the design proceeds. In the early stage, the primary objective is the reduction of uncertainty. Uncertainty leads to great difficulties in the decomposition of a task, and the allocation of tasks to a specialized design team. If uncertainty is the primary characteristic of the domain, then the integrative process is relatively ineffective. Consequently, the task team should comprise a relatively small number of generic agents. If complexity is the primary characteristic, then problem solving can be managed effectively in a larger, more differentiated and specialized team. Traditionally, uncertainty is managed through a process of sequential stages of decision-making activities, often the responsibility of different specialized agents in the team. Complexity is managed through the decomposition of a design problem into discrete, separate, and manageable subproblems for a mixture of parallel and sequential processing. Problem decomposition is employed prior to task segmentation and allocation until all remaining problem elements can be managed and solved with limited human/computer cognitive ability.

Recognizing the hierarchical nature of both individual design problems and the overall design process is the key to better understanding, and effective management, of the uncertainty and complexity of the subproblems, the global problem and the interactions between them. A design problem can be decomposed into several subproblems which are relatively independent and simple. Accordingly, the design task is divided into sequential levels and parallel activities. An alternative way is to break a problem into subproblems similar to those in hierarchical decomposition, but they are arranged as graph elements of a network graph. The interactions between the parent, uncle, and sibling subproblems, which describe a problem and its internal processes, always exist and can be treated as constraints. Constrained reasoning on the basis of the theory of commitments is intensively discussed by Stefik [21] and implemented in the MOLGEN expert system. Related research in mechanical design has been reported [22].

Schemes by which a problem is decomposed provide important structures in guiding the design process, maintaining integrity and consistency, and resolving conflicts.

Integrity and Consistency Maintenance and Conflict Resolution

In general, information and knowledge must maintain integrity and consistency as design progresses, although local and temporary inconsistency is both allowable and necessary to enable design iteration. Integrity implies the maintenance of functionally related information and knowledge. Consistency is the maintenance of the equivalence of redundant information and knowledge. Integrity and consistency are two significant issues for a knowledge base management system.

As mentioned before, it is usual to divide a system in terms of functions or structures or both. Any product design should start with the recognition of its functional requirements derived from knowledge of its intended operating environment. The functional specifications are matched onto their implementation space implicitly or explicitly or both. The functional and structural integrity and consistency must be maintained during this process. However, conflicts often appear since there is usually no one–one correspondence between functions and their implementations. Different functional attributes can be implemented by the same structures. For example, functional attribute f_1 can be replaced by structures s_1 and s_2, and f_2 by s_1 and s_3. For the function set $F = (f_1, f_2)$, the system must be able to decide what the structure set should be like,

$$S_1 = (s_1, s_2, s_3)$$
$$\text{or} \quad S_2 = (s_1, s_2, s_1, s_3)$$
or others

Considering a slightly more complicated example, two functional attributes f_1 and f_2 can be implemented by structure sets S_1 and S_2 respectively. There is an elementary structure s_{11} in S_1, and s_{21} in S_2. Suppose that it has been known that s_{11} and s_{21} are in conflict with each other according to the stored knowledge, and unfortunately there is no way of avoiding these two conflicting structures in the final solution. The system should be capable of utilizing hierarchical knowledge at more refined levels of structural implementation to resolve the conflicts. Conflicting structural implementations may result from conflicting functional requirements. To detect conflicts, the system must check the consistency of functional requirements. Some of the issues which conflict resolution must address include: conflict detection, identification and location; and the definition of resolution strategies.

It is clear that the inference process of designing production machines and manufacturing systems must be capable of conflict resolution but if this mechanism is dominant in the early stages of design then creativity, which thrives on conflicting ideas, will be stifled. Consequently the relative importance of the different inferential mechanisms will change, depending on the current status of the design.

In dealing with interactions and conflicts represented as constraints, the theory of commitments can be used. The least commitment principle states that a decision must not be made until and unless all the relevant information is available or no more can be obtained. By implication, the separation of functions from their least-committed implementations enables the generation of possible solution variants and allows consideration of many alternatives without overlooking relevant solutions. On the other hand, by the early commitment principle, a design is specified structurally as early as possible in the design process. Progressively more detailed information enables the application of a wider range of both qualitative and quantitative analytical techniques. Thus more accurate estimates of system performance can be obtained. In the modular design methodology, functions and their implementations are mixed based on the opportunistic commitment principle, which suggests that in a problem domain, some functions be always implemented by one or several fixed modular structures, which are regarded as proven according to current technology.

Viewing a knowledge-based system as consisting of a database, a knowledge base, an inference engine, a knowledge application system, and some methods of interacting with the real world, integrity, consistency and compatibility must be maintained among them. An integrated knowledge base must be accessible to multiple application programs and support a developing environment. Knowledge for conceptual design and detailed design, knowledge about manufacturing design and production planning, and knowledge about utilization should be interrelated.

Different application programs involve a number of design activities which require a variety of knowledge to be represented in different schemes and various inference strategies are used. Knowledge management and inference strategies depend not only on representation schemes but also the domain knowledge itself. The knowledge attached to the same object must be represented in different ways to achieve the most suitable solution from a particular aspect in the most efficient way. The central activity of inference is that of search. Conceptual design searches within a large solution space and choices are made among discrete alternatives. Alternatives are generated incrementally, dependent on the perceived difference between the previous solution and the expected one. The change can be measured either numerically and/or symbolically.

It can be seen from the above arguments that forcing all types of knowledge to be represented by a single form may lead to a distortion in the content of the knowledge, or invalid decision. The schemes of frame, rule, procedure and logic types, are usually used to represent object, causal, process and relationship knowledge respectively. Redundant representation by multiple schemes can be avoided by a dynamic elicitation strategy. That is, dynamic exchange in representation schemes without recompiling any schemes or consulting other schemes.

Multiple application programs, knowledge representation schemes and inference strategies must maintain compatibility within the system as a whole. Different approaches can be employed in solving a same domain problem. The knowledge in the same domain can be structured in several schemes. The knowledge in different domains can be represented by the same formalism. The knowledge represented in the same scheme can be employed by different inference strategies, or the knowledge represented by different schemes can be managed by a single inference strategy. Hybrid representation schemes and unified inference strategies seem to be necessary to design problem solving. For the above reasons, the use of the knowledge by the system must be compatible with states, contents and forms of the knowledge.

Generic Design Tasks

Engineering design, as problem solving, comprises general decision-making activities and tasks, which may need to be processed in parallel or series. It is essential to identify and incorporate generic design activities and tasks in the knowledge base, in a similar way as design agents are employed in the AIR-CYL system. Having broken a problem into an algebraic structure, a set of specialists are organized conceptually isomorphic to the complexity scheme. Each specialist has a set of operations defined to deal with the attributes and their relations to others.

Analysis and synthesis are two of the generally recognized design operations. The first phase of computer application to engineering design is analysis, which is now best understood. Design synthesis takes place simultaneously with design analysis but with less success in practice. The general form for both design analysis and synthesis can be summarized as "applying model M to entity E with attributes A in environment S under constraints R to derive attribute value V".

- In analysis, GIVEN M, A of E, S and some of R and V,
  DERIVE the others in V
  and PREDICT the others in R.
- In synthesis, GIVEN M, S, some of A of E, some of R and some of V,
  DERIVE the others in A, R and V.

Design analysis can be integrated into a design expert system in several ways. One way is to call the accessible and relevant analysis packages to perform a detailed analysis to determine if performance constraints and goals are being satisfied. Another way is to treat analytical knowledge as constraints in the knowledge base.

There are a number of well-known techniques for numeric quantitative design both in analysis and synthesis. However they demand that the problem be well defined and structured. On the other hand, design at the higher levels (such as conceptual design) can only be performed qualitatively by symbolic processing on strategies, which is domain dependent and heuristic. The quantitative techniques must be compatible with and guided by qualitative strategies so that an accurate solution can be reached.

By either analytic or synthetic approaches, design can be globally divided into several sequential stages, at each of which attention is paid to certain parts of the problem. The process stages are generally determined by levels of the problem complexity hierarchy. Early stages are concerned with principal requirements and key constraints, while the late stages deal with more detailed parameters.

Bearing in mind the generic design tasks and activities, it is possible to define more sophisticated design procedures which are sequences of decision-making activities and tasks. Different domain problems require different precedence in these sequences. Similar tasks and activities or procedures can be used in different stages for solving one or several different problems. Knowledge should be clearly organized and easily identified for different stages or in the knowledge base. A knowledge base management system must recognize the situations and control the process.

Having recognized the sequential aspect of design, which is necessary in order to reduce ambiguity, or because the later activities are dependent on the decisions of the earlier activities, it is necessary to recognize the need for parallelism in design problem solving [23]. Human designer teams in the real world are highly parallel systems. This is the natural consequence of the complexity of design problems themselves. Quality of design, expenditure of resources, and project duration constitute the multi-dimension of company preferences. The cognitive capability, information processing and storage of both human operators and computers are limited. These three characteristics of design problem complexity, company preferences and human/computer cognitive ability result in the creation of highly specialized and parallel design organizations.

Responsibility for parallel design tasks and activities may be allocated to different agents in a differentiated design team. Each agent in the team is specialized in, and responsible for, one or more subproblems or aspects of the problem. Specialization improves design quality and creates diverse points of view. Different individuals may emphasize different aspects of the problem when presented with identical problem situations. Therefore, they are able to apply their own skills and expertise to appropriate aspects of decision making, or to cooperate with and help others when they encounter difficulties. Management of design is a process of integrating diverse sources of knowledge and points of view in an environment characterized by high levels of ambiguity. By analogy, it is both possible and desirable to implement a group decision-making methodology in knowledge based systems for design problem solving. Much work remains as the complexity of the problem is not clearly defined.

Geometric Reasoning

As has been discussed previously, functions and features are closely related to the geometry of mechanical products. Functional design entails the analysis of functions of specific geometrical parts constituting the product to achieve a synthesized performance. Manufacturing and assembly features such as slots, shoulders etc. can be recognized and extracted from a geometrical model with the addition of certain kinds of knowledge. In essence, product models are basically geometry-based for design, feature-based for manufacturing and function-based for analysis. In general, the selection and early design of machine tools are dominated by the geometry of workpieces. A design system for machine tools must carry out geometric reasoning about dimensions, shapes, forms, spatial relations, etc., of both workpieces and machine components.

Taking the effect on machine tool design as an example, workpieces are preferably modelled in terms of manufacturing features including geometric form features and technological features. Geometric form features are groups of geometric entities that define attributes of nominal dimensions, shapes, and relative positions etc. It is important to recognize the major form and shape (called primary geometric features) of a workpiece. Technological features are necessary deviations of geometric form features. Dimensional, shape and positional tolerances, surface finishes, and part materials are three of the major technological features. In this connection, geometric reasoning in machine design has the special task of dealing with the relationships between geometric features and their functional purposes.

In addition to the above, geometric reasoning is one of the major processes in machine design. A machine tool as a geometric object is assembled from a number of components which are in turn geometric objects. Design is sequentially divided into several stages. Computer graphics requires strictly exact details about the geometries of an object. This is not satisfied at early stages. Therefore, qualitative geometric-reasoning techniques are preferable. For example, dimensions are defined by a range of values instead of exact values. These value ranges are refined to exact values as the design proceeds to later stages for which, consequently, computer displays are suitable.

Proposal Generation and Solution Evaluation

The problem specification mechanism has a problem space which contains the domain problems of concern. Within it, a point can be identified by a set of specifications. Details are added to the point by the complexity decomposition mechanism. Meanwhile, a solution space is generated by the proposal generation mechanism. More points are added to and detailed in the space by the design deduction mechanism. Reasonable solution points in the space are selected for further consideration. Usually, a number of solutions, which satisfy the main requirements and key constraints but differ slightly in certain attributes, are generated during the design process. Different criteria can be defined such as possibility, feasibility, acceptability and optimality for the evaluation and assessment of design solutions. Knowledge derived during evaluation of design solutions should be incorporated in the knowledge base as a solution filtering mechanism, under several modules such as acceptance module and optimization module. The acceptance module eliminates nonfeasible solutions to generate a list of realizable ones for further analysis. The aim of the optimization module is to find out the best solution among the possible/acceptable solutions based on the resources available.

Backtracking strategies are often used in knowledge base management systems, since they form an integral part of the inference process. They may be used recursively to decide when and where backtracking is needed, explain why it is used, and how it is performed. The cost in time and design resources is often very high, rising with the extent of backtracking, as a design is progressively decomposed into subproblems for parallel processing. Backtracking strategies commonly used are chronological and dependency-directed backtracking. Chronological back-tracking reverts to the last decision point if any failure is found at the current point. It is simple and systematic but inefficient. On the other hand, dependency-directed backtracking [24] intelligently goes back to the last decision point which has contributed to the failure or contradiction. This is carried out according to the records which are created during any constrained reasoning to remember the sources upon which the inference depended. Once an inference source has been found to be contradictory, all dependent inferences and decisions are retracted.

In addition to the fact that a number of mechanisms mentioned above employ backtracking, other mechanisms are available in special implementations of backtracking strategies. They include failure handling, design iteration, design recursion, design modification and design optimization. Design is backtracked when one or more requirements or constraints fails to be satisfied. Design iteration is used to obtain alternative problem solutions by criteria defined in solution evaluation such as possibility, feasibility or acceptability. Design recursion is applied when the same procedure can be employed repeatedly as a cycle until a solution is found or a failure is met. Design modification improves a design proposal by modifying the values of some of its attributes without any change in its general form. Design optimization establishes the best design plan or tunes a design proposal to the level of optimality. Redesign must be carried out if no acceptable solutions to a design problem can be found during a design course.

Intelligent Integrated Modelling and Design of Machine Tools

A machine tool can be modelled in terms of functions, structures, and/or both of them. For example, a machine tool, as a hardware implementation of the generic function "machining", is modelled by a piece of software as a processor with an input and output, which executes algorithms. Machine designers should also take the production aspect into account as shown in Fig. 8.2, so that integrity in its utilization environment and production environment is maintained.

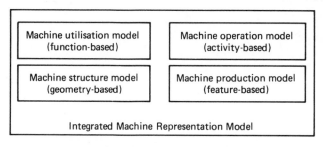

Fig. 8.2 Integrated machine representation model.

Topological Machine Representation Model

A topological model for representing machine tools, initiated by Ito [25–29], has been developed [30], based on an investigation into various existing methodologies used in the machine design literature [31]. Incompatibility still exists in such a composite model but a compromise can be achieved by defining appropriate intelligent environments. Functions, structures, activities and features, within a machine tool, can be abstracted as objects or entities. They can be arranged as graph elements (vertices and edges) in network graphs intuitively, and represented internally by symbolic notation as object frames. The relational knowledge is illustrated as edges of the graphs and represented by pointers of the frames. Topological graphs constructed this way share global and local similarities, because of the mutual dependency between entities. The key features of a graphical machine representation model are described in detail elsewhere [32] and will therefore only be illustrated here by the example shown in Fig. 8.3.

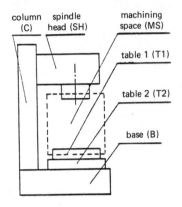

(a) A vertical machining centre

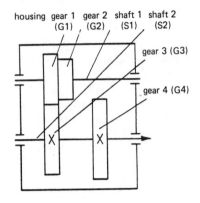

(b) A simplified spindle head

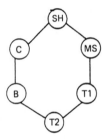

(c) Graphical model of machining centre (a)

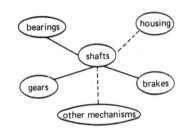

(d) Generic structure of spindle head (b)

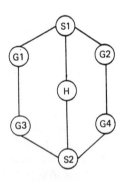

(e) Graphical model of spindle head (b)

(f) Hierarchical model of machining centre and spindle head

Fig. 8.3 An example of graphical machine representation model.

Towards Intelligent Modelling and Design

Similar to AIR-CYL approach, a team of design agents each of which is specialized in, and responsible for, one subproblem or aspect is established according to the problem complexity decomposition. One perspective is that of aspects in machine design including functions, structures, performances, cost, maintenance, reliability, production, assembly, etc., as shown in Fig. 8.4. Expertise on these aspects is covered by design agents in the team. They are responsible for evaluating design proposals, supervising the design process, and maintaining integrity. Simple and direct collection of the opinions and decisions of all the agents in such a design team does not constitute the solution to the total problem. For this, the machine design problem should be examined from a different perspective.

The perspective obtained by decomposing a machine tool into physical structures suggests an isomorphic organization to the structure of the design teams. For example, a lathe is typically composed of a spindle head, a base, two tables and a column, as shown in Fig. 8.5. Each agent in the team proposes a solution to the subproblem, which is under its responsibility and specialization. An integration of the subsolutions to the subproblems intuitively constitutes the design proposal to the total

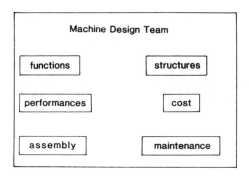

Fig. 8.4 Problem complexity decomposition.

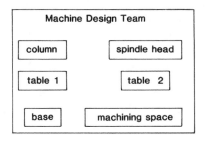

Fig. 8.5 Isomorphic organization.

problem. However, this proposal is not guaranteed to be good or reasonable. Supervision and assistance during the course of design and subsolution integration from the first team is needed.

Neither of the above design teams guarantees to generate a good solution to the machine design. Therefore, a metaphor analogous to human design organizations, for cooperative group decision making by multiple design agents and multiple design teams, is required. The whole design decision-making process is the cooperative effort of these teams of design agents, and the integration of their points of view. A team of design agents executes actual decision making. There may also be another team for giving advice during the design. A chair agent, usually the user in most systems, is required to organize the teams, allocate the tasks to the members, and conduct the whole design process. A collection of executive and advisory design teams and design agents contains the information about the designed machine, and how it is designed, as well as what has been used. Design information and knowledge can be distributed and represented in agents, teams, and their common knowledge/data base in the forms of rules, frames, etc.

The metaphor described above is far from being a fully automated knowledge-based design tool for industrial problems. However, this is a challenging step towards design problem solving by analogy to the cooperative approach used in human engineering communities.

Knowledge-Based Design of Machine Tools

The process of machine design can be generally divided into four main stages or phases: preliminary design, concept design, engineering design and detailed design.

Preliminary Design – Requirements Processing

Any decision must entail a stage of requirements processing termed the preliminary design stage. During the initial requirements stage the design problem is defined, detailed and formulated. With the definition of input requirements, general strategies of problem solving can be recognized. The tasks of the preliminary design are globally represented as a flow chart shown in Fig. 8.6. However, local parallelism is allowed (not indicated in the chart).

Input to the preliminary design is very rough and entails diverse requirements, which are described symbolically and/or numerically. They can be read into the system from a disk file or by interviewing the user. In both cases, they are recorded in an internally equivalent format.

After all the requirements have been read, the system starts carrying out requirements analysis. First of all, the requirements have to be

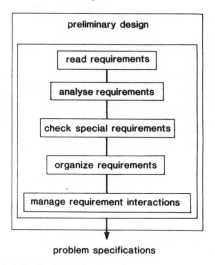

Fig. 8.6 Preliminary design – requirements processing.

checked to ensure that they fall within the scope of the system capability, as required by every design agent involved in the design teams. Any special requirements must be processed, by the user or the system, after acquisition of appropriate special knowledge, otherwise an invalid decision will be made. Alternatively the system may be unable to reach a decision at all.

Different design agents are concerned with different categories of requirements. The categorization can be carried out by posting a relevant requirement to an agent if the relationship "who is concerned with what" is known in advance. On the other hand, a design agent is able to decide whether it has an interest in a specific requirement, simply by checking whether the requirement is contained in its concern set in the knowledge base. If no record is found in its concern set, the system assumes that the design agent is indifferent to this requirement, that is, the requirement has no effect on its decision making. A particular requirement may be of concern to several agents.

The effects of two or more requirements, when they appear concurrently, are usually different from those when they are considered individually. Interactions (interdependence and conflicts) are involved. For a specific agent, two requirements may seem to be related. But for another agent they do not interact at all. In dealing with requirements interactions, cooperation between the agents is usually necessary. Preliminary design is a preparatory stage which are usually undertaken by human expert designers instead of computer systems.

Conceptual Design

Preliminary design results in the preparation of a set of design specifications which are well organized and formulated. Conceptual design is simply the interpretation of the specifications, to generate machine overall configurations, which consist of generic structures. The conceptual design of machine tools starts with functional design, proceeds into structural design, and switches between them as shown in Fig. 8.7.

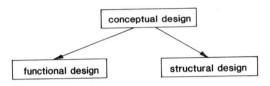

Fig. 8.7 Conceptual design.

Functional Design. The first task of functional design is to transform the requirements, under the category of machining features, into appropriate machining functions. It is feature-driven. For example, given a feature under certain environment, there is a set of candidate machining functions for choosing.

The machining functions for all machining features are matched to primary functions. For each machining function, a set of candidate primary functions is suitable. For the whole set of machining functions, there are a number of sets of primary functional combinations. They are nonordered sets, and therefore have to be arranged in certain patterns.

Structural Design. The generation of functional patterns is usually mixed with the structural design. At this step, structural modules or primitive structures are first selected in order to replace the primary functions. Some of the primary functions have to be replaced by the relations (motions or assembly relationships) between structures. Having done this, there are usually several ways of arranging the selected functions and structures into machine representation patterns.

The output of the conceptual design is the release of machine representation patterns which are represented symbolically as formulae. It is generally required that such a result be displayed on the screen for the user to carry out an intuitive evaluation for later design. However, computer graphics require strictly exact details about the geometry of an object. This is not satisfied at the conceptual level.

Once the overall configuration of a machine tool is defined, the design proceeds into lower design levels by a similar approach to that used in AIR-CYL.

Conclusions

It is possible for designers to analyse a machine tool as a product of a manufacturing system and simultaneously to include its production aspects, as hardware equipment in its utilization environment in order to maintain the general integration of manufacturing systems.

The work reported here is not related to any specific machine tool builders or purchasers, since different machine manufacturers have different approaches to the design and integration of manufacturing technology. Furthermore, different users have different requirements for specific functional or physical arrangements in the system. Consequently company specific technological environments must be included in design systems. This kind of flexibility has to be balanced by other criteria such as cost constraints. Machining operations involved are turning, milling, drilling and some kinds of their combination as in machining centres. To date the system under development must be considered as a prototype, since insufficient domain knowledge has been acquired and only a limited number of factors are taken into account in order to simplify and demonstrate potential of the methodology. However, it is possible to incorporate the potential to accommodate expertise and skills in the domain, if human experts are available. For this purpose, the general aspects of specification and management of knowledge base specially for mechanical design have been discussed on the basis of the literature in this field.

A cooperative metaphor has been proposed which simulates a human design team, although much challenging work remains for further research. Such an enhanced system will prove powerful in machine design and in the modelling of the related decision making.

References

1. Smith GW, Wang M. Modelling CIM systems Part 2: the generic functions. Comput Integr Manuf Syst 1988; 1: pp 169–178
2. Smith GW, Wang M. Modelling CIM systems Part 1: methodologies. Comput Integr Manuf Syst 1988; 1: 13–17
3. Checkland PB. Systems thinking, systems practice. John Wiley, 1984
4. Yeomans PH. Improving quality and productivity in systems modelling using the IDEFO methodologies. MicroMatch Ltd, 1985
5. Kim S. Mathematical foundations of manufacturing science. PhD Thesis, MIT, 1985
6. Seliger G et al. Descriptive methods for computer-integrated manufacturing and assembly. Int J Robotics Comput Integr Manuf 1987; 3: 15–21
7. Kilmartin BR, Leonard R. Selecting advanced machine tools by a systems approach based on key machined components. Proc Inst Mech Eng 1983; 197B: 261–269
8. Kusiak A. Production equipment requirements problems. Int J Prod Res 1987; 25: 319–325

9. McGlennon JM, Cassidy G, Browne J. ROBOSPEC: a prototype expert system for robot selection. In: Kusiak A (ed) Artificial intelligence: computer integrated manufacture. IFS, Bedford, 1987

10. Pham DT, Yeo SH. A knowledge based system for robot gripper selection. Int J Mach Tools Manuf 1988; 28: 301–315

11. Dixon JR, Simmons MK. An architecture for the application of artificial intelligence to design. In: Proceedings ACM/IEEE 21st annual design automation conference, 1984; pp 634–640

12. Ferreira PM, Kochhar B, Liu CR, Chandru V. AIFIX: an expert system approach to fixture design. In: Liu CR, Chang TC, Komanduri R (eds) Computer-aided/intelligent process planning. ASME, New York, 1985, pp 73–82

13. Fisher EL, Nof SY. FADES: knowledge-based facility design. In: Annual of International Industrial Engineering, Conference Proceedings, Chicago, IL, 1984, pp 74–82

14. Brown DC. Expert systems for design problem solving using design refinement with plan selection and redesign. PhD Thesis, Ohio State University, 1984

15. Mittal S, Dym CL, Morjaria M. PRIDE: an expert system for the design of paper handling systems. IEEE Comput 1986; 19: 102–114

16. Suzuki H, Kimura F, Sata T. Variational product design by constraint propagation and satisfaction in product modelling. Ann CIRP 1986; 35: 75–78

17. Ulrich K, Seering W. Computation and conceptual design. Int J Robotics Comput Integr Manuf 1988; 4: 309–315

18. Burbidge JL. Group technology in the engineering industry. Mechanical Engineering Publications, 1979

19. Gallagher CC, Knight WA. Group technology methods in manufacture. Ellis Horwood, Chichester, 1986

20. Chandrasekaran B. Generic tasks in knowledge based reasoning: characterizing and designing expert systems at the right level of abstraction. In: Proceedings IEEE international conference on AI applications, 1985

21. Stefik M. Planning with constraints (MOLGEN: Part 1). Artif Intell 1981; 16: 111–140

22. Sata T, Kimura F, Suzuki H, Fujita T. Designing machine assembly structure using geometric constraints in product modelling. Ann CIRP, 1985; 34: 169–172

23. Kornfeld WA, Hewitt CE. The scientific community metaphor. IEEE Trans Syst Man Cybernet 1981; 11: 24–33

24. Stallman R, Sussman G. Forward reasoning and dependency-directed backtracking in a system for computer-aided circuit analysis. Artif Intell 1977; 9: 135–196

25. Ito Y, Saito Y. Computer-aided draughting system "ALODS" for machine tool structures. Proceedings 22nd international machine tool design and research conference 1981; pp 69–76

26. Ito Y, Shinno H. Structural description and similarity evaluation of the structural configuration in machine tools. Int J Mach Tool Des Res 1982; 22: 97–110

27. Ito Y. Description of machine tools and its applications – CAD system for machine tools structures. Bull Jpn Soc Precision Eng 1984; 18: 178–185

28. Ito Y, Shinno H. Generating method for structural configuration of machine tools. Trans JSME 1984; 50: 213

29. Ito Y, Shinno H. A proposed generating method for the structural configuration of machine tools. ASMEJ 1984; 84-WA: Prod-22

30. Huang GQ, Brandon JA. Topological representations for machine tool structures. In: Davies BJ (ed) 27th machine tool design and research conference, Manchester. Macmillan, 1988, pp 173–178

31. Huang GQ, Brandon JA. An investigation into machine representation models for machine tool design. In: Conference factory 2000. Institution of Electronic and Radio Engineers, Churchill College Cambridge, pp 331–336

32. Huang GQ, Brandon JA. Machine tool analysis and synthesis based on the graphical machine representation model, In: Worthington B (ed), Advances in manufacturing technology, vol 3. Kogan Page, pp 100–104

Process Design

This section contains four chapters. The first chapter, by Inui and Kimura, describes an intelligent system incorporating product modelling techniques for designing process plans for machining prismatic parts and bending sheet metal components. The second chapter, by Milacic, proposes the use of the theories of formal grammars and automata to acquire and represent knowledge for expert process design systems. The chapter also outlines two such systems, one for designing process plans for machining rotational parts and the other, for producing plans for prismatic components. The third chapter, by Ito and Shinno, discusses the philosophy to be adopted in the development of the next generation of AI-based systems for machining process design. The last chapter in this section, by Wright et al., reports on the progress achieved by the authors' team in building an AI system that integrates machining process design and workholding element configuration.

## Chapter 9

Design of Machining Processes with Dynamic Manipulation of Product Models

M. Inui and F. Kimura

Introduction

Process planning is recognized as a critical bridge between designing and manufacturing stages [1]. In this process, manufacturing information required for realizing the product is determined, based on the complete machine part definition. Therefore, computerized process planning is said to be a key technology for constructing computer-integrated manufacturing (CIM) systems. Recently many types of automated process planning systems have been constructed. Expert system techniques are introduced for such systems in order to utilize expert production engineers' knowledge for generating process plans. However, such "expert" process-planning systems still have some problems in achieving higher levels of automation.

One problem is the difficulty in preparing enough information about complicated products. For process plan generation, process-planning systems usually require the characteristic description of products such as form features, dimensions and tolerances. In many cases, current systems do not incorporate the concept of integration with the product design stage, so system users must prepare such product information.

Process planning includes a wide range of manufacturing preparation stages from the blank material designing to the determination of machining parameters. In order to attain total automation, process-planning systems have to provide many capabilities for evaluating various kinds of constraints appearing in those stages. Most of the constraints are closely related to machining operations and their effects on the workpiece. However, current systems lack the ability of simulating the workpiece change according to the operation, so some process-planning constraints are difficult to evaluate.

To solve these problems, we propose a new process-planning system with advanced product modelling techniques. In manual practice, product information is normally defined in the form of machine drawings and utilized in many manufacturing stages. Instead of machine drawings, so called product models [2] in computers are used for representing such product information. Product design and process-planning stages are able to be integrated with this product model. Some procedures are also prepared for manipulating the product information, for example, creating and eliminating form features. By using such procedures, two types of simulation, those of prismatic part machining and sheet metal part bending, are implemented and incorporated in the proposed process-planning system. These simulators manipulate the product information according to the machining (bending) operations, and the simulation results are effectively used for evaluating various constraints in process planning.

The organization of this chapter is as follows. In the next section, a brief overview is given of automated process-planning research. In the third section, the basic representation framework of product modelling is explained. As an example of product modelling, definition and manipulation of form features are discussed in the fourth section. In the fifth section, as another example, definition and manipulation of sheet metal parts are discussed. Sheet metal parts have a very distinct feature from the prismatic parts, however, they can be handled in the same product-modelling framework. The structure and basic strategy of our product-modelling-based process-planning system XMAPP are explained in the sixth section. As the first application of XMAPP, automatic process planning of prismatic part machining is discussed in the seventh section. In the eighth section, as another application of our system, automatic process planning of sheet metal part bending is discussed.

An Overview of Automated Process-Planning Systems

Variant Process Planning and Generative Process Planning [3]

The importance of process-planning automation is well understood and much research work has already been done. Through these activities, computerized process planning can be classified into two major categories, which are variant process planning and generative process planning.

Variant process planning is a process plan database retrieval method based upon the similarity among machine parts. As a preparation, existing machine parts are coded according to the characteristic shape and technological specification and classified into some part families. For each family, a corresponding standard process plan is manually defined in consideration of the product description and stored in the database with identification code of the family. Once this standard process plan database is completed, any new incoming part can be coded in the same manner. The database search procedure finds out the most similar family to which the new part should belong, and the standard plan of the family can be referred to. When the new part has partly different specifications from the expected part of the family, a certain amount of modification must be manually done on the referred plan.

On the other hand, generative process planning synthesizes the machining process information to create a process plan for a new machine part automatically. During the preparatory stage, the knowledge about the manufacturing process is captured, classified and coded as computer programs. When a machine part is given for process planning, the system imitates a human process planner's decision-making process by using such programs on the product descriptions and determines necessary machining information. The generative process-planning system could generate proper plans without human intervention. However, process-planning methods are, as yet, not standardized or well organized for implementing complete programs. In reality, production engineers are required to use a lot of practical experience concerning the manufacturing process to accomplish their task.

In order to solve this problem, recently, many experimental process-planning systems using *expert system* techniques have been developed. An expert system is a program which represents the human experts' problem-solving knowledge as declarative rules, and make inferences with them to deduce the solution of the problem. By using this rule-based inference technique, it is expected to capture expert process-planning knowledge in computers. Almost all the process-planning systems currently under research use expert system techniques.

Process planning can be understood as the transformation of the complete machine part definition into the manufacturing process information suitable for modifying blank material into the finished part. Machining operations have a close relationship to such transformation, so they must be properly represented in terms of rules and used in the inference mechanism of the expert system. In the following discussion, some expert process-planning systems are briefly reviewed to show how such a transformation is realized.

Expert Generative Process-Planning Systems

The GARI system by Descotte and Latombe [4] is one of the earliest systems to introduce the expert system technique for process planning. In GARI, a complete machine part is represented as a set of form features, which represent characteristic shapes of the machine part such as holes and grooves. In the same framework, technological details such as dimensions, tolerances and surface roughness are also described. This form feature-based representation is commonly used in other process-planning systems, as will be explained. GARI first creates an initial and loose process plan by assigning roughing cuts and finishing cuts on such form features. Then the system applies various rules concerning machining sequences on the initial plan and successively restricts it to being suitable for actual machining. So, the transformation of the product information to the machining process is mainly attained in the initial plan preparation. In this sense, GARI does not have an explicit model of machining operations.

Several process-planning systems [5], designed in the same framework as GARI, use more efficient reasoning techniques. For example, HI-MAPP system by Berenji and Khoshnevis [6] introduces an efficient hierarchical planning technique to reduce the amount of backtracking.

TOM, developed at the University of Tokyo [7], is an automated process-planning system specialized for hole-machining processes. This system generates a process plan by searching backward from the complete state of the machine part to the initial blank state of the material. In order to realize this strategy, TOM represents the knowledge of the hole-machining operation as production rules, whose antecedent parts represent the state of the hole to be realized by the operation, and consequent parts describe the prestate of the workpiece which is suitable for executing such an operation.

The SIP system by Nau and Gray [8] also uses the backward search strategy; however, their knowledge representation framework is not production rules as adopted by TOM, but frames in which the knowledge about the machining process can be organized in a taxonomical hierarchy.

The STOPP system by Choi and Barash [9] also generates a process plan in the backward search manner. STOPP introduces two concepts named elementary machine surface (EMS) and simple process cycle (SPC) for process planning of the machine part with holes and slots. An EMS can be understood as the surface generated by a unit-machining operation represented by an SPC. Thus, in STOPP, process planning is recognized as a series of mappings from an EMS to an SPC, which generates a new EMS corresponding to the state before executing the SPC. The system continues this process until an EMS corresponding to the blank state is obtained.

The XPLANE system developed at the Twente University of Technology [10] also uses the backward search method that starts with a completely machined form feature with the solid model representation. It applies a series of solid-modelling operations corresponding to inverse-machining operations that fill up the feature until its blank state is achieved. XPLANE is different from other systems, because it uses the complete geometry of the solid model for describing form features; however, the mechanism for modifying the solid model according to the inverse operation is not clear from the available literature.

As discussed, all of the systems introduce some special inference model to implement the machining operation knowledge in the expert system's framework. However, their models are simple and limited to evaluating the specific process-planning constraints (especially concerning operation planning), so it is difficult to evaluate other process-planning constraints in the same model. For example, almost all the systems adopt the form feature-based workpiece representation with no precise geometric models, so it is difficult to detect collision of the tool with the workpiece in process planning. In order to overcome this problem, we propose to introduce advanced product-modelling techniques for process-planning automation.

Product Models

A "product model" is a computer-internal model which contains a wide range of engineering information concerning the products, which is necessary for designing, process planning and other manufacturing activities. In process planning, we especially need product models of workpieces, tools and the product itself. In this section, our concept of product models and a framework for representing them are briefly described.

Product Model Concept

The designing and manufacturing process is generally regarded as an information-processing process, in which the information necessary for machine production is generated, retrieved, modified and utilized. This information contains various kinds of basic engineering knowledge concerning design objects and manufacturing environments (such as geometrical shape; form features; dimensions; tolerances; assemblies and so on), and concerning designing and manufacturing tasks.

In advanced CIM systems, this information must be treated by computers in an integrated manner, and we propose product models to describe design objects and manufacturing environments. Briefly speaking, product models are generated in designing processes and are utilized in manufacturing preparation processes with the knowledge about the manufacturing tasks (Fig. 9.1). Because it is, however, a difficult problem to represent all this knowledge, we restrict the scope of product models to constraint conditions imposed on physical properties of the product.

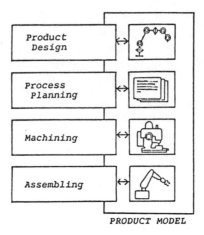

PRODUCT MODEL

Fig. 9.1 Role of product models.

Product Modelling Framework [11]

The basic idea of the framework is that the product information is treated as constraints on properties of products. A typical example is the dimensions in machine drawings. A main function of dimensions is to nominally define a product's shape. This means that dimensions impose geometric constraints on the product shape. For instance, they constrain a distance or an angle between two faces, a radius value of a cylindrical hole, and so on. Many of the important concepts composed in product modelling, in addition to dimensions, can be treated in terms of constraints, or more specifically in terms of geometrical shapes and geometric constraints imposed on them.

In order to represent the geometry of products in computers, we use a solid-modelling technique. We represent these geometrical constraints as relationships between geometric elements, such as faces, edges, and vertices of the solid model. These geometric constraints can be represented formally by using first-order predicate logic. The flexibility and universality of first-order predicate logic is well known, hence we developed our product-modelling framework concept as combination of first-order predicate logic [12] and an object-oriented approach [13]. The basic system elements used in describing models are entities, relationships and attributes of entities. They are based on first-order predicate logic and respectively represented as constants, ground literals and value functions of entities. We employed these concepts along with some manipulation functions as COMET/DB (COnceptual Modelling Experimental Tools/Database) [11] which utilizes a Prolog-like system for handling relationships, and an object-oriented language, Flavor, for dealing with the entity attributes. Our solid modelling system GEOMAP-III [14] is combined with COMET/DB in such a way that geometric elements of solid models are treated as entities and geometric properties of the elements are treated as attributes of these entities. Then we describe geometric constraints according to entity relationships.

Definition and Manipulation of Form Features [15]

In this section, as the first example of product modelling, the definition and manipulation of form features are explained. Through the inspection of current machining activities, we found that the process plan is formed based on the consideration of many factors including form features, dimensions, surface roughness, tolerances, machining specifications and so on. Of those factors, form features [16] have the most important meanings. Figure 9.2 shows some examples of form features.

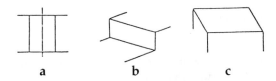

Fig. 9.2 Examples of form features: **a** through hole; **b** step; **c** flat face.

A machining operation is considered as an operation to generate a form feature in a workpiece. So it is possible to recognize which form feature remains to be machined by comparing the form features of the workpiece being machined with those of the completed machine part. Process-planning engineers mainly select preferred machining operations according to the form features. Furthermore, the order of executing operations is decided based on relations between form features. Therefore, proper definition and manipulation of form features are indispensable for realizing the process-planning automation of prismatic part machining.

Definition of Form Features

We represent form features in our product modelling framework, where they are defined by form feature types and component face relations.

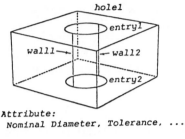

```
Attribute:
 Nominal Diameter, Tolerance, ...

Form Feature Type:
 through_hole(hole1)

Component Face Relation:
 entry_face(hole1 , entry1)
 entry_face(hole1 , entry2)
 wall_face(hole1 , wall1)
 wall_face(hole1 , wall2)
```

Fig. 9.3 Form feature definition.

Form Feature Types

For instance, a through hole feature (Fig. 9.3), is described in the database by the following logical formula:

$through\_hole(hole1)$

where *hole1* is an entity name and *through_hole* is a predicate to specify form feature type. An implication, for instance, "the through hole feature is a kind of the hole feature", can be defined by the following formula;

$hole(*x) \leftarrow through\_hole(*x)$

The symbol "←" means logical implication and "*x" denotes a variable. When it is desired to give the through hole feature its nominal diameter and other values such as tolerances, these values are treated as attributes of the form feature entity.

Component Face Relations

A form feature is considered as a set of face elements of a solid model as shown in Fig. 9.3. We represent the relationships between the form feature and its component face elements by some binary predicates. For the form feature shown in the figure, we use binary predicates *entry_face* and *wall_ face* as follows:

$entry\_face(hole1, entry1), entry\_face(hole1, entry2)$
$wall\_face(hole1, wall1), wall\_face(hole1, wall2)$

The symbols *entry1, entry2, wall1* and *wall2* are face element names of the solid model. In this product-modelling system, both the form feature types and the component face relations are defined as primitive relations [17].

Relations between Form Features

In the same framework, we represent relations between form features. For instance, we express the relation of a through hole feature connected to another hole feature's wall face as the binary predicate *branch_connect*. This *branch_connect* relation regulates the machining order of two hole features. Figure 9.4 shows that *through1* and *blind1* have a *branch_connect* relation. The system has the following primitive relations in the database:

> *through_hole(through1)*
> *wall_face(blind1,face1)*
> *entry_face(through1,face1)*

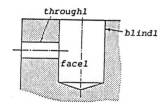

branch_connect(through1 , blind1)

[Primitive Relation]
through_hole(through1)
wall_face(blind1 , face1)
entry_face(through1 , face1)

Fig. 9.4 Branch connect relation.

but we need the relation

> *branch_connect(through1,blind1)*

for determining the machining order of *through1* and *blind1*. To derive this relation, we apply the following rule:

> *branch_connect(*x,*y)* ← *through_hole(*x)* & *entry_face(*x,*f)* & *wall_face(*y,*f)*

where & means logical conjunction. In our product-modelling system, all the derivable relations are thus defined using rules. Those relations are called defined relations [17].

Manipulation of Form Features

In order to evaluate the constraint concerning the machining operation, a machining operation simulator must be realized for process-planning systems. Most of the workpiece modifications required for the simulator are concerning form features, because of their close relation to the machining operation. Therefore, we defined procedures for creating/eliminating form features to cope with this requirement. The modification of the solid model description of the product model is done by Boolean set operations and local operations implemented in GEOMAP-III. These operations are prepared to create and eliminate the geometrical shape of form features, which are corresponding to some machining operations. For example, the drilling operation is simulated by a local operation to make a hole shape. In addition to drilling, the system supports local operations for grooving and milling, but other operations can be easily added. Here we explain procedures for creating and eliminating form features with an example (Fig. 9.5).

Form Feature Creation

According to the request of creating a step feature, the system makes the geometrical shape of the new step feature *step1* by the local operation of GEOMAP-III. Then some primitive relations for the step feature:

> *step(step1)*
> *entry_face(step1,face1)*
> *wall_face(step1,face5)*
> *bottom_face(step1,face6)*
> *side_end_face(step1,face4)*
> *open_end_face(step1,face2)*
> *open_end_face(step1,face3)*

are asserted with face elements of the solid model. Attribute values of the created form feature and its component face elements, such as surface roughness, are then specified.

Generated form features often change some description of other existing primitive relations. In this example, the primitive relation

> *entry_face(hole1,face1)*

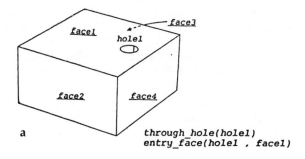

through_hole(hole1)
entry_face(hole1 , face1)

a

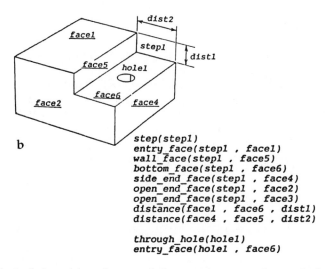

b

step(step1)
entry_face(step1 , face1)
wall_face(step1 , face5)
bottom_face(step1 , face6)
side_end_face(step1 , face4)
open_end_face(step1 , face2)
open_end_face(step1 , face3)
distance(face1 , face6 , dist1)
distance(face4 , face5 , dist2)

through_hole(hole1)
entry_face(hole1 , face6)

Fig. 9.5 Manipulation of form feature: **a** before creating a step feature; **b** after creating a step feature.

must be replaced with

entry_face(hole1,face6)

For every primitive relation, we prepare the predicate to check the consistency of the description by using the topology of the solid model. For instance, to check the consistency of the *entry-face* relation,

consistent_entry_face(*feature,*face)

is used. In this case, the system, at first, adds the relation

entry_face(hole1,face6)

into the product model database, when *face6* is newly created. Then it
applies this *consistent_entry_face* predicate and eliminates the inconsistent
one (in this case, *entry_face(hole1,face1)*) from the database.

Finally, according to the location specifications for the created form
feature, some dimensions are set as the form of the geometric constraint
on the face elements. In this example,

> *distance(face1,face6,dist1)*
> *distance(face4,face5,dist2)*

are asserted in the database.

Form Feature Elimination

According to the elimination of the step feature *step1*, face elements *face5*
and *face6* must be eliminated. Some face eliminations can be regarded
as merging of face elements. In this case, *step1*'s bottom face *face6* is
regarded as being merged by the entry face *face1*. The system first selects
all the primitive relations which relate to each face element to be merged.
Then it creates new relations which are almost the same as the selected
relations, but all the appearance of the face to be merged in the relations
is replaced by the merging face. In this example, for all the relations
concerning *face6*:

> *distance(face1,face6,dist1)*
> *bottom_face(step1,face6)*
> *entry_face(hole1,face6)*

the following relations (*face6* is replaced by *face1*) are created and added
in the product model database,

> *distance(face1,face1,dist1)*
> *bottom_face(step1,face1)*
> *entry_face(hole1,face1)*

Then, the geometrical shape of the *step1* and *step1* itself are eliminated.
According to the elimination of *step1* and component face elements, *face5*
and *face6*, the product-modelling system automatically eliminates the
following relations concerning the eliminated entities,

> *step(step1)*
> *entry_face(step1,face1)*
> *wall_face(step1,face5)*
> *bottom_face(step1,face1)*
> *bottom_face(step1,face6)*

side_end_face(step1,face4)
open_end_face(step1,face2)
open_end_face(step1,face3)
distance(face1,face6,dist1)
distance(face4,face5,dist2)
entry_face(hole1,face6)

Finally, an inconsistent dimension specification:

distance(face1,face1,dist1)

is detected and eliminated from the database.

According to the change of primitive relations, all the other defined relations, which can be derived from primitive relations, are changed.

Definition and Manipulation of Sheet Metal Parts [18]

As another example of product modelling, the definition and manipulation of sheet metal parts are discussed in this section. In manual activities, process-planning engineers often consider the effect of bending, and its reverse operation, expanding. For instance, in order to determine the blank's shape, the sheet metal part must be developed in consideration of the geometrical modification of bends. Therefore, a sheet metal part product model must be designed, with which bending and expanding can be accurately simulated.

Definition of Sheet Metal Parts

Metal forming has the distinct characteristic of stressing the metal at localized areas only, and in the case of bending, this localized stress occurs only along the cylindrical area of the bending radius on which the forming force is applied (Fig. 9.6). The remaining metal is not stressed during bending, and its shape and contour remain unchanged [19].

In bending, the sheet metal is stressed in tension on one surface and in compression on the other. When a cross-section is made through the stressed area (Fig. 9.6), the line of zero stress is called the neutral axis, therefore, the neutral axis is a true representation of the original blank length [19]. We approximate the sheet metal modification as the rotation around an imaginary axis Ax, which is determined as shown in Fig. 9.7a. When the rotational angle ϕ is $\psi = 180° - \omega$, then a "bent" condition is

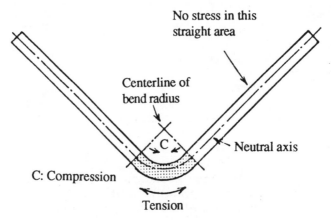

Fig. 9.6 Stresses during bending.

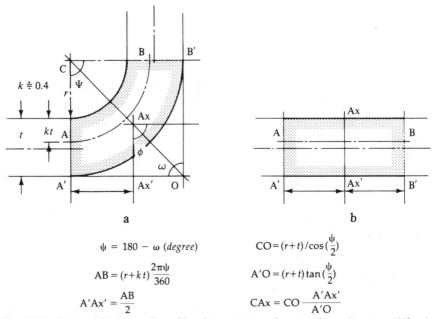

$$\psi = 180 - \omega \; (degree)$$

$$AB = (r+kt)\frac{2\pi\psi}{360}$$

$$A'Ax' = \frac{AB}{2}$$

$$CO = (r+t)/\cos\left(\frac{\psi}{2}\right)$$

$$A'O = (r+t)\tan\left(\frac{\psi}{2}\right)$$

$$CAx = CO\,\frac{A'Ax'}{A'O}$$

Fig. 9.7 Definition of the centreline of bending rotation: **a** bent condition ($\phi = \psi = 180° - \omega$); **b** expanded condition ($\phi = 0°$).

represented, and when ϕ is 0°, an "expanded" condition is represented (Fig. 9.7b). In this expanded condition, the distance between the line AA′ and BB′ is the same as the length of the neutral axis AB.

It is therefore possible to represent the sheet metal part in computers as the set of stiff and thin plates representing the unchanged region, and

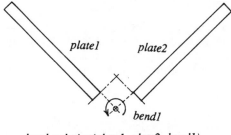

bend_relation(plate1, plate2, bend1)

Fig. 9.8 Definition of sheet metal part product model with plates and bends.

bending axes representing imaginary centrelines of bending rotation. Plates are connected to each other by bending axes. In the following discussion, such bending axes are called bends (Fig. 9.8). We represent these plates, bends and their mutual relationships in the framework of product modelling.

Plate and Bend Definition

Each component plate of the sheet metal part is created as a thin plate like solid model and registered as an entity. Then it is described in the product model database by the following logical formula:

   plate(plate1)

where *plate1* is an entity name and *plate* is a predicate to specify the plate entity in the database. Thickness of the part is treated as an attribute of the plate entity. A bend is also created as an entity and described in the database in the same manner. Bending angle, radius and length are defined as attributes of the bend entity.

Bend Relation

A relationship between two plates and one bend which connects these plates is represented by a predicate, *bend_relation*. For instance (Fig. 9.8), the *bend_relation* between the plate on the left side (*plate1*), the plate on the right side (*plate2*) and *bend1* is described in the database as follows:

   bend_relation(plate1,plate2,bend1)

Manipulation of Sheet Metal Model in Bending

Based on the definition of the sheet metal part product model, bending is simulated as the transformation of plates according to the bending axes. Here we explain the homogeneous transformation used in the simulation and describe some of the simulation procedures that have been developed.

Homogeneous Transformation [20]

In this discussion, point vectors are denoted by lower case, bold face characters (e.g. **v, u,** . . .) and coordinate frames by upper case, bold face characters (e.g. **H, C,** . . .).

Vectors. A point vector $\mathbf{v} = a\mathbf{i}+b\mathbf{j}+c\mathbf{k}$ where **i, j** and **k** are unit vectors along the x, y and z coordinate axes respectively, is represented in homogeneous coordinates as a matrix:

$$\mathbf{v} = [x,y,z,w]^T$$

where $a = x/w$, $b = y/w$, $c = z/w$. The vector at the origin is represented as $[0,0,0,n]^T$, and vectors of the form $[a,b,c,0]^T$ are used to represent directions.

Transformations. A transformation of the space **H** is a 4×4 matrix and can represent translation and rotation. Given a point **u**, its transformation **v** is represented by the matrix product: $\mathbf{v} = \mathbf{Hu}$. The transformation **H** corresponding to a translation by a vector $a\mathbf{i}+b\mathbf{j}+c\mathbf{k}$ is

$$\mathbf{H} = \begin{bmatrix} 1 & 0 & 0 & a \\ 0 & 1 & 0 & b \\ 0 & 0 & 1 & c \\ 0 & 0 & 0 & 1 \end{bmatrix}$$

The transformation corresponding to rotation about the x axis by an angle ω is:

$$\mathbf{Rot}(x,\omega) = \begin{bmatrix} 1 & 0 & 0 & 0 \\ 0 & \cos\omega & -\sin\omega & 0 \\ 0 & \sin\omega & \cos\omega & 0 \\ 0 & 0 & 0 & 1 \end{bmatrix}$$

Inverse Transformation. In general, given a transformation with elements

$$
T = \begin{bmatrix} n_x & o_x & a_x & p_x \\ n_y & o_y & a_y & p_y \\ n_z & o_z & a_z & p_z \\ 0 & 0 & 0 & 1 \end{bmatrix}
$$

then the inverse is:

$$
T^{-1} = \begin{bmatrix} n_x & n_y & n_z & -\mathbf{pn} \\ o_x & o_y & o_z & -\mathbf{po} \\ a_x & a_y & a_z & -\mathbf{pa} \\ 0 & 0 & 0 & 1 \end{bmatrix}
$$

where $\mathbf{p}$, $\mathbf{n}$, $\mathbf{o}$ and $\mathbf{a}$ are the four-column vector and $\mathbf{pn}$ means the vector dot product of $\mathbf{p}$ and $\mathbf{n}$.

Coordinate Frames. We can interpret the elements of the homogeneous transformation matrix as four vectors describing a coordinate frame. For instance, the transformation matrix $\mathbf{C} = [\mathbf{a,b,c,o}]$ shows a coordinate frame which has three axis directions $\mathbf{a,b,c}$ and the position of the origin $\mathbf{o}$ of a coordinate frame rotated and translated away from the reference coordinate frame. When a vector $\mathbf{u}$ is transformed to $\mathbf{v}$ by the homogeneous matrix $\mathbf{C}$, with respect to the reference coordinate frame ($\mathbf{v} = \mathbf{Cu}$), then $\mathbf{u}$ is considered as a vector described in the coordinate frame $\mathbf{C}$ and $\mathbf{v}$ is the same vector described in the reference coordinate frame.

Transformation Equations. Consider two descriptions of the position of the point $\mathbf{P}$;

$$\mathbf{P} = \mathbf{A}^A\mathbf{B}^B\mathbf{C} \text{ and } \mathbf{P} = \mathbf{D}^D\mathbf{E}$$

where $^A\mathbf{B}$ means that $\mathbf{B}$ is the transformation matrix with respect to $\mathbf{A}$. As both positions are the same, equating the two descriptions,

$$\mathbf{A}^A\mathbf{B}^B\mathbf{C} = \mathbf{D}^D\mathbf{E}$$

To solve this equation for $^A\mathbf{B}$, we must premultiply by $\mathbf{A}^{-1}$ and postmultiply by $^B\mathbf{C}^{-1}$ to obtain

$$^A\mathbf{B} = \mathbf{A}^{-1}\mathbf{D}^D\mathbf{E}^B\mathbf{C}^{-1}$$

Coordinate Frames of Sheet Metal Parts

All the component plates of the sheet metal part model are defined to have their own coordinate frames (Fig. 9.9). Each of them is assigned the homogeneous transformation matrix corresponding to its frame with respect to the reference coordinate frame. For instance, when a matrix P_1 is assigned for *plate1*, a vector $\mathbf{u}$ is transformed to $\mathbf{v}$ with respect to the reference coordinate frame as follows;

$$\mathbf{v} = \mathbf{P_1 u}$$

All the bends are also defined to have their own coordinate frames. Each bend connects two plates, and has two homogeneous transformation matrices with respect to the coordinate frames of its right and left side plates. For instance, when a *bend_relation*;

bend_relation(plate1,plate2,bend1)

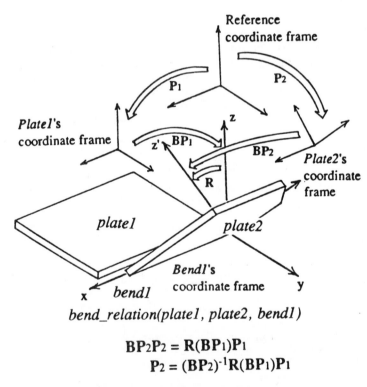

bend_relation(plate1, plate2, bend1)

$$\mathbf{BP_2P_2 = R(BP_1)P_1}$$
$$\mathbf{P_2 = (BP_2)^{-1}R(BP_1)P_1}$$

Fig. 9.9 Coordinate frames of a sheet metal part.

is described in the database, *bend1* has two transformation matrices $\mathbf{BP}_1$ and $\mathbf{BP}_2$ with respect to the coordinate frames of *plate1* and *plate2*. Bending rotation matrix $\mathbf{R}$ corresponding to the bending operation is also specified for the *bend*. This matrix represents the modification of the *bend's* coordinate frame in a bent condition with respect to the coordinate frame in an expanded condition. This transformation is the rotation around the imaginary bending axis as explained.

Definition of Bending Rotation Matrix

In order to simulate an actual modification according to bending, the bending rotation coordinate frame must be changed in consideration of following factors.

Springback. The metal nearest the neutral axis has been stressed within the elastic limit. Hence, the forming force having been removed, this portion tries to return to its original shape and causes some degree of "springback" [19]. This springback is simulated by decreasing the degree of rotation of the bending transformation matrix.

Manufacturing Errors. Products manufactured by bending operations inherently contain some effect of manufacturing errors, which are caused by false adjustment of bending machines, false setting of workpieces and so on. These errors can be simulated as three types of transformation of the bending rotation coordinate frame. As shown in Fig. 9.10, manufacturing error caused by false adjustment of bending machines, punches and dies is simulated as the displacement in the direction of the y axis of the bending rotation coordinate frame. Bending error caused by false setting of the workpieces is simulated as the rotation around

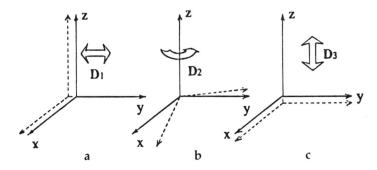

Fig. 9.10 Modification of *bend's* coordinate frame for simulating manufacturing errors: **a** false adjustment of bending machines, punches and dies; **b** false setting of workpieces; **c** difference in sheet thickness and forming force.

the z axis, and an error caused by the difference in sheet thickness and forming force is simulated as the displacement in the direction of the z axis. Matrices $\mathbf{D}_1$, $\mathbf{D}_2$ and $\mathbf{D}_3$ are prepared to transform the bending rotation coordinate frame according to these manufacturing errors. The modified bending rotation matrix $\mathbf{R}^{new}$ is given by the following equation:

$$\mathbf{R}^{new} = \mathbf{D}_3^{-1}\mathbf{D}_2^{-1}\mathbf{D}_1^{-1}\mathbf{R}\,\mathbf{D}_1\mathbf{D}_2\mathbf{D}_3$$

Bending Simulation Procedure (Fig. 9.9)

Here we explain the procedure for simulating sheet metal part modification by the bending operation with an example. When the system determines to execute the bending operation, it selects one component plate, *plate1*, as reference. Then it changes the transformation matrix $\mathbf{P}_1$ of *plate1*, in order to place this plate in the proper location with respect to the reference coordinate frame. According to this displacement of *plate1*, another side plate, *plate2*, must be placed in a new location by modifying the transformation matrix $\mathbf{P}_2$ of *plate2*. By using the transformation equation shown in the figure, $\mathbf{P}_2$ is derived as follows:

$$\mathbf{P}_2 = (\mathbf{BP}_2)^{-1}\mathbf{R}(\mathbf{BP}_1)\mathbf{P}_1$$

The system traces all the component plates of the sheet metal part by using the *bend_relation*, and modifies their transformation matrices one by one according to the procedure explained above, in order to place them in the proper location with respect to the reference coordinate frame.

Product Model-Based Process-Planning System

Based on the product model manipulation capabilities, we developed a process-planning system called XMAPP (eXperimental Model based Assistant for Process Planning), which can generate the process plan of prismatic part machining and sheet metal part bending.

Structure of XMAPP System

Figure 11 illustrates the structure of XMAPP system. The system is constructed in Common Lisp environment and running on a Symbolics 3640 Lisp machine. The system can be divided into three major modules:

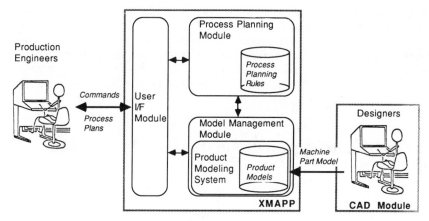

Fig. 9.11 Structure of XMAPP system.

user interface module, process-planning module and product model management module.

The user interface module provides graphical output, multi-window management and menu interaction functions. This module is developed dependent on the Lisp machine's window system.

The process-planning module is built as a knowledge-based expert system, which prepares a specially tuned knowledge base for process planning of specific tasks such as prismatic part machining and sheet metal part bending. All the knowledge utilized in process planning is represented as production rules and stored in the knowledge base.

The product model management module is built based on the product-modelling system. In addition to the basic product-modelling functions, it also prepares various functions for manipulating product models as discussed in the previous sections.

Basic Process-Planning Strategy

As shown in Fig. 9.11, for using XMAPP, a complete machine part (or sheet metal part) product model must be prepared in the product model database. This model is specified and provided by the CAD module in terms of CIM systems.

When a user requests to start process planning, XMAPP repeats trying production rules for the specific process-planning task by referring to the product information stored in the product model database until all the necessary results are successfully generated. When some product model modifications are required, for example to evaluate the effect of the machining operation, the process-planning module requests the model management module to modify the corresponding product information, and the model management module applies some necessary product model manipulation functions to the model to fulfil the request. In process planning, the system records the process of model modifications. Based on this record, a manufacturing sequence required to realize the product is also generated. In following sections, detail plan generation strategy of the specific task and its application examples are discussed.

Process Planning of Prismatic Part Machining [15]

Our first application of XMAPP was process planning of prismatic part machining. The basic strategy is the backward search starting from the complete state of the machine part to the initial blank state. As already reviewed in the previous section, this method is well used in many conventional process-planning systems. However, in addition to the process plan, XMAPP can generate the blank material product model with the complete geometry of the solid model as the result of process planning.

The importance of blank material designing is well understood, as properly designed blank material is effective in reducing machining cost and time. It is very difficult to realize this blank material designing capability in the conventional expert system framework, because such systems prepare no functions to manipulate the product information, especially geometrical shape, dynamically in problem solving.

This capability is realized in the XMAPP system by using the form feature manipulation functions of the model management module. Here we consider casting as the blank material production method, because it is common especially in the medium batch production of machine parts.

Backward Search with Inverse Machining Operations

We assume that the machine part can be recognized as the result of some necessary machining operations on the blank material. So it may be possible to obtain the blank material product model by executing a series of product model modifications, each of which corresponds to the inversion of the required machining operation on the machine part product model.

The system generates the blank material product model by following two steps. Machining operations are usually executed in order to generate form features. So the system, at first, selects the machining operations which are thought to be suitable to generate form features of the machine part by using rules for machining operation selection. Then it inversely

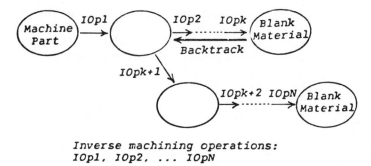

Inverse machining operations:
IOp1, IOp2, ... IOpN

Fig. 9.12 Prismatic part machining process planning with inverse machining operations.

executes the selected machining operations for the machine part product model in proper order, and modifies the product information (geometrical shape, form features, dimensions and face roughness), until all the applicable inverse operations are executed (Fig. 9.12). Finally the machine part product model is changed to the blank material product model.

When an inverse operation is invoked, the system records the corresponding machining operation in the special database. When the blank material model is obtained, the system refers to the recorded operations in the database. The inverse sequence of the recorded machining operations corresponds to the required machining sequence for manufacturing.

Inverse Machining Operation Application Rules

To control the execution sequence of inverse operations, inverse operation application rules are used. An example of a rule is:

```
RULE (inverse,
    ⟨if⟩ ((selected(*FT),
    specified_machining(*FTs, groove, end_milling,*QTY),
    member_of(*FT,*FTs),
    quality?(*FT,rough),
    entry(*FT,*EF),flat_face(*EF),
    or(quality?(*EF,rough), quality?(*EF,fine)))),
    ⟨then⟩((request(ModelManager, :eliminate, (nil groove *FTs unma-
    chined nil)))))
```

If an end-milling operation is selected for the groove form feature, and the form feature's condition is roughly machined, and its entry side form feature is flat face which is roughly or finely machined, then request the model management module to eliminate the specified groove form feature.

The rule is applied according to the state of the machine part (workpiece) product model. The ⟨if⟩ part of the rule contains the precondition that must be satisfied before a certain inverse operation can be executed. The ⟨then⟩ part represents the series of actions to be executed one by one when this rule is invoked. In this case, the model management module eliminates the machine part's groove features and changes dimensions according to the request.

Process-Planning Example

We applied the XMAPP to generation of a machining operation plan of a prismatic part as shown in Fig. 9.13a. Figure 9.13b shows the corresponding machine part product model. In Fig. 9.13a, some form feature names such as FF1, HL1 and so on are specified. This complete machine part model is prepared in the database in advance. Figure 9.14 shows the blank material product model automatically generated by the system. Figure 9.15 shows a part of the machining operation plan generated by the system automatically and the workpiece orientation in this machining phase. Symbols appearing in parentheses in the figure are form feature names which are shown in Fig. 9.13a.

Process Planning of Sheet Metal Part Bending [18]

In this section, process planning of sheet metal part bending is discussed. Sheet metal parts are manufactured by applying a series of forming operations, such as bending, flanging and hemming on blanks. In those forming operations, bending is the most commonly used, especially in order to obtain rigidity and to obtain a part of desired shape to perform a certain function [19]. In the trial manufacturing stage, or in the job shop type of production, bends are made manually with bending machines, punches and V-dies (Fig. 9.16). Process planning of this bending operation, which includes the determination of bending sequence, selection of bending tools and generation of numerically controlling data, is important to produce high-quality products without manufacturing errors.

Process Planning with Bending Simulations (Fig. 9.17)

When a user requests to start process planning, the system first extracts some dimensions on which critical tolerances are specified. Then it requests the model management module to execute the expanding

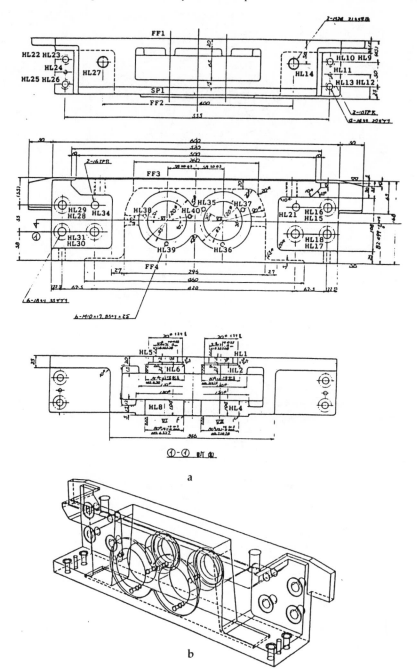

Fig. 9.13 An example prismatic part: **a** drawing of the machine part; **b** machine part product model.

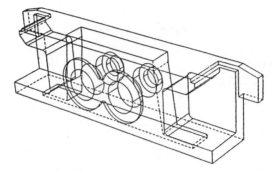

Fig. 9.14 Designed blank material product model.

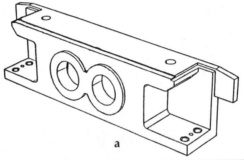

a

```
Phase 2  Workpiece: Y+ Direction
            Machine: HORIZONTAL_MACHINING_CENTER
            Machining Reference Faces: (FF4)
Op. 2.1 : ROUGH FACE_MILLING => (FF1) FLAT_FACE
Op. 2.2 : ROUGH FACE_MILLING => (FF2) FLAT_FACE
Op. 2.3 : CENTER_DRILLING => (HL16 HL18 HL20 HL29 HL31 HL33) HOLE
Op. 2.4 : CENTER_DRILLING => (HL21 HL34) HOLE
Op. 2.5 : FIN DRILLING => (HL16 HL18 HL20 HL29 HL31 HL33) HOLE
Op. 2.6 : FIN DRILLING => (HL21 HL34) HOLE
Op. 2.7 : SINKING => (HL15 HL17 HL19 HL28 HL30 HL32) HOLE
Op. 2.8 : ROUGH BORING => (HL1 HL5) HOLE
Op. 2.9 : ROUGH BORING => (HL2 HL6) HOLE
Op. 2.10 : ROUGH BORING => (HL4 HL8) HOLE
Op. 2.11 : FIN BORING => (HL1 HL5) HOLE
Op. 2.12 : FIN BORING => (HL2 HL6) HOLE
Op. 2.13 : FIN BORING => (HL4 HL8) HOLE
Op. 2.14 : FINE BORING => (HL1 HL5) HOLE
Op. 2.15 : FINE BORING => (HL2 HL6) HOLE
Op. 2.16 : FINE BORING => (HL4 HL8) HOLE
Op. 2.17 : FIN FACE_MILLING => (FF1) FLAT_FACE
Op. 2.18 : FIN FACE_MILLING => (FF2) FLAT_FACE
Op. 2.19 : CENTER_DRILLING => (HL35 HL36 HL37 HL38 HL39 HL40) HOLE
Op. 2.20 : FIN DRILLING => (HL41 HL42 HL43 HL44 HL45 HL46) HOLE
Op. 2.21 : M10 TAPPING => (HL35 HL36 HL37 HL38 HL39 HL40) HOLE
Op. 2.22 : CHAMFERING => (CH1 CH2 CH3 CH4) CHAMFER
```

b

Fig. 9.15 Generated process plan and corresponding workpiece orientation: a workpiece
orientation; b process plan.

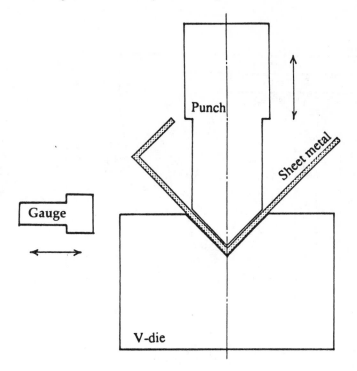

Fig. 9.16 Bending with a punch and a V-die.

operations for the sheet metal part model and change it to the blank model. In this expanding phase, required tools for bending (punches and V-dies) satisfying the bending conditions such as bending angle, radius, length and material of the part, are selected.

Then, by using the determined blank model, the process-planning module generates the bending sequence and other numerically controlling data. It is possible to formulate bending operation planning as a simulation of the series of bending, which modifies the workpiece product model state from an initial blank to a final product. We implemented this formulation with the strategy of the forward production system [21]. The process-planning module starts generating the bending sequence from the initial state of the workpiece model. It repeats the selection of the bending part, determination of the workpiece positioning method and continually requests the model management module to simulate the determined bending on the workpiece model until it reaches the final product state. During this bending simulation, constraints concerning tool collisions and tolerances are evaluated. The series of simulated bending is adopted as the bending operation plan. Finally, some process plan modifications are done to attain more efficient workpiece handling and tool changing.

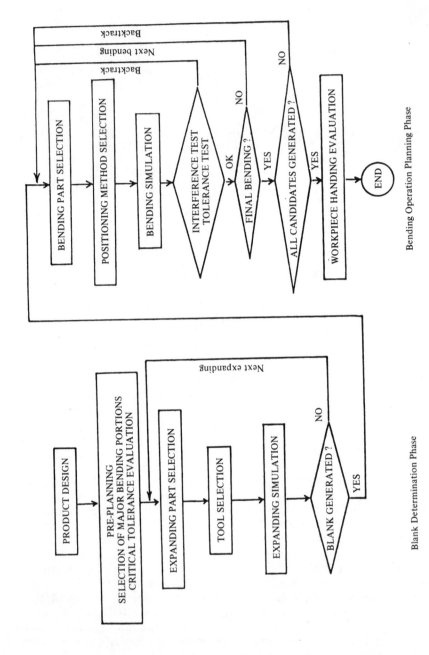

Fig. 9.17 A processing flow for bending process planning.

Evaluation of Process-Planning Constraints

By using the sheet metal part product models and bending simulation procedures, the system evaluates process-planning constraints concerning tool collisions and tolerances.

Tool collision

Fatal collisions between bending tools and the workpiece must be avoided. Collisions between the component plate and the tool shape solid models are detected by using the interference detection function prepared in the solid-modelling system GEOMAP-III. This collision detection function is executed in the bending operation planning phase. When the material is sufficiently elastic, some tool collisions, with the plates located parallel to the bending axis, is permitted. We prepare the table for this allowable collision, and the system refers to it during the collision detection.

Tolerance

Constraints concerning tolerances are checked in the bending operation planning phase. In order to check that the workpiece to be manufactured satisfies the tolerances specified on the sheet metal part, the difference between the design specifications and the dimensions of the corresponding plates of the workpiece model after bending simulation, must be evaluated. Currently speaking, the model management module simulates three types of manufacturing errors: errors in the displacement along the y axis, rotation around the z axis, and displacement along the z axis of bending rotation coordinate frame. Each manufacturing error has some range between maximum and minimum value. Threfore a total of eight ($=2^3$) kinds of bending rotation matrix exist. When the system determines to execute a series of bending operations, the model management module tries these eight bending simulations for each bend, and checks the distance between the plates, on which the tolerance is specified. If the distance exceeds the tolerance range, the system cancels the determined operations.

Process-Planning Example

We applied the system for generating a bending operation plan of a sheet metal part shown in Fig. 9.18. As shown in the figure, a tolerance is specified on the dimension between two plates. The system determines the blank, and generates the bending process plan automatically. Figure

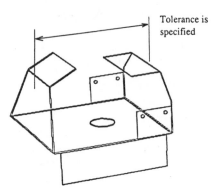

Tolerance is
specified

Fig. 9.18 An example sheet metal part.

9.19a–g shows the modification of the workpiece model according to the generated plan. The sheet metal model specified by an arrow in Fig. 9.19a is the blank part. In a–g, the lower sheet metal model shows the state before bending, and the upper one shows the state after bending. The message shown in the lower right in the figure is the specification of tools and the numerically controlling data for the bending machine.

Conclusion

In this chapter we have described the automatic process-planning system XMAPP, which is developed based on the concept of integration with product models. XMAPP has the capability of automatic blank material designing, process planning of prismatic part machining and sheet metal part bending. Process-planning systems are required to evaluate various kinds of constraints concerning manufacturing processes. Most of the constraints are closely related to machining operations and their effects on the workpiece. In order to evaluate such constraints, a machining operation simulator is implemented and incorporated in our process-planning system. XMAPP represents process-planning methods and constraints as production rules, and applies them to product models. These rules not only use product models for referring to the product information, but also invoke the machining operation simulator to modify the product models dynamically as needed in the process of problem solving.

Only the conceptual framework of the process-planning automation based on product models has been discussed. Details have not been given about the particular process-planning knowledge concerning

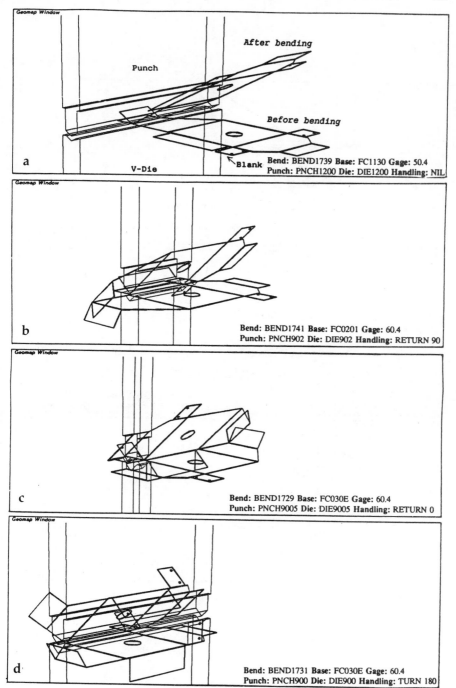

Fig. 9.19 Generated bending operation sequence.

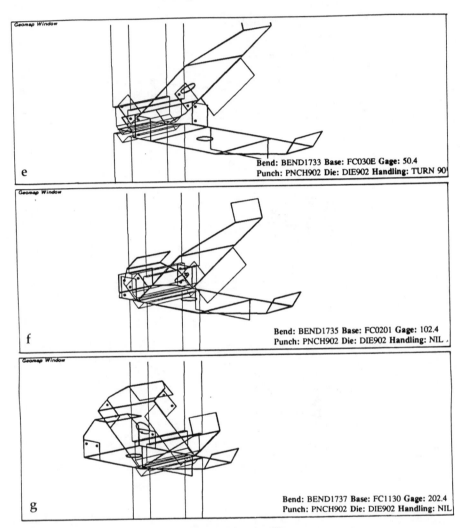

Bend: BEND1733 Base: FC030E Gage: 50.4
Punch: PNCH902 Die: DIE902 Handling: TURN 90°

e

Bend: BEND1735 Base: FC0201 Gage: 102.4
Punch: PNCH902 Die: DIE902 Handling: NIL

f

Bend: BEND1737 Base: FC1130 Gage: 202.4
Punch: PNCH902 Die: DIE902 Handling: NIL

g

Fig. 9.19 (*continued*).

the specific process-planning task. However, the above framework is considered to be useful for practical applications if such knowledge is available.

Acknowledgements

This research work was partly funded by the Development Project for Product Modelling System organized by the Japan Society of Precision Engineering. Special thanks are due to the industrial members of this project for offering us precious information about actual process-planning practices.

References

1. Chang TC, Wysk RA. An introduction to automated process planning systems. Prentice-Hall, Englewood Cliffs, 1985, pp 18–22
2. Lillehagen FM, Dokken T. Towards a methodology for constructing product modeling database in CAD. In: File structure and data bases for CAD. North-Holland, Amsterdam, 1982, pp 59–91
3. Wysk RA et al. Automated process planning systems – an overview of ten years of activities. Preprints of 1st CIRP working seminar on computer aided process planning, 1985, pp 13–18
4. Descotte Y, Latombe J-C. Making compromises among antagonist constraints in a planner. Artif Intell, 1985; 27: 183–217
5. Tsang JP. The Propel process planner, Proceedings CIRP manufacturing seminar for process planning, Penn State, 1987, pp 71–75
6. Berenji HR, Khoshnevis B. Use of artificial intelligence in automated process planning. Comput Mech Eng, 1986; September: 47–55
7. Matsushima K et al. The integration of CAD and CAM by application of artificial intelligence techniques, Ann CIRP 1982; 31: 329–332
8. Nau DS, Gray M. SIPS: an application of hierarchical knowledge clustering to process planning. Proceedings ASME symposium on knowledge-based expert systems for manufacturing, ASME WAM 1986, PED vol 24, pp 219–225
9. Choi BK, Barash MM. STOPP: An approach to CADCAM integration. Comput-Aided Des, 1985; 17: 162–168
10. van't Erve AH, Kals HJJ. XPLANE, a generative computer aided process planning system for part manufacturing. Ann CIRP 1986; 35: 325–329
11. Kimura F et al. A uniform approach to dimensioning and tolerancing in product modeling. Preprints CAPE'86 1986: 1: 166–178
12. Gallaire H, Minker J. Logic and data bases. Plenum Press, New York, 1978
13. Weinreb D, Moon D. Flavors: message passing in the Lisp machine. AI memo 602, MIT Artificial Intelligence Laboratory, 1981
14. Kimura F. GEOMAP-III, designing solids with free-form surfaces. IEEE Comput Graphics Applic 1984; 4: 58–72
15. Inui M et al. Extending process planning capabilities with dynamic manipulation of

product models, Proceedings CIRP manufacturing seminar for process planning 1987, Penn State, pp 273–280

16. CAM-I Inc. Product definition data interface. System specification document, SS560120100, DR-84-GM-03, CAM-I Inc. 1984

17. Kowalski R. Logic for problem solving. North-Holland, Amsterdam, 1979, pp 133–146

18. Inui M et al. Automatic process planning for sheet metal parts with bending simulation, Proceedings ASME symposium on intelligent and integrated manufacturing analysis and synthesis, ASME WAM 1987, PED vol. 25, pp 245–258

19. Eary DF, Reed EA. Techniques of pressworking: sheet metal, Prentice-Hall, Englewood Cliffs, NJ, 1958, pp 56–76

20. Paul RP. Robot manipulators: mathematics, programming, and control. MIT Press, Cambridge, MA, 1981, pp 9–40

21. Nilsson NJ. Principle of artificial intelligence. Tioga, Palo Alto, CA, 1980, pp 281–287

Chapter 10

Theoretical Approach to Knowledge Acquisition and Knowledge Representation in CAPP Expert Systems

V. R. Milacic

Introduction

Developments in artificial intelligence (AI) and knowledge engineering have led to task-oriented products and the so-called expert systems. Different methods of building expert systems have been proposed [1]. The task-oriented approach offers expert systems in very different areas of human activities of which those in engineering are of special interest at the present time.

Computer-aided process planning is one of the most important manufacturing engineering fields to be reconsidered from many different aspects. This has been successfully done by Ham [2]. In the section "Suggestions to the Future Directions", Ham has paid special attention to the potential role of AI techniques in integrated planning stating that "in the future AI would become the key technology in generating, representing, integrating and utilizing the planning intelligence that is essential to the future of computer-integrated manufacturing."

If this is so, it means that a new framework or even an adequate theory has to be developed for intellectual activities in manufacturing engineering, including process planning. Some of the already developed expert systems for process planning and computer-aided design and manufacture show very strong application limitations, mainly because they are built on the basis of the analogy with expert systems in medicine and other fields. This means that it is necessary to introduce broader and deeper research activities in different scientific and engineering domains.

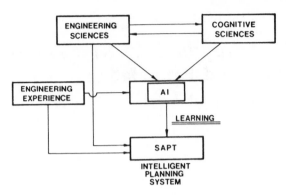

Fig. 10.1 New concept for creating intelligent process-planning system.

In Fig. 10.1 a new concept to create intelligent planning systems is proposed. The main elements of the concept are:

- Engineering sciences including manufacturing
- Cognitive sciences
- Engineering experience (skill)
- Artificial intelligence
- Intelligent planning system as one of the possible outputs of the whole structure

The long-range research project "Intelligent Manufacturing Systems" [3–4] directed by the author is dealing with this concept. At this point we shall consider some learning aspects for acquiring knowledge in the expert system and inference process.

Some practical research results are applied in the building of the SAPT knowledge-based expert system.

Learning Mechanism as a Backbone of the Expert System

If we start with the general definition given by H. Simon [5]:

> Learning denotes changes in the system that are adaptive in the sense that they enable the system to do the same task or tasks drawn from the same population more efficiently and more effectively the next time,

and continue with a more specific definition:

Learning process includes the acquisition of declarative knowledge, the development of motor and cognitive skills through instruction or practice, the organisation of new knowledge into general, effective representations and discovery of new facts and theories through observation and experimentation [6]

then the set of key words which are essential to recognize a strong relation with an expert system's structure becomes evident.

Our discussion is concentrated on learning mechanisms as a cognitive invariant in humans, in order to acquire facts, skills and more abstract concepts. Two main aspects of learning are:

- Knowledge acquisition as a conscious process whose result is the creation of new synthetic knowledge structures and neural models
- Skill refinement as a subconscious process by reason of repeated practice

A process planner creates process plans by inferring at both the conscious and subconscious levels.

In the building of an expert system for process planning the problem is faced of how to combine knowledge acquisition and skill refinement in order to obtain an adequate solution for a given task. There exist two extreme approaches in the building of expert systems. In the first approach the skill refinement and empirical knowledge defined by heuristics serve as the basis for the expert system. A much higher knowledge-based concept of expert systems includes machine learning.

Nowadays machine learning is predominantly a field of interest to psychologists and artificial intelligence researchers. A short historical sketch of machine learning is shown in Fig. 10.2. During almost 50 years

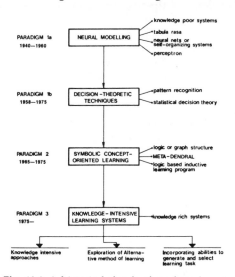

Fig. 10.2 A historical sketch of machine learning.

there have been three main paradigms moving from knowledge-poor systems to knowledge-rich systems. The first paradigm of machine learning deals with neural modelling and decision-theoretic techniques. The main characteristics are perceptrons and statistical decision theory development. The next paradigm is symbolic concept-oriented learning. The logic-based inductive learning program and META-DENDRAL are practical features of the second period of development. At present the paradigm is concentrated upon knowledge-intensive learning systems, including the exploration of alternative methods of learning and incorporating the abilities to generate and select the learning task.

Process planning is a data-rich and knowledge-poor engineering domain. This is the bottleneck in the development of adequate knowledge-based expert systems. It thus means that the learning paradigm plays an important role in the elimination of the above mentioned limitation. The amount of inference involved in learning is a function of the learning strategy adopted. Generally speaking there are the following learning strategies:

- Rote learning
- Learning from instruction
- Learning by analogy
- Learning from examples
- Learning from observation and discovery

The learning mechanism has been developed between the teacher and the learner or, in our case, between the expert or an environment on the one hand, and the expert system or the user of an expert system on the other hand. The inference performance between the teacher and the learner is reciprocal, and as already mentioned, depends on the type of learning strategy. So, rote learning presupposes maximum inference of the teacher and non inference of the learner. However, in the case of learning from observation and discovery the opposite relationship applies. This correlation is shown in Fig. 10.3 giving approximate quantification of inference performance contributions between the teacher and the learner subject to different learning strategies.

The learning system may acquire rules of behaviour, description of physical objects, problem solving heuristics, classification taxonomies over a sample, etc. However, the knowledge acquired could be considered as a function of the representation of that knowledge. According to Carbonell and Michalski [6] there are the following types of acquired knowledge:

- Parameters in algebraic expressions
- Decision trees
- Formal grammars
- Production rules
- Formal logic-based expressions and related formalisms

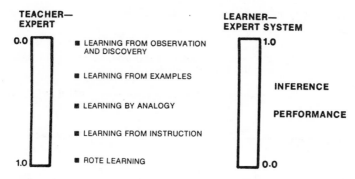

Fig. 10.3 Learning redistribution between expert and expert system.

- Graphs and networks
- Frames and schemas
- Computer programs and other procedural encodings
- Taxonomies
- Multi-representation

The quoted list of types of acquired knowledge is rather long and offers broad scope for research. The learning strategies and acquired knowledge could also be coupled to a domain of application. So, a three-tuple is formed:

⟨LS,KE,DA⟩

where:

- LS   learning strategies
- KE   knowledge acquisition
- DA   domain of application

A very brief overview shows that there exists quasi-morphism between all three elements of the given structure. Some plausible relations between these elements are shown in Fig. 10.4.

At this point we will focus on the formal grammar approach of applying the theory of automata with production rules to build expert systems in engineering design, and particularly in process planning [7]. The formal grammar approach is a rather complex linguistic concept which is related to regular expression, finite-state automata, context-free grammar and production rules.

Engineering knowledge has very strong logical latent structure. The theory of automata gives a framework for discovering this structure and developing new knowledge on the basis of learning from examples.

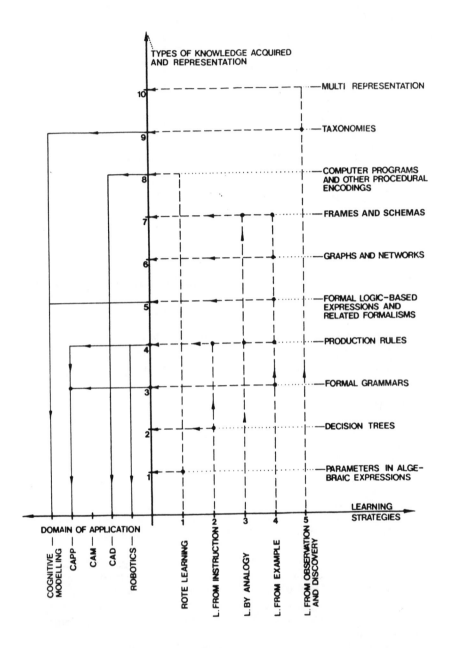

Fig. 10.4 Plausible relations between learning strategies, types of knowledge acquired and representation, and domain of application.

Theory of Automata – Knowledge Acquisition Method

Knowledge in production engineering seems predominantly to be empirical. However, there is a strong logical foundation in decision making for different manufacturing tasks including process planning.

Engineering logic is a priori oriented towards machine logic which covers the functional and physical behaviour of engineering products. This means that the cognitive process is deeply oriented towards machine logic. It seems that engineering reasoning is rather rigid and inappropriate for discovering new concepts and machines. However, if abstract machines are considered, the situation is quite different in the sense of impoverishment of the knowledge structure. The main feature of the new concept is to organize in a formal way a broad chunk of knowledge. Many years of research by the author's group confirm the advantages of applying the concept of languages, and corresponding grammars and automata for solving a given problem. The transition process is:

This is rather complex and diversified. The offered structure of transition is not only complex but with a very rich content of knowledge as well.

The second postulate is that knowledge parsing provides sufficient knowledge chunks for solving a given problem. Some practical aspects of machines, languages and computation theory are applied for process planning knowledge acquisition.

Let us take a primitive model of two intermediate states of knowledge:

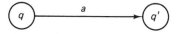

This yields the relation

$$q \longrightarrow aq'$$

However, the given structure is a basic one. It means that the initial state of knowledge (q) is defined, and the final state of knowledge (q') determined by input (a). According to the nature and the number of inputs there are different types of automata. It is possible to distinguish three general types of automata: finite state machine (FSM) transducer, stochastic automata and fuzzy automata. The important differences in all three types of automata are related to the number and definition of inputs and their values. The three basic automata structures are shown below.

The finite state automaton (FSA) transducer is defined as:

$$M = \langle Q, q_0, \Sigma, \Phi, f, g \rangle$$

where:

Q	nonempty finite set of states
$q_0 \in Q$	initial state
Σ	nonempty finite set of input symbols
Φ	nonempty finite set of output symbols
f	state transition function

$$f: Q \times \Sigma \rightarrow Q$$

g	output function

$$g: Q \times \Sigma \rightarrow \Phi$$

It is obvious that with the given definition we have one input between two nodes. The input is defined as:

$$f(q, a) = q'$$

and the output as:

$$g(q, a) = b$$

so that the connection between two nodes is

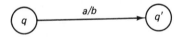

The stochastic automaton is used to weaken the relation between the predecessor and successor nodes, i.e.:

$$SA = \langle Q, q_0, \Sigma, V, \delta, f \rangle$$

where:

Q	finite nonempty set of states
$q_0 \in Q$	initial state
Σ	finite nonempty set of inputs or instructions
V	(0, 1) is the valuation space
δ	transition function

$$\delta: Q \times \Sigma \times Q \rightarrow V$$

f the final state determination function

$$f: Q \to V$$

The nature of stochastic automaton is characterized by the probabilistic approach which is defined by

$$\sum_{q' \in Q} (q, a, q') = 1$$

$$\delta\,(q, a, q') = x$$

and

$$f(q) = 1$$

This means that the given relation is now

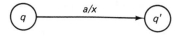

A further step for weakening of the relation between the predecessor and the successor is to introduce a fuzzy automaton:

$$FA = \langle Q,\ q_0,\ \Sigma,\ V,\ \delta,\ f \rangle$$

The notation is identical as in the stochastic automata. The difference could be noted in changing probability into possibility concept. This means that in the fuzzy automaton the max–min function is extended to a function

$$\delta': Q \times \Sigma^* \times Q \to V$$

For the input string a, δ' is

$$\delta'(q, a, q') = \begin{cases} 1 & \text{if } q = q' \\ 0 & \text{otherwise} \end{cases}$$

For input string $a_1, a_2 \ldots a_n \in \Sigma^*$ the delta function is

$$\delta'\,(q, a_1, a_2 \ldots a_n, q') = \bigvee_{q_1, \ldots q_{n-1} \in Q} [\delta(q, a_1, q_2) \wedge \delta(q_1, a_2, q_2) \wedge \ldots$$
$$\ldots \delta(q_{n-1}, a_n, q')]$$

The notation V means that only the maximum is taken from all possible collections of $n-1$ states in Q. Practically this means that $\Sigma(q, a, q')$ differs from 1.

The automata concept has been introduced in order to build a corresponding basic structure for knowledge generating and acquiring in the whole process-planning domain. It is evident that different pieces of knowledge for process plan generation are of different nature.

Process planning, roughly speaking, could be defined according to three axes dealing with workpieces, manufacturing process and tooling. The structure of workpieces according to geometrical and technological recognition can be considered as a deterministic structure defined by the FSM concept. This means that the knowledge has adequate structure to generate corresponding contents for solving a process-planning problem. By applying production rules it is possible to formalize preliminary defined knowledge and offer some kind of inductive reasoning for generation of new knowledge. This means that the establishing of knowledge framework opens the possibility for adding of additional knowledge contents. Developed structure for knowledge formalization is the basic structure used to enhance the existing knowledge.

However, different kinds of knowledge require adequate corresponding approaches. For example, the tooling in process planning appears as a very complex and, at the same time, plausibly easy to define knowledge base. It is evident that the correlation between a tool and type forms has three main relations. The first is that a given tool can generate a single type form (or feature). It has a highly deterministic nature (one-to-one mapping). The second case is when one tool can generate different type forms. This is a one-to-many mapping, or the introduction of or-logic. In this case it is clear that we have to introduce the stochastic approach to solve a given problem. This is the reason for proposing the stochastic automata structure.

Finally, in the case where the set of tools has to generate different type forms, this means that it will be necessary to apply the fuzzy concept in solving a given problem.

The three main automata structures are applicable for process planning knowledge generation. It is not necessary to go further into details to explain the nature of the knowledge and of the problem solving.

The real life demands increase the complexity for a given problem of a process plan. This means that one-dimensional and multi-dimensional structures require different approaches. The two main structures are evident. The first structure deals with rotational parts, and the second with prismatic parts.

SAPT – Knowledge-based Expert System

The SAPT knowledge-based expert system is a part of the Designer® expert system dealing with process planning. There are two different expert systems covering rotational and prismatic parts.

SAPT-R Expert System

The SAPT-R expert system is a process-planning system for rotational parts based on a hybrid concept combining group technology and type (feature) technology concepts in order to bridge rotational and prismatic parts process-planning approaches. The second characteristic of the SAPT-R system is that the theory of automata has been applied to generate and acquire knowledge about the production rules.

Process planning for rotational parts could be considered as a parsing string of external and internal type forms which can be divided into three parts: left, right and middle. If the three subsets of type forms are combined it is possible to generate a part family (Fig. 10.5). A geometrical automaton for a given example of workpieces can generate a whole family of workpieces with the same geometrical contents.

The same approach can be used for process-planning recognition, as well as for tooling recognition for a given part family. In order to generate realistic solutions the M-paradigm has been developed [8].

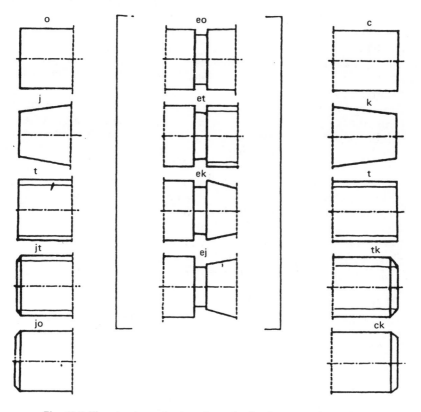

Fig. 10.5 The structure of external standardized rotational part forms.

The concept offered has two main advantages:

- to represent the knowledge in a formalized and compact manner
- by corresponding automata
- to acquire new knowledge by the taxonomy of the set of automata

This means that with the use of a compact representation model it is possible to generate extensive corresponding geometric and technological knowledge for process planning.

A very general knowledge concept is not sufficiently precise to offer adequate structure for real life reasoning. In other words, it is necessary to include logical and process planning skill features into the inference engine. The M-paradigm provides some of these concepts.

The structure of the M-paradigm consists of seven postulates, which are as follows:

P1. The string description of a part which belongs to a given part family defines the contents of all symbols from the starting to the final symbol for external and internal elementary forms (EFs).

P2. All existing combinations of EFs are searched and memorized into strings, dividing into external and internal substrings, expanding up to third-order combinations (maximum of three letters of the alphabet).

P3. The transitional EF (e-groove) is neighbouring with two other EFs which cannot be transitional according to geometrical and technological logic.
    Comment: in the case of second-order combinations the members with the e-groove symbol have been omitted. For third-order combinations this symbol is retained in the case when the middle member is transitional. It is clear that an alphabet could have more than one symbol for a transitional EF.

P4. The substrings defined under P3 are necessary and sufficient to determine productions in formal language grammar. Productions are directly generated from substrings. Nonterminal symbols are common for the same terminal symbols.

P5. The connection between grammatical productions describing external and internal EFs is obtained with LAMBDA productions. The LAMBDA productions have been defined for all nonterminal symbols chained with productions of terminal symbols which are final for outer EFs.

P6. The connection between grammatical productions describing internal EFs and final nonterminal grammar symbol (T) is obtained with LAMBDA productions. The connections are defined productions for all nonterminal and terminal symbols which are final elements for internal EFs.

P7. LAMBDA productions automatically generate connections between the initial and final nonterminal symbols for internal EFs.

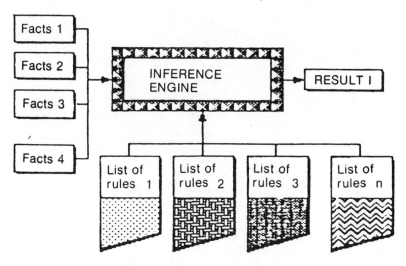

Fig. 10.6 SAPT expert system shell.

The structure for geometric and technological recognition is used to evaluate corresponding process plans. Finally, the inference engine has a shell able to manipulate separate production rules lists covering the process planning field (Fig. 10.6) (WP → TF → TTS → TO → TOH → MT → FIX).

The entire structure of process-planning knowledge is divided into blocks of knowledge organized into sublists of rules. The meta-graph of process planning has the following nodes as nonterminal structure:

- WP   workpiece
- TF   type forms (geometric and technological features)
- TTS   type technological sequence
- TO   tool-cutting part
- TOH   tool-holding part
- MT   machine tool
- FIX   fixturing

This means that the knowledge base is structured into seven chunks of knowledge connected with adequate inputs. The expert system shell has control rules with different purposes, like: strategic rules, control rules for selecting adequate rules for problem solving, and rules for evaluation of some facts. The proposed inference engine concept results in a simpler inference procedure for selecting appropriate knowledge for solving a given problem.

The SAPT-R expert system shell is written in Lisp coding and the hierarchical structure is given in Fig. 10.7.

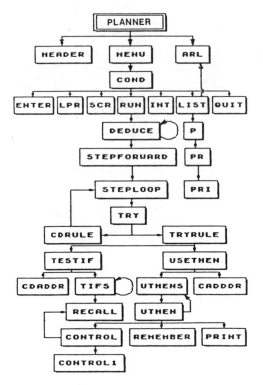

Fig. 10.7 SAPT expert system – hierarchical Lisp tree functions.

The inference engine is triggered by DEDUCE starting with TEST IF and then USE THEN rule chaining.

The SAPT-expert Lisp hierarchy tree has the following subroutines:

- TECHNOLOG – the highest-level function dealing with strategic concepts like group technology or type (feature) technology.
- HEADER – gives basic parameters of the system.
- ARL – prints available list of rules using auxiliary functions like LIST.P, PR, PRI.
- LIST – prints available list of rules.
- INT – selects list of rules for sequential use in inference procedure.
- ENTER – inputs the initial factors for inference process.
- SCR, LPR – gets inference results on screen or line printer.
- RUN – starts inference procedure.
- DEDUCE – the highest level of inference.
- STEPFORWARD – selected list of rules and inferring from first rule by STEPLOOP, TRY and TRYRULE.

- TESTIF – analyses IF part of production rule and compares with list of facts CDADDR, TIFS.
- RECALL – discovers facts in IF part by comparison.
- USE THEN – triggers when all facts have all symbols in IF part.
- UTHENS, UTHEN, REMEMBER – the THEN part of the rule is added as a result of triggering in the list of facts.
- CONTROL (STRAT)
- CONTROL 1 (STRAT1) – called in order to check that there are no more control functions.

The given procedure is repeated for all production rules in acting list of rules. The procedure then moves to higher level loop for the next list of rules.

Production rules of FSM are compiled into control rules of the expert system shell. An example of production FSM rule is taken as: S → cA, then the corresponding control rule of the expert system shell is:

```
(RULE S1
(IF (S) (C))
(THEN (Exec) (TRANS ('s) ('c) ('A))))
```

Lisp function TRANS deals with the contents of the list of facts. In the case of LAMBDA it has been used as LTRANS. These functions have three parameters which correspond to elements of production rules. The "raw" structure of grammatical rules is refined by the given M-paradigm, and by the Lisp interpreter triggering a special mechanism in the inference engine by EXEC instruction for IF-THEN rules. The structures of the expert system shell and LISP-functions used in both cases are given in Fig. 10.7 and in Appendix 1.

The procedure of building a corresponding automaton without and with the M-paradigm is now detailed. If one considers the test group of parts in Fig. 10.8(a), it is evident that there exist two groups of elementary forms:

External EFs		*Internal EFs*	
c	cylinder	h	hole
k	cone external right	g	thread internal
t	thread external	f	groove internal
e	groove external		
j	cone external left		

The part description could start from right to left for external EFs and from left to right for internal EFs. In order to describe a given set of parts it is possible to build the following strings:

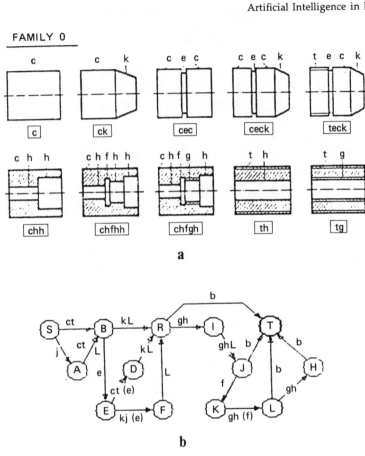

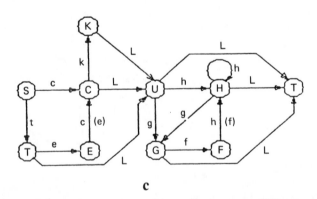

Fig. 10.8 a Test group; b FSM for family test group; c FSM – applied M-paradigm; d Family
1; e FSM for family 1; f family 2; g FSM for family 2.

FAMILY 1

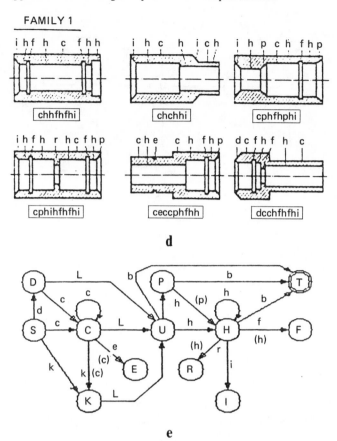

d

e

FAMILY 2

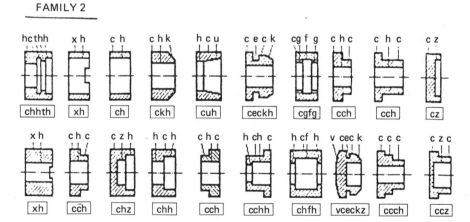

f

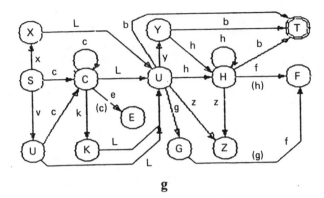

g

Fig. 10.8 (continued.)

1. cb	4. tgb	7. teckb	10. ceckb
2. ckb	5. cecb	8. chhfhb	
3. thb	6. chhb	9. chgfhb	

where b – blank.

The context free grammar is:

$$G = (V, T, P, S)$$

where

    V nonterminal symbols S,A,B,E,D,F,R,I,J,K,L,M,T

    T terminal symbols c,t,j,e,k,g,h,f

    P production rules:

S→cB	S→tB	S→jA	A→tB	A→cB	A→λB	B→kR
B→eE	B→λR	E→kF	E→jF	E→cD	E→tD	F→eE
F→λR	D→kR	D→λR	R→gI	R→hI	I→gJ	I→hJ
J→bT	J→fK	K→hL	K→gL	L→fK	L→hM	I→λJ
L→gM	L→bT	M→bT				

$S = S'$ where $S \in V$.

The corresponding finite state automaton is used to organize and formalize the chunk of knowledge. For the technical solution of a given problem, the TPARS-software for the PDP 11/34 for the operating system RSX 11M has been used. Applying the complex structure one can generate a finite state automaton for the family test group (Fig. 10.8b).

However, for further generalization the M-paradigm is applied. The formal description is not adequate in comparison with real technological demands. There are two main points.

The creation of an automaton for a given set of parts is predominantly determined by the engineer and his feeling about how to generate the necessary process-planning data. This means that the production rule procedure is more empirical than exact. Two engineers can develop different expert concepts in order to solve a given problem. The second limitation is related to the number of parts and heterogeneous features generating demands for richer knowledge structure than that defined by a limited number of set members.

This is the main reason for the introduction of the already mentioned M-paradigm. The M-paradigm is defined by seven postulates which give a complete picture of the recognized demands.

The automaton is generated step by step going through the postulates of the M-paradigm. Following the given postulates one can obtain a more realistic FSM (Fig. 10.8c). Some of the important steps for building this automaton are explained below.

The starting (c,t) and finite (c,k,t) zones of a given part are generally defined with the first postulate for the test group of parts considering the external description. The next step is to define the middle area of a part, by applying the second postulate. Selecting the second- and third-order combinations it is possible to create substrings for the external description as:

ck, ce, ec, te, cec, eck, tec

for the test group of parts. It is clear that the second-order combinations are included in the third-order combinations. It is evident that the third-order combinations which have a groove as the middle member of the string are valid. More precise definitions of knowledge including geometrical and some technological features are recognized and included to build the real picture of a part. The fourth postulate separates some strings:

ck, cec, tec

in order to develop a corrected version of the initial basic structure of the parts family shown in Fig. 10.5. This means that the left side is defined with c and t; the right side with c, k and t; and the middle with combinations of ck, cec and tec strings.

So, the structure analyses for external elementary forms of the family test group have been defined using the first three M-paradigm postulates. The three substrings are sufficient to introduce the next postulate to generate productions for the grammar of external EFs taking into account each substring. For the first $C \rightarrow kK$ substring ck there are two production rules reading: $S \rightarrow cC$ and $C \rightarrow kK$ where S is a nonterminal starting symbol. This is the knowledge content of the first substring. For the next substring cec additional production rules are: $C \rightarrow eE$ and $E \rightarrow cC$. And finally, for the third substring tec it is evident that new grammatical rules are generated as: $S \rightarrow tT$ and $T \rightarrow eE$.

All necessary components are available to build the first part of FSM with:

- nonterminals – S, C, E, K, T
- terminals – c, t, e, k

which is shown in the left part of Fig. 8c.

The next step is to link external and internal features for a given part by applying LAMBDA production or empty string (the fifth postulate), by generating U-nonterminal. In this manner, the left part of Fig. 10.8c is completed.

A similar procedure is applied for internal features description using the sixth postulate. The substrings of the second-order combinations for internal EFs are: hh, hf, fh, hg, and gf; while the substrings of the third-order combinations are: hhf, hfh, hgf and gfh. The accepted production rules are:

$$U \rightarrow hH, H \rightarrow hH, H \rightarrow gG, H \rightarrow fF, F \rightarrow hH$$
$$V \rightarrow gG, G \rightarrow fF$$

Finally, according to the seventh postulate, the production rule is $U \rightarrow T$. There are all the necessary production rules to build the right side of the FSM structure and put it together with the first part. The FSM for the family test group using the M-paradigm is shown in Fig. 10.8c.

The M-paradigm is a grammatical inference procedure, the result of which is recognized.

In Fig. 10.8d, f, two families of parts based on the group technology concept with corresponding automata (Fig. 10.8e, g) are presented. We now have all relevant elements to explain how the SAPT system works for geometrical recognition by generating productions of part family 1 (Fig. 10.8d). The first step is to list input facts including the strings of the part family with their length L and the position where internal elements U start. The input facts are specified as follows:

External EFs

c   cylinder
e   groove – external
d   bevel – external left
k   cone – external right
e   transition external EF

Internal EFs

h   hole
f   groove – internal
p   bevel – internal right
i   bevel – internal left
r   the smallest internal diameter
fr   transition inner EFs

strings of part family:

Part no.	*String*	U	L
1	chhfhfhi	2	8
2	ckchhi	4	6
3	cphfhphi	2	8
4	cphfhrhfhi	2	10
5	ceccphfhh	5	9
6	dcchrhfhi	4	9

The next step is evaluation of the second- and third-order substrings for external and internal elementary forms:

- External EFs substrings   ck, kc, ce, ec,cc, dc, ckc, cec, ecc,dcc, phi, fhr, hrh, rhf, fhh
- Internal EFs substrings   hh, hf, fh, hi, ph, hp, hr, rh, hhf, hfh, fhf, fhi, hhi, phf, fhp, hph

The finite input dealing with starting and terminating alphabet symbols in the strings for external and internal elementary forms are as follows:

- Starting symbols for external forms: cd
- Terminating symbols for external forms: c
- Starting symbols for internal forms: h, p
- Terminating symbols for internal forms: ih

Applying the M-paradigm, productions are generated:

S→cC	C→kK	S→kK	K→cC
C→cC	S→dD	D→cC	C→eE
E→cC	U→hH	H→hH	H→iI
U→pP	P→hH	H→pP	H→fF
F→hH	H→rR	R→hH	C→λU
D→λU	K→λU	H→bT	P→bT
I→bT	U→bT		

from which direct evaluation of FSM given in Fig. 10.8e could be made.

The structure of production rules is shown for the first production rule:

```
(RULE S1
 (IF (S) (C))
  (THEN (Exec) (TRANS) ('S) ('c) ('C))))
```

which can be used for all production rules.

This is the list of production rules (AUTORULES) which are used in the inference engine in order to solve a given problem. An example of the work of the inference engine is given in Appendix 2 for selected rotational parts.

The combinatorial experiment for geometrical recognition using the automaton in Fig. 10.8e gives the following results:

Class	Pass	Rejected
1	2	1
2	6	12
3	18	72
4	50	418
5	132	2280
6	338	12010
7	846	62046
8	2084	316984, etc.

where class is defined by number of elementary forms for a given part.

The explained procedure has been used to build FSM for Family 2 (Fig. 10.8g).

Filtering through automata for different sets of part families has been determined as giving the real set of parts which belong to the family defined by the initial set of parts family (Fig. 10.9). The pattern recognition procedure (Fig. 10.10) includes the classification of a given part in corresponding part family group, or generates new knowledge for building of a new part family. It is important here to stress that the starting set of parts includes much more knowledge. In other words, the starting set of parts for a given part family is the initial knowledge which is enhanced by proposed corresponding automaton. In Fig. 10.11 the portion of the enhanced test part family is shown, which is obtained by generating real part family from a given test part family 0 (Fig. 10.8a).

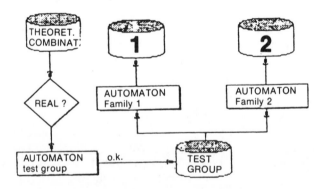

Fig. 10.9 Filtering through automata.

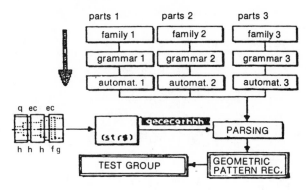

Fig. 10.10 Pattern recognition procedure.

Geometrical pattern recognition is already defined. The next step is to generate technology pattern recognition in a similar manner by introducing a technological automaton for a given part family. The procedure is similar with geometrical pattern recognition. The integral screen contents are shown in Fig. 10.12. The first window shows the part drawing. The second window gives the automaton structure for technological recognition which builds basic rules for process planning. The grammar concept and corresponding automaton are used to obtain the second filtering step in order to generate the process plan for a given part. The third window contains sets of production rules for geometrical and technological recognition. At the bottom are explicitly given the strings which describe geometrical and technological features of a given part.

Test results for geometrical and technological automata efficiency are shown in Fig. 10.13. Using the filtering procedure by applying the geometrical automaton in the first step has resulted in a plot below the theoretical one.

The next step is to apply the technological automaton for geometrically accepted parts which also reject some of them. This produces the lowest plot. The given procedure can be used for all part families in order to cover the theoretical area of sets of part families.

SAPT-P Expert System

The prismatic part process planning is more type (feature)-oriented than the group technology approach. This means that automata theory could be extended to space parsing of strings with extensive transition characteristics from rotational to prismatic part families.

The first assumption is to generalize the feature (type) concept from rotational to prismatic parts. The generalization includes increased number of type forms, space location of type forms with defined position

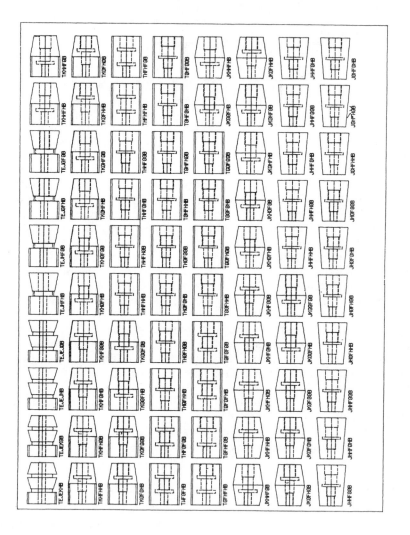

Fig. 10.11 Portion of enhanced test part family 0 using proposed automaton.

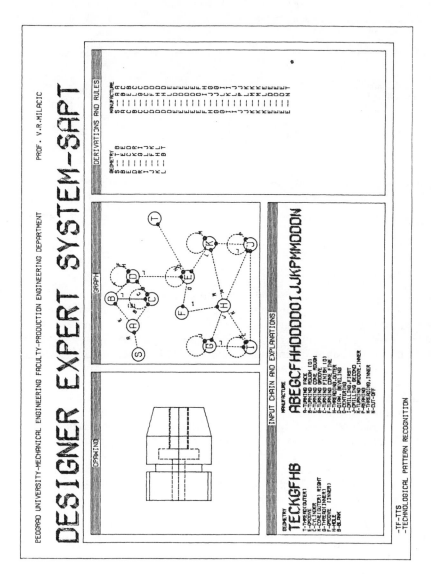

Fig. 10.12 Screen display for geometrical and technological recognition.

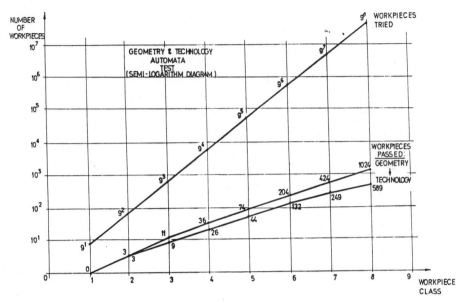

Fig. 10.13 Tried and accepted part family by automata.

accuracy, and extensive variety of tools for the provision of given manufacturing process. This means that besides isomorphic and homeomorphic behaviour there also exists quasi-morphism among the given entities of the manufacturing space. The given complexity generates the need for specific concepts and a much larger pool of organized knowledge. At this stage SAPT-R is an intensively rule-based and machine tool-oriented system whereas SAPT-P is developed as a prototype for machining centres.

The four main nodes for the manufacturing graph are: workpiece, process plan, tools and tool holders. Each workpiece (WP) has a multi-level hierarchical structure defined by complete sets of type forms (CTF), groups of type forms (GTF), type forms (features TF), and elementary forms (EF). For the first three nodes the interconnection is obtained along three axes (Fig. 10.14). In the space defined in such a way iso-, homeo- and quasi-morphisms could be applied in order to define interfacing of these three entities.

The hierarchical structures for all three groups of entities could be considered as the basic structures and should vary subject to the complexity of a given workpiece family. Extensive research results are reported in [8–10]. Here the basic structure in all three planes will be shown and special attention will be paid to the coupling between planes and to the tool plane in particular.

On the workpiece plane the TF contains the basic knowledge. Higher levels could be considered as a kind of meta-knowledge, and at the same

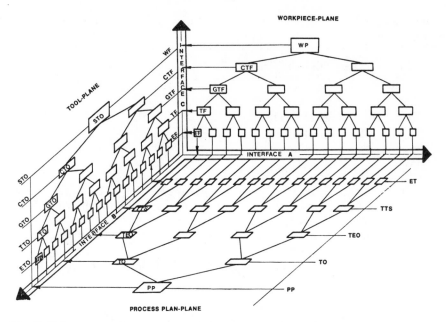

Fig. 10.14 Interconnection among workpiece–process plan–tool planes as a hierarchical structure.

time adding some knowledge. However, the EF is the basic unit (primitive) of knowledge for a given workpiece. The knowledge about type forms could be expressed as:

⟨TF⟩: : = TYPEFORM ⟨name⟩ IS ⟨elementary form⟩ WITH
⟨elementary form⟩ [, ⟨elementary form⟩];

It is clear that type form is defined by elementary forms, like:

⟨EF⟩: : = ELEMENTARY FORM ⟨name⟩ HAS ATTRIBUTES
⟨attribute⟩ [,⟨attribute⟩];
⟨ATR⟩: : = FOR ATTRIBUTE ⟨attribute⟩ QUESTION IS ⟨question⟩
TYPE OF VALUE IS ⟨value⟩;
⟨VALUE⟩: : = SET (⟨element⟩ [,⟨element⟩]|⟨number⟩

where, for example:

FOR ATTRIBUTE *diameter*
QUESTION IS *diameter*
TYPE OF VALUE IS *number*.

At the process plan (PP) plane the hierarchical structure has been defined as:

$$PP \rightarrow TO \rightarrow TEO \rightarrow TTS \rightarrow ET$$

where:

- TO   technological operation
- TEO   technological elementary operation
- TTS   type technological sequence
- ET   elementary transformation

Each of these entities is precisely defined. The TTS is the basic knowledge in order to evaluate the process plan for a given workpiece. The basic unit of knowledge is defined as elementary transformation (ET). In a similar manner as for the workpiece the definition of TTS is:

⟨TTS⟩: : = ELEMENTARY TRANSFORMATION FOR ⟨type form⟩ ARE

       ⟨elementary transformation⟩ ⟨elementary transformation⟩.

⟨ET⟩: : = ELEMENTARY TRANSFORMATION FOR ⟨elementary form⟩

       IS ⟨elementary transformation⟩

The higher levels like TEO and TO are some sort of meta-knowledge with additional new knowledge at these levels.

The INTERFACE-A enables connection between these two planes (WP–PP) among corresponding hierarchical levels. The basic interface is between elementary forms (EF) and elementary transformations (ET). A special group of production rules has been developed as follows:

⟨PRODUCTION RULE FOR ET⟩: : = IF ⟨elementary form⟩

       : & condition

       THEN ⟨elementary transformation⟩

where:

    ⟨condition⟩: : = ⟨fact⟩ ⟨relation⟩ ⟨fact⟩
    ⟨fact⟩: : = ⟨attribute⟩ ⟨value⟩
    ⟨relations⟩: : = ·= | = ⟨ | ⟨ ⟩ = | ⟩

The tool plane has the following hierarchical structure:

$$STO \rightarrow CTO \rightarrow GTO \rightarrow TTO \rightarrow ETO$$

where:

- STO   tool system
- CTO   complete set of tools
- GTO   group of tools
- TTO   type tool
- ETO   elementary tool

Again the basic tool structure is defined by type tool structure (TTO), and the primitive structure is given as elementary tool (ETO). A tool has a complex knowledge structure. In general a tool as physical structure has: cutting element (r), tool holder interface (n), and tool holder–machine connection (d). The number of elements varies widely.

There are three grouping strategies of type tools. Location-based grouping means that all different machining operations are running on a one side, one part, or one part family basis. In the case of a machining centre this represents a strong requirement. Auxiliary time is a minimum. Type forms-based grouping means that the similarity among a set of type forms defines sets of tools. This is a strong requirement for tool planning and control, but a weak one for machining. Tool change is very frequent. Finally, manufacturing process-based grouping is a weak requirement for workpiece manufacturing, but a strong one for tool manufacture. One example of tool structures for drilling and milling is shown in Fig. 10.15.

Tool selection highly depends on workpiece and process plan planes via interface B and via interface C, as well as on the combination of both interfaces. Tool structure based on type forms (interface C) could be performed as forward and backward chaining, as well as depth-first and breadth-first searching (Fig. 10.16).

The production rule for the mutual use of B and C interfaces in the definition of elementary tool is:

$\langle$ETO$\rangle$: : = IF $\langle$elementary form$\rangle$ AND $\langle$elementary
transformation$\rangle$ & $\langle$statement$\rangle$
THEN $\langle$elementary tool$\rangle$ MARK $\langle$grade$\rangle$;
$\langle$statement$\rangle$: : = $\langle$condition$\rangle$ $\langle$arithmetic expression$\rangle$
$\langle$arithmetic expression$\rangle$: : = $\langle$variable$\rangle$ IS $\langle$fact$\rangle$ $\langle$operation$\rangle$ $\langle$fact$\rangle$
$\langle$operation$\rangle$: : = + | − | * | /
$\langle$grade$\rangle$: : = 1 | 2 | 3

For example, the production rule for hole drilling is:

IF hole AND drilling   & diameter = $\langle$70
& diameter 5 IS *5
& diameter 5 = $\langle$ length
THEN r2 MARK 2

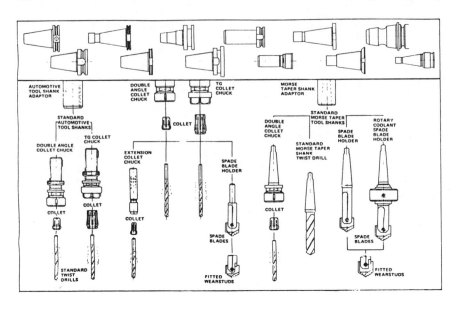

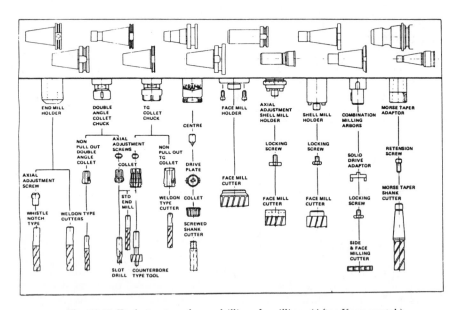

Fig. 10.15 Tool structure for: **a** drilling; **b** milling. (After Kennametal.)

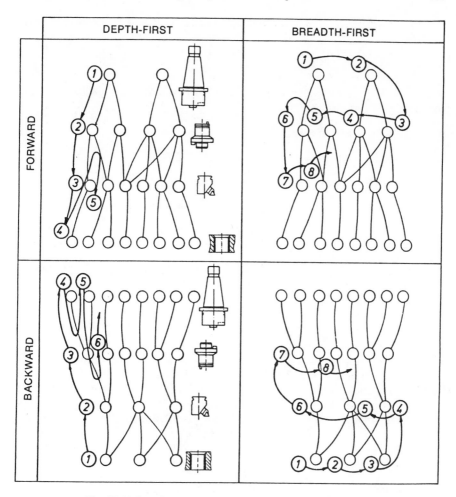

Fig. 10.16 Searching strategies for tool structure generation.

The knowledge also covers cutting elements and cutting conditions using specially formed tables of facts for machining centre process planning. Two kinds of knowledge are organized in the dedicated dictionaries and categorical coupling rules. The dictionary for elementary forms contains 71 entities, the dictionary for elementary transformations has 261 entities (with 38 basic ones) and the dictionary for tools covers 163 entities. The categorical coupling rules relate to the physical structure of tools, coupling tools with machine tools as well as coupling tools with workpiece and process plan structure. The coupling between cutting parts and tool holders and expansions has 331 rules, as well as between expansions and other expansions and tool holders 100 more rules.

Coupling functions (C) defined by interfaces B and C are also given:

- $ET \xrightarrow{C} ETO$ (cutting element) – 165 rules for forward chaining (interface B)
- $ETO \xrightarrow{C} ET$ (elementary transformation) 152 rules for backward chaining (interface B)
- $TTO \xrightarrow{C} TTS$ (interface C)
- $TTO \xrightarrow{C} TF$ with two elements (interface C)
- $TTO \xrightarrow{C} TF$ with three elements (interface C)
- $TTO \xrightarrow{C} TF$ with maximum number of elements (interface C)

The rules are systematically organized for use. However, the expert system has been built for machining centres and mainly for workpieces for machine tool builders.

Let us explain the structure given in the example of type form TF-15 (Fig. 10.17a). The type technological sequence is defined with the following elementary transformations:

$$TTS \ (TF \ 15) \rightarrow \langle ET217, ET163, ET1, ET2, ET12, ET28 \rangle$$

The six elementary sequence transformations could be obtained by six elementary tools defined by 16 different alternatives (Fig. 10.17b). This means that the variability block of knowledge for a given example is a rather complex matrix from which different adequate tool sets could be derived. Figure 10.17c shows one of the possibilities.

The already given tool plane and tree structure of tools for the practical example, is shown in Fig. 10.18. Using fuzzy automata this knowledge could be organized in a similar way to the FSM concept for the SAPT-R expert system.

The formal representation of tool knowledge for the given type forms (TF 15 and TF 82) based on locational grouping could be expressed as:

$$TTO \ 15 \ n2 \ ([r9, r1, r15, d5, r67, d20, r38, d10])$$

.

.

.

$$TTO \ 82 \ n2 \ ([r31, d10]) \ for \ plane \ machining$$

.

.

.

At the level of GTO, one formal expression, as an example is given:

$$GT \ 03 \ ([r35, d11, r9, r1, r15, d5, r67, d20, r38, d10])$$

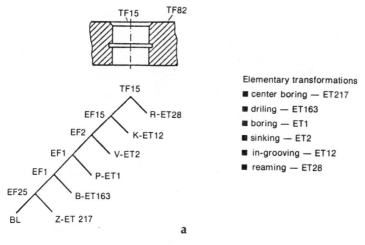

Elementary transformations
■ center boring — ET217
■ driling — ET163
■ boring — ET1
■ sinking — ET2
■ in-grooving — ET12
■ reaming — ET28

a

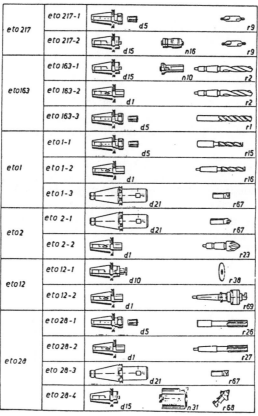

b

Fig. 10.17 Tool selection for given example: **a** TF 15 and TTS 15; **b** tool definition for TTS; **c** (overleaf) example of selected tool set for given example.

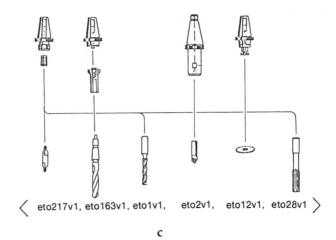

$$\langle \text{ eto217v1, eto163v1, eto1v1,} \quad \text{eto2v1,} \quad \text{eto12v1, eto28v1 } \rangle$$

c

Fig. 10.17 (continued).

at the level of CTO:

CTO 1 ([r31, r9, r1, r15, d5, r67, d20, r38, d10])

and finally at the level of STO:

STO 1 ([r38, r9, r1, r15, r67, d20, r31, d10, r35, d11, r55, r49, d5])

Using a similar type of presentation, knowledge could be formalized for tools grouping based on type forms and manufacturing processes. The meanings of symbolic expressions are given in Figs 10.17b and 10.18.
    The prototype of SAPT-P expert system is developed in Prolog coding with the following knowledge base:

● TF – production rules	58
● EF – attributes	63
● Production rules EF–ET	32
● Coupling rules EF–ET	134
● Production rules as elementary path	24
● Production rules for cutting element of tools (TO)	287
● Production rules for tool holders	124

    The systematic classification of vocabulary and corresponding production rules with built in inference mechanism have been used for experimental assessment of SAPT-P for a selected family of workpieces manufactured on machining centres for a collaborating industrial company.

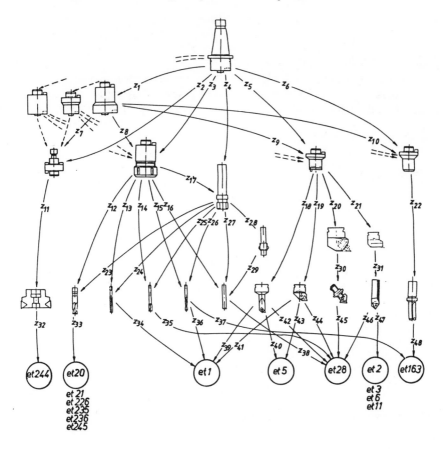

Fig. 10.18 Graph of tool structure for given example.

Conclusion

After five years' research and development the author and his team have developed a theoretical approach and two prototype expert systems for process planning in the conceptual domain.

The SAPT expert systems for rotational and prismatic parts were tested in Lisp and Prolog coding in order to compare their suitability to the knowledge engineer and for the acquisition of manufacturing knowledge. However, we have not come to the conclusion because it is still necessary to investigate the concept of shells for process planning expert systems.

However, the knowledge base for process planning and some reasoning strategies including learning features have already been developed and experimentally proved in the working environment. It is evident that a severe limitation is appearing when the number of rules rapidly increases and the reasoning process becomes unacceptably time consuming. This means that if the system is segmented into subsystems for specific machines the developed concept could be applicable.

At this stage the SAPT-P expert system is tested for a machining centre using personal computer facilities. The expanding local area network concept will include SAPT-P in a larger workshop environment.

Appendix 1

```
TEH
(lambda nil)
     (progn nil)
          (HEADER)
          (ARL arlist)
          (setq urlist (list (quote stratrules)))

HEADER
(lambda nil)
     (progn nil)
          (cls)
          (sprint (quote MECHANICAL_ENGINEERING_FACULTY
))
          (sprint)

ARL
(lambda (arlist))
     (progn (rlist))
          (sprint (quote Available_rulelists_are:))
          (setq rlist arlist)
          (while (not (null rlist)))
               (sprint (cons (quote ) (car rlist)))
               (setq rlist (cdr rlist))

MENU
(lambda nil)
     (progn nil)
          (sprint (quote Make_a_choice:))
          (sprint (quote =========))
          (sprint (quote R = run))
          (sprint (quote E = enter))
          (sprint (quote L-list-rulelists))
          (sprint (quote P = printer))
          (sprint (quote S = screen))
```

```
                  (sprint (quote I = select-rulelists))
                  (sprint (quote Q = quit))
                  (COND)

          COND
          (lambda nil)
              (progn (user)
                  (setq user (read))
                  (cond ((equal user (quote R))   (RUN)))
                        ((equal user (quote E))   (ENTER))
                        ((equal user (quote L))   (LIST))
                        ((equal user (quote P))   (LPR))
                        ((equal user (quote S))   (SCR))
                        ((equal user (quote I))   (INT))
                        ((equal user (quote Q))   (QUIT))
                        (t (sprint (quote INCORRECT_ENTRY)))
                  (MENU)

          ENTER
          (lambda nil)
              (progn nil)
                  (sprint (quote Enter_facts:))
                  (setq factg (read))

          RUN
          (lambda nil)
              (progn (factgg facts urlistt rules))
                  (setq factgg factg)
                  (while (not (null factgg)))
                      (setq facts (car factgg))
                      (setq urlistt urlist)
                      (while (not (null urlist)))
                          (setq rules (eval (car urlistt)))
                          (sprint (car urlistt))
                          (sprint (quote =========))
                          (DEDUCE)
                          (setq urlistt (cdr urlistt))
                      (setq factgg (cdr factgg))
                  (sprint (quote End-of_run))

          LIST
          (lambda nil)
              (prog nil)
                  (ARL arlist)
                  (print (quote Rulelist_name?))
                  (P (eval (read)))
```

```
LPR
(lambda nil)
    (progn nil)
        (de sprint (s) (lprint s))
        (print (quote All_outputs_to_printer))

SCR
(lambda nil)
    (progn nil)
        (de sprint (s) (print s))
        (print (quote All_outputs_to_screen))

INT
(lambda nil)
    (progn nil)
        (sprint (quote Enter_rulelist_names:))
        (setq urlist (read))

QUIT
(lambda nil)
    (progn nil)
        (sprin (quote Quit_requested))
        (exit)

P
(lambda (rules))
    (progn (rulelist))
        (setq rulelist rules)
        (PR rulelist)

PR
(lambda (rulelist))
    (cond ((null rulelist) (sprint (quote ==========
    ==))))
        (t (PRI rulelist)

PRI
(lambda (rulelist)
    (progn (rule))
        (setq rule (car rulelist))
        (sprint (quote ===========))
        (sprint (cons (car rule) (cadr rule)))
        (sprint (car (cddr rule)))
        (sprint (cdr (cddr rule)))
        (setq rulelist (cdr rulelist))
        (PR rulelist)

DEDUCE
(lambda nil)
    (progn (progress))
```

```
                    (cond ((STEPFORWARD) (setq progress (DEDUCE))))
                    (t progress)
STEPFORWARD
(lambda nil)
      (progn (rulelist))
        (setq rulelist rules)
        (STEPLOOP rulelist)

STEPLOOP
(lambda (rulelist))
      (cond ((null rulelist) (quote nil)))
          (t (TRY rulelist)

TRY
(lambda (rulelist))
      (cond ((TRYRULE (car rulelist)) t))
          (t (CDRULE rulelist))

TRYRULE
(lambda (rule))
    (and (TESTIF rule) (USETHEN rule))

CDRULE
(lambda (rulelist))
      (progn (rlist))
          (setq rlist (cdr rulelist))
          (STEPLOOP rlist)

TESTIF
(lambda (rule))
      (progn (ifs)
          (setq ifs (CDADDR rule))
          (TIFS ifs)

USETHEN
(lambda (rule))
      (progn (thens success))
        (setq thens (cdr (CADDDR rule)))
        (UTHENS rule thens success)

CDADDR
(lambda (I))
    (cdr (cadr (cdr 1)))

TIFS
(lambda (ifs))
      (progn (ifsh))
          (setq ifsh ifs)
          (cond ((null ifsh) t))
            ((RECALL (car ifsh)) (TIFS (cdr ifsh)))
```

RECALL
(lambda (fact))
    (cond ((member fact facts) fact))
        (t nil)

CADDDR
(lambda (k))
    (cadr (cddr k))

UTHENS
(lambda (rule then success))
    (cond ((null then) success))
        (t (UTHEN rule then success)))

UTHEN
(lambda (rule then success))
    (progn (thensh successh))
        (setq thensh then)
        (setq successh success)
        (cond ((equal (car thensh) (list (quote   ))) (STRAT (cdr
        thensh))))
            ((REMEMBER (car thensh)) (setq successh (PRINT rule
thensh successh)))
        (setq thensh (cdr thensh))
        (UTHENS   RULE THENSH SUCCESSH)

REMEMBER
(lambda (new)
    (cond ((member new facts) nil))
        (t (setq facts (cons new facts)) new)

STRAT
(lambda (thensh))
    (progn nil)
        (cond ((not (RECALL (car thensh))) (STRAT1 thensh)))

TRANS
(lambda 9startnode transaction destnode))
    (progn nil)
        (setq string (cdr string))
        (setq factg (SUBST destnode startnode factg))
        (setq factg (SUBST (car string) transaction factg))
        (setq facts (car factg))
        (print (cons startnode (cons (quote) (cons transaction
        destnode))))
        (DEDUCE)

LTRANS
(lambda (startnode destnode))
    (progn nil)

```
                    (setq factg (SUBST destnode startnode factg))
                    (setq facts (car factg))
                    (print (cons startnode (cons (quote   ) (cons (quote lambda)
                    destnode))))
                    (DEDUCE)
STRAT1
(lambda (thensh))
       (progn nil)
              (sprint (quote Rule_used:))
              (eval (car thensh))
```

Appendix 2

For a given part family 1 (Fig. 10.8d) the automaton (Fig. 10.8e) has been automatically generated, at the same time as the list of production rules. The list of some of production rules in Lisp editor is entered.

```
(Rule Si
       (IF          (S((c))
       (THEN        (Exec) (TRANS ( S) ( c) ( C))))

(Rule C1
       (IF          (C) (k))
       (THEN        (Exec) (TRANS ( C) ( k) ( K))))

(Rule S2
       (IF          (S) (k))
       (THEN        (Exec) (TRANS ( S) ( k) ( K))))

(Rule K1
       (IF          (K) (c))
       (THEN        (Exec) (TRANS ( K) ( c) ( C))))

(Rule C2
       (IF          (C) (c))
       (THEN        (Exec) (TRANS ( C) ( c) ( C))))

(Rule S3
       (IF          ( S) ( d))
       (THEN        (Exec) (TRANS ( S) ( d) ( D))))

(Rule D1
       (IF          (D) (c))
       (THEN        (Exec) (TRANS ( D) ( c) ( C))))
```

```
(Rule  C3
    (IF            (C) (b))
    (THEN          (Exec) (TRANS ( C) ( e) ( E))))
```

....

....

```
(RULE  K2
    (IF            (K))
    (THEN          (Exec) (LTRANS ( K) ( U))))

(RULE  H5
    (IF            (H) (b))
    (THEN          (Exec) (TRANS ( H) ( b) ( T))))

(RULE  P2
    (IF            (P) (b))
    (THEN          (Exec) (TRANS ( P) ( b) ( T))))

(RULE  I1
    (IF            (I) (b))
    (THEN          (Exec) (TRANS ( I) ( b) ( T))))

(RULE  U3
    (IF            (U) (b))
    (THEN          (Exec) (TRANS ( U) ( b) ( T))))
```

This list of production rules is available as AUTORULES for the SAPT-expert user. The process planner gets a net rotational part (Fig. A1) to create the process plan by checking. If the given part belongs to the automaton of the existing part family, it will be accepted. Otherwise it will be rejected. The string of the new rotational part is dccchfhb. The SAPT-expert system with instruction TEH offers on the screen the following contents:

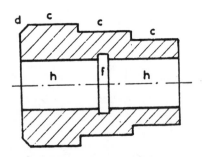

Fig. 10.A1 New rotational part.

TEH
  MECHANICAL_ENGINEERING_FACULTY
      SAPT_EXPERT

Available_rulelists_are:
        (autorules)   shown all above listed rules

Make_a_choice:
==========

        R=run   –start inference engine
        E=enter   –initial facts string of part
        L=list_rulelists   –searching automatically rule list
        P=printer   –output results in any stage of the
                                            work
        S=screen   –output results in any stage of the
                                            work
        I=select_rulelists   –manual selection of rules
        Q=quit   –accepted (T) or rejected part if T is
                                            not the last symbol

E   –first step
    dcchfhb

R – start inference engine

autorules
==========

Rule_used:
(S ⟩⟩⟩ d.D)
Rule_used:
(D ⟩⟩⟩ c.C)
Rule_used:
(C ⟩⟩⟩ c.C)
Rule_used:
(C ⟩⟩⟩ c.C)
Rule_used:
(C ⟩⟩⟩ lambda.U)
Rule_used:
(U ⟩⟩⟩ h.H)
Rule_used:
(H ⟩⟩⟩ f.F)
Rule_used:
(F ⟩⟩⟩h.H)
Rule_used:
(H ⟨⟨⟨b.T)
End_of run

Q
Quit_requested

The inference engine selects the first variable block (by definition it is S) and first terminal symbol (in our case it is d), and tries to find the corresponding rule from the list of autorules. This is rule S3. The process continues until the final T variable block is ready, otherwise the part is rejected.

References

1. Hayes-Roth et al. Building expert systems. Addison Wesley, London, 1983
2. Ham I. CAPP: present and future. CIRP Ann Manuf Technol 2, 1988; 37: no. 2
3. Milacic RV (ed). Intelligent manufacturing systems I. Elsevier, Amsterdam, 1988
4. Milacic RV (ed). Intelligent manufacturing systems II. Elsevier, Amsterdam, 1988
5. Simon H. Why should machines learn. In: Machine learning. Springer-Verlag, Berlin, Heidelberg, New York, 1984
6. Carbonell J, Michalski R. An overview of machine learning. In: Machine learning. Springer-Verlag, Berlin, Heidelberg, New York, 1984
7. Milacic RV. Manufacturing systems design theory. Mechanical Engineering Faculty, Beograd University,Beograd, 1987
8. Urosevic M. SAPT – Expert system. MSc thesis, Beograd University, 1988
9. Veljovic A. Technological knowledge base. PhD thesis, Beograd University, 1988
10. Putnik G. Tool knowledge base. MSc thesis, Beograd University, 1988

## Chapter 11

Knowledge-Based Computer-Aided Process Planning

Y. Ito and H. Shinno

Introduction

Process planning is one of the most important procedures in production because of its interfacing function between the design and the manufacturing procedures. The manufacturing procedure consists mainly of three core activities, i.e., machining, assembly and inspection. Accordingly, there are various kinds of process planning depending on each activity; however, research and development are so far concentrated on process planning for machining. This trend may be attributed to the fact that machining is deeply concerned with part manufacturing, which is fundamental to producing the goods, appliances, equipment, devices and so on required by society. In this chapter, thus, process planning for machining is described.

Although the scope of the description will be limited to machining, i.e., to process planning for the machine tool-based industry, process planning is very tedious work and requires complicated information processing. Thus, with advances in computer technology, its automated version, i.e. computer-aided process planning (CAPP) has, in due course, become one of the major objectives in research and development activities. In this case, CAPP can be interpreted as being the interfacing or linkage function between CAD and CAM. There has been much effort to develop effective and efficient CAPP over the world.

At present, CAPP can be classified into three types, i.e., variant (decision table), semi-generative and generative as shown in Fig. 11.1. Each has been vigorously improved in terms of its performance, quality and capability since the beginning of the 1970s (e.g. [1]). However, even after more than 20 years of development, it is still believed that CAPP has not yet sufficient capability to be fully of practical use. As shown in Fig. 11.2, it appears that most companies employ simple computerized

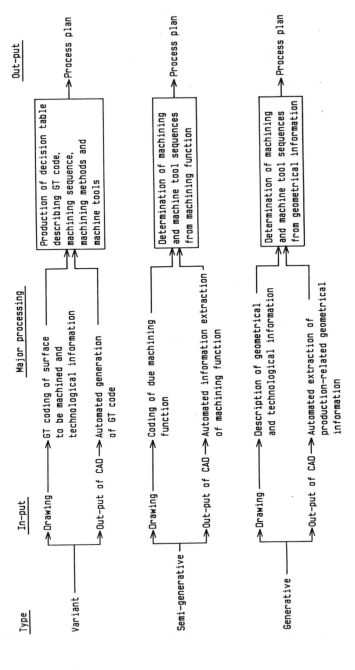

Fig. 11.1 Existing CAPP systems (by Iwata).

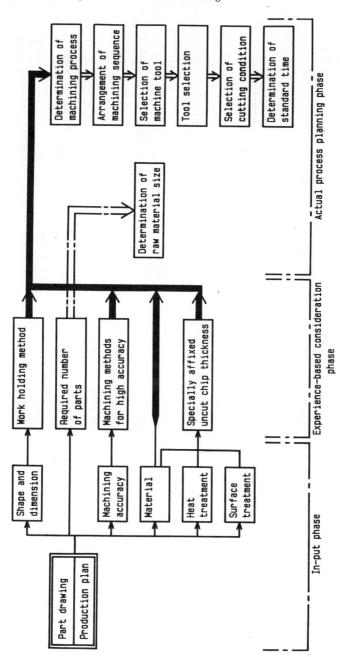

Fig. 11.2 Simple computerized process planning employed by industry (in the case of axisymmetrical parts).

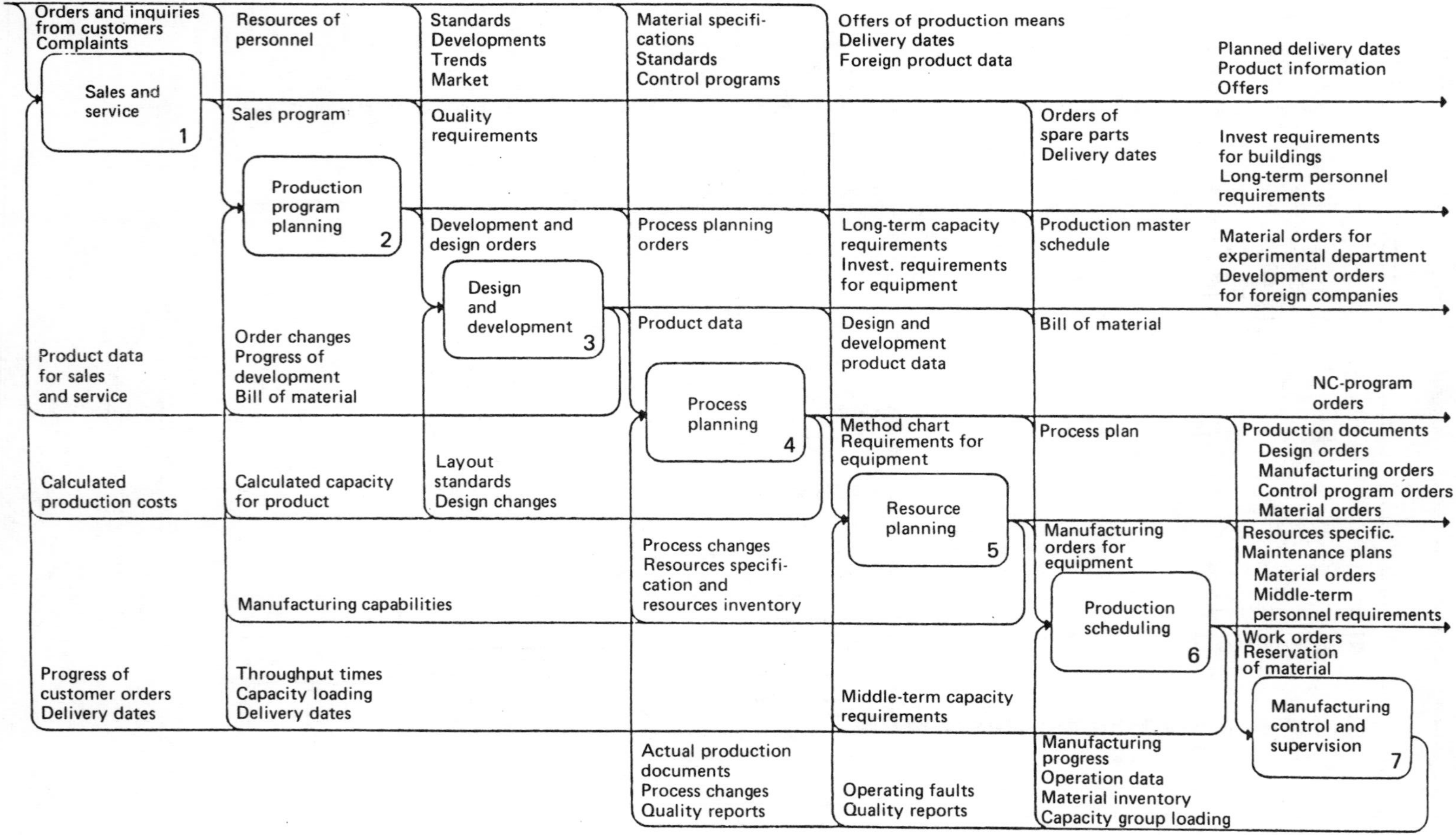

Fig. 11.3 CIM concept proposed by Fraunhofer-IPK Berlin and VW-Gedas.

process planning. This trend may be derived from the following characteristic features of process planning:

1. Process planning plays a role of intermediary between the design and manufacturing of the factory floor level. In design, however, the geometric-related information should be processed, whereas manufacturing requires production-related information. In general, both kinds of information are not compatible with each other, but stand completely opposite.

2. Process planning can be characterized by its experience-oriented nature, also having considerable dependence on production facilities and production personnel's ways of thinking. It is not unusual that two process engineers belonging to the same company produce different process plans to the same part drawing.

In addition, CAPP has recently increased its importance as the computer-integrated manufacturing (CIM) concept has become a major trend in production technology. CIM itself is expected to be a kernel function of factory (or flexible) automation (FA) of the next generation. Although the exact definition of CIM is not fixed as yet, one of the major interpretations is that CIM possesses an integrated software function, mainly consisting of CAD, CAM and material requirement planning (MRP) as shown in Fig. 11.3 [2]. Of these, CAD and CAM are still believed to play an important role, and without having a qualified CAD/CAM interface, i.e., CAPP, the CIM concept could not be realized.

In consideration of such situations, and also of the characteristic features of process planning, there are now two potential ways to enhance the present CAPP: one involves using the product model (e.g. [3]) and another is based upon artificial intelligence (AI). In the former case, the essential idea is to evolve the geometric models so far used in CAD, i.e., primitive, wire-frame and surface models, to one simultaneously having the geometrical and manufacturing information, i.e., the product model. By this, the establishment of an automatic interface between CAD and CAM is achieved to some extent. The product model certainly has potential; however, it also presents the following problems:

1. Due to having much information, the model is to be very complex.

2. In principle, there is not a one-to-one relationship between the geometrical specification and the method of generating a part. Even in the production of the same part, various machining methods are available, and it is sometimes difficult to say which method is the most suitable.

As a result, AI-based CAPP may be considered to be superior to that using the product model, and concerned researchers and engineers have been very interested in CAPP with AI function, i.e., knowledge-based CAPP, since AI technology became a topic in the production field at the beginning of the 1980s. This chapter deals with some knowledge-based CAPP systems of the next generation, the characteristic features of which are based on deep understanding of the thinking patterns (i.e. way of thinking, decision making and use of flair) of experienced engineers, and which take into consideration the essential phase in the production of the part drawing, i.e., the interfacing function of the part drawing itself. Most of the knowledge-based CAPP systems so far developed are those simply incorporating expert systems, and for reasons of poor consideration of the engineer's thinking patterns, they may be regarded to be systems of the first generation. In addition, this chapter may contribute to a better understanding of the importance of the analysis of thinking patterns of experienced process engineers. Without having such information, real knowledge-based CAPP is not likely to become established.

Survey of Previous Work

Regarding the research and development of knowledge-based CAPP, at present, the major activity is concentrated on CAPP with expert systems,

Table 1. Organizations having investigated or currently developing CAPP of expert systems type.

Country	Organization
Denmark	Technical University of Denmark
Japan	Chuoh University
	Hokkaido University
	Kobe University
	NEC Co.
	Tokyo Institute of Technology
	University of Tokyo
Norway	Senter for Industriforskning
	Technical University of Norway
The Netherlands	Twente Institute of Technology
UK	Systime Co.
	University of Aston in Birmingham
	University of Edinburgh/Coventry Polytechnic
	University of Leeds/GEC
	University of Manchester Institute of Science and Technology
USA	Allied Co.
	Metcut Research Associates
	Ingersoll Milling Machine Co.
	MIT
	MRA Inc.
	Pennsylvania State University
	University of Illinois
Yugoslavia	University of Ljubljana

i.e., a variant of knowledge-based CAPP. An example of organizations where expert systems have been or are being applied to CAPP is shown in Table 1. This table has been obtained from a literature survey and on-the-spot investigation. It can be suggested that most of the CAPP programs with expert systems employ, in general, production rules or frames to represent the explicit or apparent knowledge of the experienced process engineer, simultaneously maintaining the information-processing flow such as shown in Fig. 11.2. For instance, Alting et al. [4] have used such an architecture, as shown in Fig. 11.4. Thus, they appear to be at the first stage of utilizing the engineer's knowledge; however, the necessity is to have more insight into the engineer's thinking patterns.

From such, a recognition of what CAPP of the future (i.e., knowledge-based CAPP of the next generation) could be like using AI technology is understood to a large extent. However, it is pointed out that there have been few such activities on CAPP because of difficulties in obtaining the necessary basic data. Such data include the thinking patterns of the experienced engineer rationally represented with those suitable for computerized processing, and also the information implicitly described on the part drawing concerning directly the machining, assembly and inspection procedures.

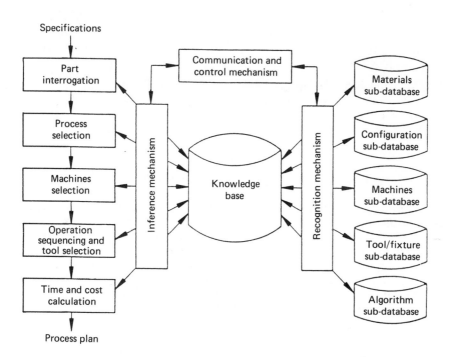

Fig. 11.4 Example of architecture of CAPP with expert systems.

Knowledge-Based CAPP of the Next Generation

Regarding knowledge-based CAPP of the next generation, at present, three types have been proposed:

1. *Thinking way-based CAPP (TW-CAPP)*. This type of idea is based on the effective and efficient implant of the experienced engineer's thinking patterns into the computer environment [5, 6]. As is easily imagined, CAPP with expert systems could not be used for the development of TW-CAPP, and other different approaches are required to establish it.

2. *Flair-based CAPP (FA-CAPP)*. In manual process planning, the engineer concerned produces the process plan after only glancing at the part drawing. This implies that he/she employs flair and experience. FA-CAPP is an attempt at positively utilizing the flair of an experienced engineer.

3. *CAPP of modified expert systems type (CAPP-MEX)*. The intention of this type is to transfer the production procedure described on the part drawing directly to the computer environment, because the part drawing itself is essentially an interface between the design and the manufacturing procedures. When producing the part drawing, for instance, the designer is required not only to deal with geometrical information, but also to have enough knowledge of how the parts are generated, assembled and inspected.

In the development of these CAPP systems, thus, the basic necessities are to have enough understanding of the thinking patterns of the process engineer (for the first two types of CAPP), and also of the implicit information described on the part drawing (for the third type of CAPP).

Thinking Patterns of Experienced Process Engineers

In general, AI technology consists of knowledge representation, knowledge utilization, knowledge acquisition and man–machine interfacing. Of these, knowledge acquisition is the most difficult, because of its experience-dependent character. The investigation of the thinking patterns of process engineers belongs to the subject of knowledge acquisition, but it is not easy because of human-oriented problems, for instance, the uncertainty and ambiguity in decision making. Due to such difficulties and uncertainties and also to the individual-related characteristic features, the research results so far obtained cannot be, in some cases, guaranteed to be valid. In brief, at present, there are no evaluation or justification methods regarding both the methodology and research results of knowledge acquisition.

Figure 11.5 is a flow chart of process planning reported by Iwata and Fukuda [7]. They have tried to analyse the thinking patterns of an engineer, especially to clarify the procedure whereby knowledge is used by the engineer. Through many interviews with experienced engineers, as shown in Fig. 11.5 by the hatched lines, they have suggested several procedures suitable for incorporation in expert systems. In due course, they have then proceeded to the development of a prototype program; however, the implant of the thinking pattern is not enough as yet, remaining in the stage of the improvement of CAPP with expert systems. Figure 11.6 shows a simplified result of another approach reported by Chen and Ito [5]. In this case, they have tried to describe the thinking patterns of the experienced engineer using a directed graph, where points (vertices) and arrowed lines (directed edges) represent the core items for conducting the decision making and information flow necessary to achieve the objective core item in process planning respectively. For reasons of difficulty and in order to obtain reasonable results, they have, at first, determined several core items, and then carried out interview investigations. Through a series of interviews, Chen and Ito have pointed out several interesting findings as follows:

1. The individual-dependent character of process planning can be verified by comparing the directed graph. For instance, Figs 11.7 and 11.8 represent the thinking patterns of two different engineers, and it is easy to understand such a characteristic feature.

2. Even though there are various thinking patterns, they can be arranged and simplified as shown in Fig. 11.6 by choosing the most important core items and directed edges. This simplified graph can be characterized by the closed loop within it, also implying the existence of the strong mutual interrelation among the core items. This loop can, furthermore, clearly demonstrate why process planning has considerable uncertainty.

3. The thinking pattern of the process engineer is very unstable, and it changes easily, especially when he has a concrete input, i.e., a part drawing. Figure 11.9 is a thinking pattern of the process engineer who exhibited the thinking pattern of Fig. 11.7 when not having the part drawing. In this graph, it is very interesting that the number of closed loops is considerably reduced, and that the core item MT has disappeared. Probably, the process engineer could break the closed loop by the stimulus of an external input so as to ease the decision making and information processing. The driving force for this is not clear, but is thought to be based upon his long experience and flair.

4. It is, furthermore, noteworthy that the same process engineer presents, more or less, another plan for the same part after a certain time interval.

As described above, the thinking patterns of an experienced process engineer have much uncertainty and ambiguity, which are not compatible with computerized processing, but show the nature of process planning.

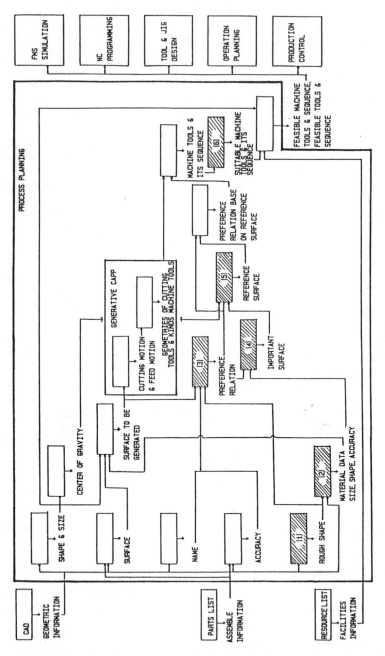

Fig. 11.5 Flow chart of process planning with expert systems (by Iwata and Fukuda).

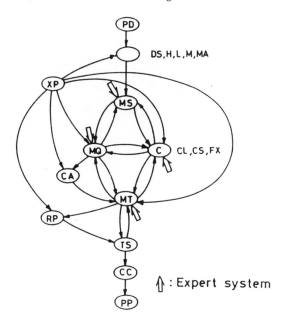

↥ : Expert system

C: Chucking	CA: Re-chucking procedure
CC: Selection of cutting condition	DS: Detailed shape and dimension
CL: Chucking Length	CS: Chucking portion
H: Heat treatment	L: Lot size
M: Material	MA: Machining accuracy
MQ: Selection of machining sequence	MS: Selection of raw material size
MT: Selection of machine tool	PD: Part drawing
PP: Process sheet	RP: Rearrangement of process plan
TS: Tool selection	XP: Experience data
FX: Jig and fixture	

Fig. 11.6 Simplified thinking pattern of experienced engineers and core items which can incorporate expert systems.

It is thus desirable to implant such thinking patterns into the computer environment, if possible. In this regard, the simplified graph of Fig. 11.6 is considered to have strong potential, provided that a rational database will be provided. Chen and Ito have thus proposed a relational database structure embodying the idea of an "index volume of hierarchical type" [6]. Figure 11.10 is a schematic representation of this idea, which maintains the closed loop characteristics, but can search for possible solutions by gradually narrowing the index volume. At present, the development of a prototype is ongoing using the software "KEE".

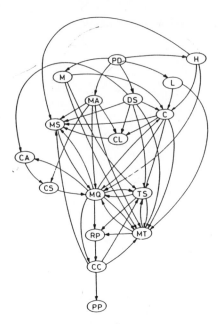

Fig. 11.7 Example of thinking pattern (1).

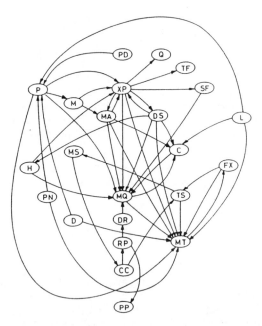

DR: Check with experience data   P: Priority
PN : Penalty   Q: Quit the job
SF : Selection of factory   TF: Task force

Fig. 11.8 Example of thinking pattern (2).

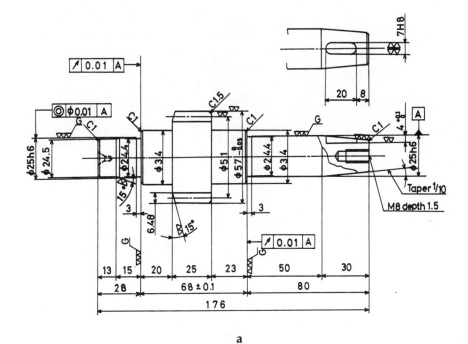

a

a: energy and time consideration

IL: In-house material list

b

Fig. 11.9 a Referred part drawing; **b** Change of thinking pattern when providing part drawing.

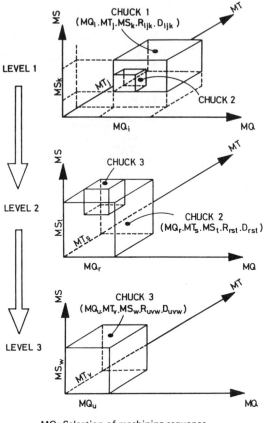

MQ: Selection of machining sequence
MT: Selection of machine tool
 D: Required data files
MS: Selection of raw material size
 R: Required rules

Fig. 11.10 A database structure of index volume of hierarchical type.

Flair of Experienced Engineers in Process Planning

As many engineers have frequently witnessed, it is not unusual that experienced process engineers can produce a suitable process plan to each part drawing without spending more than 5 min on it. This amazing ability is believed to be derived from his/her flair to a large extent and also from knowledge about parts accumulated by his/her long experience. In brief, the engineer's flair eases the process planning, and in due course there could be high feasibility of utilizing such flair to establish knowledge-based CAPP, i.e., flair-based CAPP. This type of CAPP is, of course, considered to be the most human-oriented or skill-based, and may be interpreted as a system of using the pattern recognition ability of the experienced engineer. Ito and his colleage Ihara, of Chuoh University, have been involved in research work on the development of such a CAPP program.

For the sake of its further advancement, Fig. 11.11 demonstrates the concept of FA-CAPP as proposed by them. In this concept, the kernel is which intuitive thoughts are applied by an engineer instantaneously in process planning. Table 2 contains information on the decision factors considered by experienced engineers in the machine tool industry when they produce a process plan after only glancing at the part drawing. As can be seen, most of the engineers point out the same factors, and thus it may be expected to establish knowledge-based CAPP of the flair type by combining AI with pattern recognition technology.

Knowledge-based CAPP Utilizing Implicit Information on Part Drawings

In general, process planning of the manual type is carried out using the part drawing as an input. In this case, the engineer converts the geometrical information into data for manufacturing by extracting implicit information from the part drawing. Thus, the engineer plays the roles of the converter and interpreter in information processing. In CAPP, on the contrary, the part drawing is replaced with the part model, e.g., computer-oriented part description, which is, of course, for the CAD system. The problem is that the part model for CAD has only geometrical information, but CAM requires manufacturing information. CAPP should, thus, implement the interface function between the geometrical and manufacturing information on behalf of the process engineer. Such an explanation is very easy, but its establishment is very difficult. As a result, CAPP with expert systems is now the focus of research and development activities to provide an effective remedy to this problem.

When changing the viewpoint, however, one can be aware of the essential feature of the part drawing, i.e., its having by nature the interfacing function between design and manufacturing. As widely recognized, the preferred or complete part drawing can be produced by the designer or engineer who has deep knowledge of machining, assembly and inspection, and it is, furthermore, noteworthy that the information related to such knowledge is implicitly described on the part drawing. This leads to the idea of positively using the part drawing in CAPP. Figure 11.12 is a proposal for such a concept by Ito et al., which can produce both the part drawing and process planning [8]. This is considered to be one of the modified CAPP packages with expert systems, and its core is the information conversion using the knowledge base, which involves such machining, assembly and inspection information

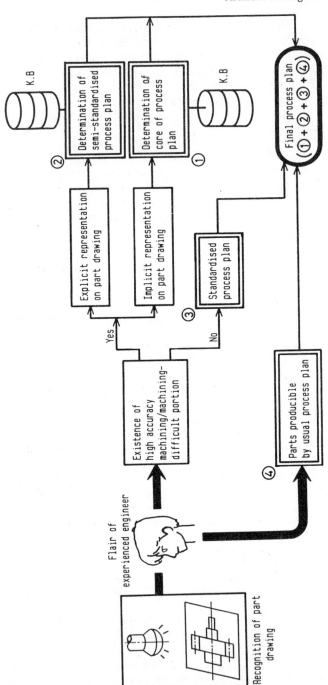

Fig. 11.11 Concept of flair-based CAPP.

Table 2. Engineer's viewpoints in process planning

Manufacturer	Engineer's viewpoints			Remarks
	1st order	2nd order	3rd order	
A	S	MA	H	Objectives: parts for prototype
				Major premise: permissible production cost
	S	H	MA	4th order: CP
B	CP	H	MX	Due to the company's environment, MA is out of viewpoints
C	S	MA	MX	
D	S	MR	MX	4th order: MA and H
E	MX	CR	S	Major premise: functional validity of part within assembly drawing
	S	MX	CP	

CP: compatibility with production facilities and worker's ability.
CR: cost reduction.
H: existence of heat treatment/reference of part material.
MA: existence of high accuracy machining portion (including necessity of jig).
MR: raw material and its shape/size.
MX: existence of machining-difficult portion/portion required of special form generating method (including necessity of jig).
S: shape/size of part along with machining feasibility.

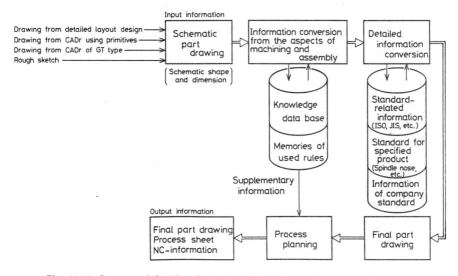

Fig. 11.12 Concept of CAPP utilizing implicit information on part drawing.

Table 3. Representative knowledge required in draughting process of part drawing

	Required knowledge
Assembly/transportation technologies	Assembly/disintegration methods of product Ease of assembly of part Adjustment method of product Ease of part transportation Machine element
Machining technology	Form-generating movement/structure/control of machine tool Attachment of machine tool Cutting/grinding tools Jig/fixture Material properties of workpiece Heat treatment/other processing methods
Others	Process control Ergonomics/safety technology Production cost Maintenance technology

necessary to produce the part drawing as shown in Table 3. In actual situations, each item within Table 3 should be divided into more detailed subitems so that the production rules are allowed to become easily available.

The proposed system can, furthermore, be characterized by the following functions.

1. When producing the part drawing, the necessary information can be released from the knowledge base, and applied to the production of the process plan after temporarily recording its usage history.

2. The primitive and datum are, as shown in Fig. 11.13, employed to represent the subpart, and a group of primitives and data enables a part to be described.

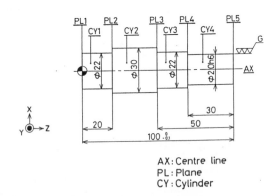

AX: Centre line
PL: Plane
CY: Cylinder

Fig. 11.13 Part description with primitives and datum.

3. Each primitive corresponds to only one machining method. This ensures the one-to-one relation between the geometrical shape and its generating method. In general, there are various machining methods for a part shape, and this causes problems in the establishment of CAPP.

4. The knowledge bases for assembly and inspection are converted into those for machining so as to ease the database management.

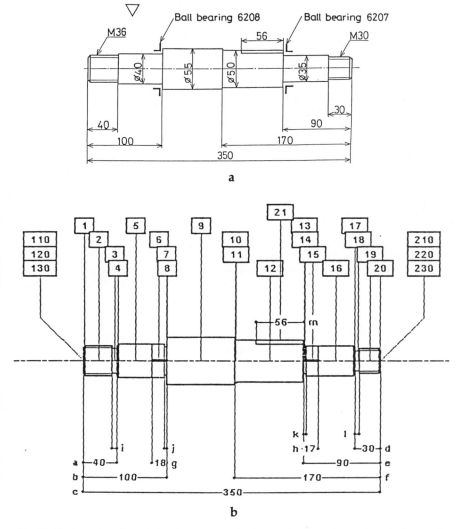

Fig. 11.14 Example of part drawing automatically produced: **a** input (rough drawing); **b** output (final drawing).

```
                  *** -------->>> PROCESS SHEET <<<-------- ***
                    ** --------->> Material Size <<----------**
    Length:            358
    Diameter           60
                    **---------->> Machining Sequence <<----------**


Machining Method No 1 LFC (facing)
Machine Tool: 'Turning Center'
Primitive of Machining: (110)
Clamp Primitive: Raw Material
      (:direction: right position: 144)
Machining Method No2 DCN (centering)
Machine Tool: 'Turning Center'
Primitive of Machining: (120)
Clamp Primitive: Raw Material
      (direction: right position: 144)
Machining Method No3 DS (spot facing)
Machine Tool: 'Turning Center'
Primitive of Machining: (130)
Clamp Primitive: Raw Material
      (direction: right position: 144)
Machining Method No4 L (turning)
Machine Tool: 'Turning Center'
Primitive of Machining: (1 2 3 4 5 6 7 8 9)
Clamp Primitive: Raw Material
      (direction: right position: 144)
Machining Method No5 L (turning)
Machine Tool: 'Turning Center'
Primitive of Machining: (1)
Clamp Primitive: Raw Material
      (direction: right position: 144)
```

Fig. 11.15 Example of process sheet automatically produced. (*Note*: objective is the part shown in Fig. 11.14 and a part of the process sheet is shown here.)

A prototype of the proposed system has been realized using the Small Talk 80 software and an AI work station, and Figs 11.14 and 11.15 are examples of the output. Owing to the capability limitation of the AI station, the radial dimension, required tolerance and accuracy of the part are given in the form of a list. At present, the prototype can manage axi-symmetrical parts only, but the basic idea is considered to be available for box-like parts.

Conclusion

Through a series of activities on advanced knowledge-based CAPP, it can be understood that CAPP with expert systems has some limitations in its capability and potential. There could possibly be another way to establish knowledge-based CAPP of the next generation or to enhance CAPP with expert systems. One such way is considered to be knowledge-based CAPP in which the main stress is put on implants of thinking patterns of the process engineer or essential roles of the part drawing into the computer environment.

The engineer's thinking patterns should be precisely clarified and deeply understood, otherwise real knowledge-based CAPP will not be established. The primary finding of this chapter is that the thinking pattern of an engineer, especially the decision making procedure, is the most difficult to represent rationally with a method suitable for the computer environment. Although facing many problems, knowledge-based CAPP incorporating the engineer's thinking pattern has surely the highest potential. In conclusion, it is thus strongly recommended that the thinking pattern of the experienced engineer should be actively investigated in future along with its rational representation and mathematical processing.

References

1. VDI-ADB. Elektronische Datenverarbeitung bei der Produktionsplanung und – steuerung V – Automatische Arbeitsplanerstellung. VDI Taschen Bücher T61/62, VDI-Verlag, Düsseldorf, 1974
2. Spur G, Mertins K, Süssenguth W. Wege zu einem unternehmensspezifischen Referenzmodell der rechnerintegrierten Fertigung. ZwF 1988; 83: 481–485
3. Kimura F, Suzuki H. Variational product design by constraint propagation and satisfaction in product modelling. Ann CIRP 1986; 35: 75–78
4. Alting L et al. XPLAN – an expert process planning system and its further development. In: Proceedings 27th international MATADOR conference, Manchester. Macmillan, 1988, pp 155–163
5. Chen MF, Ito Y. Investigation on the engineer's thinking flow in the process planning of machine tool manufacturer. In: 13th NAMRC proceedings NAMRI of SME, 1985, pp 418–422
6. Chen MF, Ito Y. A proposal of data base structure for automatic process planning with expert system. In: 14th NAMRC proceedings. NAMRI of SME, 1986, pp 523–527
7. Iwata K, Fukuda Y. Representation of know-how and its application of machining reference surface in computer aided process planning. Ann CIRP, 1986; 35: pp 321–324
8. Ito Y, Shinno H, Saito H. A proposal for CAD/CAM interface with expert systems. Robotics Comput Integr Manuf 1988; 4: 491–497

Applications of Artificial Intelligence to Part Setup and Workholding in Automated Manufacturing

P.K. Wright, P.J. Englert and C.C. Hayes

Introduction

With the eventual goals of reducing labour costs and increasing quality in small-batch manufacturing, several research institutions and industrial organizations have been developing research methods that:

1. Analyse the part setup planning actions of human machinists with the aim of creating *automated part process planning* methods at the computer numerically controlled (CNC) machine tool level.
2. Analyse part clamping and fixturing methods in milling that can then be used for designing and operating *automated workholding devices*. This research has focused on innovative hardware design, interfacing of sensors to clamping components and the development of mathematical models to approximate clamping.

Despite this ongoing research, few automated planning and/or clamping systems are currently in everyday production use. Present systems are either relegated to performing laboratory demonstrations in predictable environments, or narrow computer simulations that ignore many aspects of the physical phenomena that arise in metal-cutting and its clamping needs. A more synergistic approach that incorporates the talents of computer scientists, engineers and craftsmen is needed for automated process planning and workholding. This approach should comprise the following main elements, also shown in Fig. 12.1:

1. Continued development of models for metal-cutting and clamping processes based on first principles of solid mechanics, statics and dynamics.

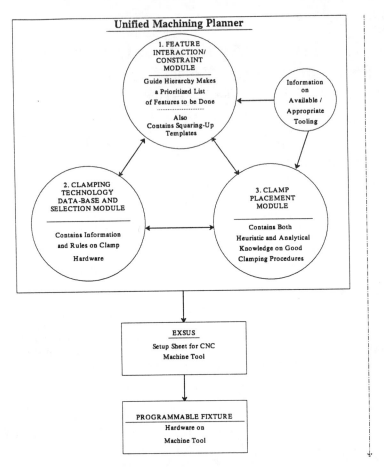

Fig. 12.1 Components of the unified system for part setup planning and workholding (Unified Machining Planner).

2. Interviews with expert machinists that expose their thought processes associated with their sensory and planning activities during setup: these interviews lead to heuristics that "fill in the gaps" in the analytical methods above.
3. The construction of programmable and flexible hardware devices that, when used during real-time setup and machining activities, test the complementary analytical and heuristic methods above.

These three activities need to be integrated and contribute towards an *intelligent, self-sustaining machine tool* for small-batch manufacturing (Fig. 12.2).

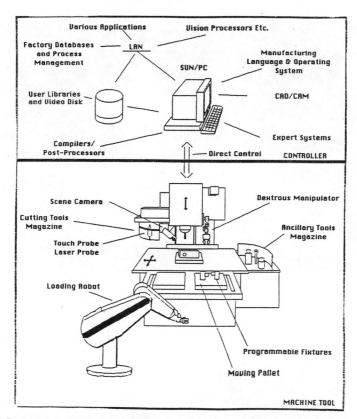

Fig. 12.2 Projected machine environment. Kitamura MyCenter 1/485 machine tool, three-axis, two-pallet; Fanuc 10M controller; Sun 3/60C colour workstation, networked to the laboratory's Ethernet; MCS Anvil-5000 cadcam software, with 3D design/draughting, solids, IGES and NC; Renishaw MP9 touch probe, with macro routines; Synermation MFNCO1 postprocessor software; user's libraries, including Anvil's users group and CarrLane.

Experimental Work

Protocol analyses with human machinists and various hardware experiments have resulted in the following prototypes:

1. An expert system for the setup planning of rectangular parts in parallel-sided vices on a three-axis milling machine.
2. Various automated fixturing devices for rectangular parts.

The first warrants no detailed discussion, simply being an automatically operable parallel-sided vice needed to support the above expert system

Fig. 12.3 Existing programmable fixture.

work. However, a second device, shown in Fig. 12.3, is a prototype automated toe-clamp fixture that serves as the basis for our future work. As we move ahead to more complex geometry parts, both the expert system for setup planning and these programmable devices will be correspondingly more complex: the planning rules (both analytical and heuristic based) and the degrees of freedom in the mechanical device must be extended.

The current laboratory environment is captured in Fig. 12.2. The equipment provides a rich design environment but also facilitates file-transfers between the Sun-based computer-aided design (CAD) system (Anvil-5000) and the Fanuc controller on the CNC machining centre. For example, touch-trigger probe routines can readily be transferred back and forth giving information on both part transformations, part rotations, and part shape [1].

Within this design environment we are also developing creative design tools that enable the programmer to browse through libraries of downstream manufacturing devices such as tools, fixtures and "standard parts" stored on an analogue videodisc recorder. These parts images can be called up in a Sun-window at the same time as the Anvil-5000 programmer is working on his design and NC programming procedures. With motivational and guiding images from the real manufacturing environment the programmer has a much greater chance of creating an NC program that leads to "the first part right the first time".

Literature Review on Part Setup and Workholding

Research in Part Setup Planning

The commercial impact of research in process planning has been demonstrated by industrial experience. For example, the advanced machining systems (AMS) project [2] at General Dynamics in Fort Worth Texas has shortened the turn around time that it takes to implement a minor design change to a part. It used to take about three weeks and it can now take as little as one day. This includes the time to make a design change to the CAD drawing, produce a new process plan, and machine the part. Despite such advances, there is room for further improvement. The majority of systems are not fully automated; humans must make many of the high-level decisions. There are also strict limitations on the type and complexity of the parts accepted. Typically these systems can only generate plans for simple parts, or parts that are very similar to ones that the system already knows how to do. A totally novel part of even modest complexity would be rejected.

Many process planners have been written. A few major examples in roughly chronological order, starting with the oldest, are: APPAS [3], CAD/CAM [4], CAPP [5], CMMP [6], CAPPSY [7], AUTAP [8], COBAPP [9], AUTOPLAN [10], AACHEN [8], AUTOCAP [11], GENPLAN [12], GARI [13], TOM [14], CAPS [15], MIPLAN [16], CMPP [17], PROPLAN [18], DCLASS [19], EXPS-E [20], CUTTECH [21], HI-MAPP [22], AMRF [23], XCUT [24], and SIPS [25]. These programs use a range of different methods, and have had varying degrees of success. Of these, AMRF, XCUT, SIPS, and CUTTECH, all solve parts of the process-planning problem that complement, rather than overlap our work. This is because we focus on planning at a higher level of setup organization, whereas these systems all plan within one setup or one feature.

- The automated manufacturing research facility (AMRF) is actually a collection of systems for automating manufacturing. It contains a process planner which plans for cutting only one side of a part at a time. The planner consists of both a plan generator, and a verifier that checks if planning was done correctly. The system can choose tools but does not prescribe fixturing, nor does it work the plan around available tools.

- XCUT [24], like AMRF, makes plans for parts that have a one-sided geometry. It decomposes features into separate cuts, uses geometry and tolerance information to choose tools. Then it groups together cuts using the same tool so that they may all be cut consecutively, saving tool changes. Lastly it chooses cut depths, and tool feeds and speeds.

- SIPS [25] is a system that selects the least cost machining operations for creating an individual feature. It uses a best first search strategy.

- CUTTECH [21] makes decisions about ordering features, choosing tools, cut depths, speeds and feeds on the basis of "machinability" data.

Both GARI and HI-MAPP work at the same level of planning as our research. In some ways the systems look the same: all work on multi-sided parts, and all start with a loosely constrained abstract initial plan that is successively refined and constrained at greater and greater levels of detail. However, neither GARI nor HI-MAPP make use of an extensive understanding of the interactions between features to help cut down search and guide planning, although they both recognize their importance.

Planners in the general area of artificial intelligence outside of the domain of manufacturing processes have recognized many of the pieces of the machining-planning processes. But, so far, none of the planners treat the chronological development of a plan in one cohesive method. Virtually all of the planners referenced in this review recognize the importance of goal interactions in planning, but their method of dealing with this problem is different from ours. Typically they do not foresee problems in the job specification and avoid them. Instead they make plans with mistakes in them and use critics to recognize and correct them after the fact. Time is wasted fixing and replanning. Hacker [26] and Noah [27] are both examples of planning using critics.

Research in Workholding

Our research on programmable and/or conformable fixtures was first reported in 1982 [28]. Others have also realized that flexible, programmable workholding devices (as a "mechanical follow-on" from the above research in planning) will improve the overall performance of automated manufacturing systems. Research scientists from the Hungarian Academy of Sciences and the Budapest Technical University [29] have developed a Prolog system that configures clamps about a workpiece using CAD information and rules based on geometrical constraints.

Asada and By [30] have implemented conformable clamps for assembly operations by using a magnetic table, a series of horizontal and vertical locators and clamps, a robot, and a CAD system all under computer control. The locators and clamps are configured by the robot using commands generated by the CAD system. In their case study, the commands are based on equations that define accessible positions for the clamps about the surface of an electric hand drill.

Ferreira et al. [31] are developing a system known as AIFIX, which decomposes the overall fixturing geometry into independent subproblems and tries to solve each subproblem toward the final goal of workholding.

A programmable vice able to accommodate varying part shapes and sizes has ben designed and fabricated [32]. The National Bureau of Standard (NBS) vice includes a variable depth jaw, opens wide enough to provide for robotic part loading, applies a controlled pressure to the part, and is able to adjust the position of the part on the jaw.

Thompson and Gandhi [33] have experimented with fluidized beds to hold unusually shaped parts. Here, the part is embedded in a vessel of fine powder which, when vibrated, behaves like a fluid, embodying the part. When the vibration is removed, the powder loosely holds the part. But, when a clamping plate is applied to the top surface of the powder, the whole assembly becomes strong enough to hold the part for machining.

Comparisons and Contrasts Between the Literature and Present Work

In our approach so far, a program called Machining Planner has been developed, comprising approximately 275 OPS5 rules. Experienced machinists have judged the program to make better plans, on average, than a journeyman with five years of experience (but of course in the limited domain of rectangular parts on a three-axis machine). Machining Planner is a high-level process planner that groups features into setups and orders setups. It does not divide features into all their component operations, plan cutter paths, or choose feeds and speeds. However, the standard Anvil-5000 NC package in our laboratory can do these tasks as can AMRF, XCUT, SIPS and CUTTECH. Thus the combined knowledge from the research results in Machining Planner and the standard features of Anvil-5000 do come close to a complete method of process planning.

Machining Planner is different from other process planners like GARI and HI-MAPP because it uses feature interactions to cut down search, and to help generate the solution. The method may be applicable to other domains where there are strong interactions between the parts of the problem.

The results from the previous studies on workholding have been encouraging. However, although they show the feasibility of some hardware innovations, they have not revealed much information on where to put clamps around an arbitrary part. Consequently, our research considers the geometrical constraints in workholding, partly from the viewpoint of traditional engineering mechanics and partly from the viewpoint of the more craft-oriented knowledge that is acquired from many years of experience. A tractable approach at first was to limit the problem domain and concentrate on rectangular stock. Now, we are expanding this work to more complex part geometries and correspondingly more complex fixtures with greater degrees of freedom.

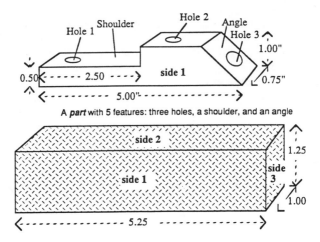

Fig. 12.4 Typical part and starting stock.

Results

To demonstrate the functionality of the current planning software, suppose the part shown in Fig. 12.4 is going to be made from the accompanying piece of stock. The part is represented in Machining Planner as a rectangular block with features subtracted from it. In this case, there are five features – three holes, an angle and a shoulder. At the current stage our planning software "understands" eight features: blind-hole, through-hole, pocket, blind-slot, through-slot, angle, shoulder and channel. Yet this is a limited library compared with the PADL-oriented research [34, 18, 35]. The program begins by querying the user about the stock size and its surface finish (machined, rolled or saw-cut): the surface finish is important and may change the squaring-up procedure described below. Next the program queries the user about the outer envelope of the part.

The "automatic" part of the program then takes over in which the features are grouped together into *setups*; the setups are ordered and the tools for each cut selected. This is the essential part of Machining Planner and its knowledge was compiled from extensive interviews with two machinists, each with 20 years' experience [36]. The main theme of the planning is to merge two simultaneous constraints, namely:

- The squaring-up process
- The feature-interaction considerations

For the part in question, the merging process is shown in Fig. 12.5. On the right-hand side the Machining Planner must create an orthogonal

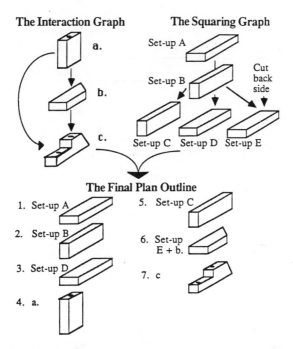

Fig. 12.5 Merging the squaring-up and feature interaction graph.

block, otherwise the features will not be placed accurately relative to each other. The squaring process always has to occur whatever the part geometry. Thus we concentrated on this in our earliest work and the various squaring-up plans are shown in Hayes and Wright [37].

It is the merging of the features into the squaring-up plan that is the challenging aspect of human craftsmanship and, consequently, of the creation of an expert system rule-set that can mirror the deep human experience. The reason this is difficult is that cutting one feature may make it very hard or impossible to cut subsequent ones. This is best explained by referring to Fig. 12.6. This shows two "bad" plans above the "good" plan. The "bad" plans are ones that a human machinist would avoid based on his/her experience: one of these plans causes too much vibration in the last step and the other causes the drill point to skid on the surface of the chamfer. The "good" plan carefully orders the setups in such a way as to avoid these problems. The way in which the setup sheet is presented to either a novice machinist or an automated machine tool is thus shown in Fig. 12.5. When the squaring-up is combined with the feature interactions, seven setups are needed to complete the part.

The software's 275 OPS5 rules contain many heuristics, learned from our two experienced machinists, that help to order the setups. In many cases the program uses built-in generalized patterns to spot the

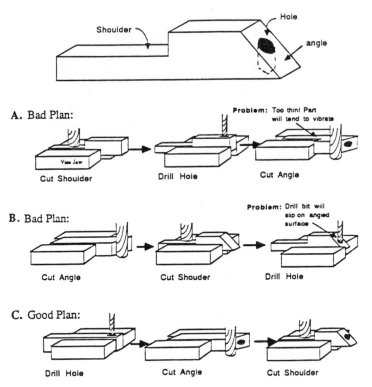

Fig. 12.6 Examples of planning showing two "bad" plans and the final "good" plan.

interactions quickly. For example, for the hole–angle interaction, the program has a pattern that matches any drilled hole or depression that enters a nonorthogonal surface. If the pattern is matched, then it puts a restriction on the plan that the hole must be cut before the nonorthogonal surface. Similarly, for each additional pattern, there is an associated operator that tells the program how to avoid the interaction. These are termed "feature interaction graphs". At present we have identified a number of these as shown below in Table 1. Evidently one important aspect of our continuing research is to expand the protocol analysis with machinists and obtain more feature interaction graphs that reflect the broad experience of machinists.

In summary it is the first two categories, which involve generating constraints, that are the ones that have the most room to grow. Productions can be added to these categories, greatly increasing the range of parts that the system can handle, even within the rectangular type of part. As we move more into toe-clamps and other fixture types, the first two categories above will expand dramatically.

Table 1. Summary of production rules in Machining Planner

Production/rule type	No. in system
Identifying feature interactions and constructing feature interaction graphs	21
Generating other machining technology constraints	50
Choosing the squaring graph	23
Merging the interaction graphs with squaring graphs	55
Generating the final plan from the merged groups	18
Entering data, entering missing data, miscellaneous	95
Choosing between alternative manufacturing methods	13

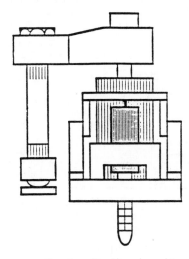

Fig. 12.7 Existing clamp cross-section (see Fig. 3) and machine-tool clamp-loading device.

As a "mechanical follow-on" to Machining Planner, we have built and used automatically operable fixturing devices. As a precise follow-on from the current planner, we needed only a hydraulically operated parallel-sided vice: the demonstration was therefore so simple it barely warrants a mention here.

A more advanced device is shown in Fig. 12.3 and it is the basis for the machining of a wider range of part geometries needing toe-clamps. The individual clamps that are used in the device are shown in Fig. 12.7. A small gripper, backed up with an RCC device, can be mounted in the machine tool spindle to load the clamps from their storage locations (at

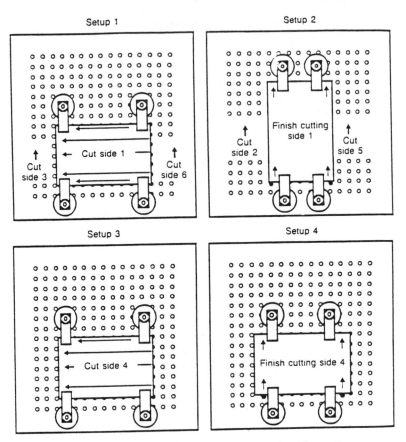

Fig. 12.8 Setups for squaring-up rectangular stock with toe-clamps.

the sides of Fig. 12.3) into any of the 99 holes shown in the baseplate. The clamps are secured into the holes when the radially expandable bushings shown at the bottom of Fig. 12.7 are automatically operated (Hazen and Wright [38] describe all the mechanical details). Thus far we have loaded the part-stock into the baseplate in Fig. 12.3 with an industrial manipulator (adjacent to the machining centre) and then closed down the clamps through the hydraulically operated system.

As a brief example it is possible to use the results from Machining Planner to decide on where to put the clamps during a squaring-up operation on a large block that must be held in toe-clamps rather than the parallel-sided vice. Thus Fig. 12.8 shows the square-up process for a rectangular block sitting on parallels on the baseplate. The parallels provide clearance between the end of the tool and the baseplate during the edge-milling operations in setups 1 and 2. The toe-clamps initially obscure the block corners in setups 1 and 2 and need to be carefully repositioned in setups 3 and 4 to allow access to these corner areas.

Future Activities

The structure of a Unified Machining Planner, currently under development and its connection to the mechanical devices is shown in Fig. 12.1. Unified Machining Planner consists of modules that each perform specific tasks but communicate with each other. These modules are:

- The *feature interaction/contraint module* that is basically an extension of the current Machining Planner but for toe-clamps and other fixtures.
- The *clamp placement module* that proposes clamp configurations about the workpiece, workpiece positions and fixture locators. The knowledge in the clamp placement module will be derived both from heuristics and from simple stress analyses.
- The *clamp selection module* that chooses, from a database of clamp and fixture units, the appropriate workholding components for a particular setup.

In the stucture we currently perceive for Unified Machining Planner, the *feature interaction module* will "lead the way" for a possible machining plan. But this plan will need to be checked out against the available clamping devices (in the clamp selection module) and the practical constraints of how the part will be clamped (the clamp and part placement module). The inner details we envisage in each of the three modules are shown in Table 2. Planning guides based on heuristics plus analyses will be listed by the categories of *necessary* (Level I), *preferred* (Level II) and *efficiency* (Level III). This ordering does not imply that Level III guides are unimportant or to be bypassed but that a hierarchy must be established when resolving conflicts in the expert system work. These conflicting courses of action might involve, for example, the choice of one type of clamp over another, or the decision to machine one feature before another.

The ultimate product of the work in Unified Machining Planner will be to produce the expert set-up sheet (EXSUS) as shown in Fig. 12.9. This is a specific set of descriptions that can be downloaded to the machine tool controller (in our case the Fanuc) and used to operate the mechanical clamps (Fig. 12.3). It is a description of the workpiece, clamps, and cutting tools sufficient to uniquely refine individual "setups" on the machine. Collectively these setups will transform the raw stock into a finished part.

Figure 12.9 displays views of how we envisage the future EXSUS program to be formatted. Two distinct production states or setups are

Table 2. Hierarchical guidelines formed by combining experts' heuristics with analytical work

Feature selection	Clamp selection	Clamp and part placement
Level I. Necessary feature machining guides A. Proper steps to achieve specified feature and surface tolerances B. Proper tools and processes to produce stated surface finishes C. Proper tooling for features D. Orthogonal, machined part sides before features are cut (templates) *Level II. Feature interaction preference guides* A. Drill holes on flat, not curved, surfaces B. Drill a hole before cutting a shoulder when the hole's axis must extend along a face of the shoulder C. Avoid situations where the cutter must be in contact with two surfaces at the same instant D. Cut any features that will make the cutting of another long, thin feature less difficult *Level III. Feature machining efficiency guides* A. Choose features that will require few cutting tool changes B. Choose features that will require few changes in clamp positions C. Reserve the cutting of angles for the latter stages of production D. Cut pockets with equal radii in the same or successive operations	*Level I. Necessary clamp envelope guides* A. Part must fit within the confines of the clamp's working envelope *Level II. Problem specific preference guides* A. Volume removal considerations B. Part stability considerations C. Clamp bending – resistance to cutting force tradeoff D. Part buckling – resistance to cutting force tradeoff E. Part vibration – production rate tradeoff F. Part deformation, access to surface and resistance to movement tradeoffs *Level III. Clamping efficiency guides* A. Clamp changeover considerations	*Level I. Necessary placement guides* A. Part location accuracy, deformation, and production rate tradeoffs B. Part-clamp interference considerations *Level II. Problem specific preference guides* A. Part bending, resistance to cutting force, and production rate tradeoffs B. Part buckling, resistance to cutting force, and production rate tradeoffs C. Part resistance to movement considerations D. Part deformation, resistance to movement, and production rate tradeoffs E. Part deformation, resistance to movement, and production rate tradeoffs *Level III. Placement efficiency guides* A. Symmetry considerations B. Part positioning efficiency considerations

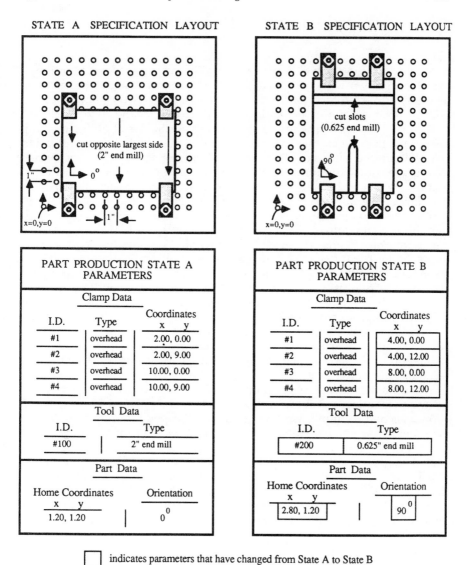

Fig. 12.9 Part production states with associated parameters for setup with overhead toe-clamps on a tooling plate.

shown for a part to be held with overhead toe-clamps to a tooling plate, and their associated parameters.

In production if something were to go wrong unexpectedly we would expect to need to TRACE back (see Fig. 12.1) through the now Unified Machining Planner and to find alternative choices for tooling, fixtures or stock.

Quantitative Analyses of Clamping

In the third column, Level II of Table 2 and in the text we have mentioned the quantitative work we plan that will contrast with the experts' heuristics from protocol analyses with machinists. Two examples of the kind of work we are investigating are mentioned below.

The first is by Mason who has studied the mechanics of object-pushing operations that are an essential component of many robotic manipulation tasks [39]. In the course of his research he developed a computer program that, given a distribution of support forces on an object, a coefficient of friction between the supports and the object, and object dimensions, determines possible rotation centres for varying angles of attack of an applied disturbance force. We have already run some simulations with this routine and shown that it may be used as a check of the validity of part-clamp machining setups [40]. The magnitude of the resultant friction forces at each of the clamp positions will act in a direction that is orthogonal to a line that connects the coordinates of a rotation centre and the clamp location. A force equilibrium equation may then be used to determine the maximum cutting force, at a given angle of attack that may be applied to the part before slip occurs. In this example, part slip occurs due to one of several tagged parameters of the flawed production state; clamp style, clamp positions and clamp forces. If the maximum clamp force is fixed, a second pass through the control dialogues will focus on alternative ways to derive an admissible part production state by changing the clamp style, clamp positions, or possibly reducing values for the machining parameters.

As a second example, we have analysed the possible movements of a plate held in a parallel-sided vice and the movement of this same plate as a cutting tool interacts with it [40, 41]. This analytical situation is shown in Fig. 12.10. For simplicity we show just the potential ways in which the cutter could move the part transversely (Fig. 12.10b) and by tilting (Fig. 12.10c). Buckling could occur across the width of the part w in Fig. 12.10a especially as part dimensions increase. To prevent part movement in Fig. 12.10b the simple relationship is

$$F_{cutter} < 2[F_{vice}][\mu_{part-vice}]$$

where μ can be estimated from experience with oily but rough surfaces [42]. The force of the vice and the cutting force can thus be compared but the vice force must be maintained below a critical level to prevent buckling. Obviously if the part width w is small, buckling is not a problem, but a check for part crushing in compression between the vice jaws should be done. Englert's recent thesis [41] shows these analyses and extends the crushing analysis to the toe-clamps where their smaller surface area creates more chance for a local indentation in the part.

Part clamped in a standard machine tool vice, subjected to cutting tool forces

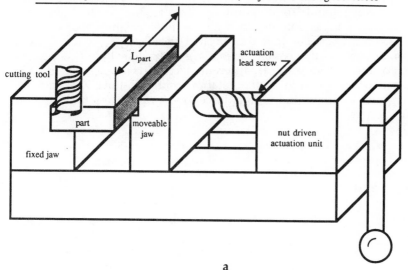

a

Simplified model of part subjected to a purely transverse force
(force vector parallel to jaws) in the vice

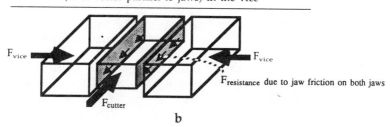

b

Simplified model of part subjected to a force that creates a moment
(force vector perpendicular to jaws) in the vice

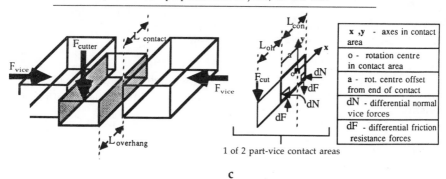

1 of 2 part-vice contact areas

c

Fig. 12.10 Schematic of part subjected to forces in a vice **a** and associated transverse force **b** and moment **c** models.

If the primary cutting force in Fig. 12.10 is downward then the part wants to rotate about some point, o, due to the applied moment $F_{cutter}(L_{overhang} + a)$ as shown (with all the nomenclature that follows) in Fig. 12.10b, c. Assume that the applied vice load F_{vice} acting in a direction normal to the contact area, is evenly distributed along the bottom portion of the area. If differential contacts of area $dx\,dy$ are constructed on either side of the rotation centre o, they will each be acted upon by a differential normal force, dN, and a differential friction force, dF, where

$$dN = \frac{F_{vice}dx}{L_{contact}}$$

and

$$dF = \mu_{part\text{-}vice}dN$$

The differential friction forces must act in opposite y axis directions so that they may form a friction couple to counteract the externally applied moment. Equilibrium must be satisfied to keep the part from slipping, and so the net sum of forces and torques acting on the entire contact area must be zero. The sums of the differential forces and moments are obtained by integrating over the regions in which their differential components act. The equilibrium equations, including both part-vice contact regions, are given by Englert and Wright [40].

The solution for the two unknowns, F_{cutter} and a, from this set of simultaneous integral equations is straightforward. Their values are

$$a = -L_{oh} + \frac{L_{con}}{2} \sqrt{4r^2 + 4r + 2}$$

and

$$F_c = 2\mu_{part\text{-}vice}F_{vice}\,[-1 - 2r + \sqrt{4r_2 + 4r + 2}]$$

where $r = L_{oh}/L_{con}$.

So to prevent a part subjected to orthogonally applied loads from slipping in the vice during machining the following relation must be upheld:

$$F_c < 2\mu_{part\text{-}vice}F_{vice}\,[-1 - 2r + \sqrt{4r^2 + 4r + 2}]$$

Conclusion

The Unified Machining Planner consists of knowledge bases related to feature interactions, clamping hardware and clamping placement. These three knowledge bases are being partly built up from heuristic rules (from protocol analyses with machinists) and partly built up from the quantitative methods shown above on buckling and frictional stability. The completed Unified Machining Planner will then specify particular setups for the machine tool in the EXSUS format shown in Fig. 12.9. Connections with a programmable fixturing device of the type shown in Fig. 12.3 will then permit real-time machining.

References

1. Greenfeld I, Hansen FB, Wright PK. Self-sustaining, open system machine tools. In: North American manufacturing research conference, 1989
2. McMahon RL. Advanced machining systems. From presentations given at AMS industrial briefing, 1987
3. Wysk RA. An automated process planning and selection program. PhD thesis, Purdue University, 1977
4. Wysk, R. Miller D, Davis R. The integration of process selection and machine requirements planning. In: Proceedings, AIIE Systems Engineering Conference, 1977
5. Link CH. CAM-I Automated process planning. Technical report, SME, 1978, paper no. MS78-213
6. Dunn MS, Mann WS. Computerized production process planning. Proceedings 15th NCS Annual Meeting and Technical Conference, 1978
7. Spur G et al. CAPPSY – A dialogue system for computer-aided manufacturing planning. In: Proceedings, 19th MTDR conference, 1978
8. Eversheim W, Holz B, Zons K. Application of automatic process planning and NC programming. In: Proceedings, Autofact West, SME, 1980
9. Phillips RH, El Gomayel JI. Computerized process planning for metal cutting. In: Proceedings Eighth North American manufacturing research conference, 1980
10. Tempelhof K. A system of computer aided process planning for machine parts. Elsevier North-Holland, New York, 1980
11. El-Midany TT, Davies BJ. AUTOCAP – a dialogue system for planning and sequence of operations. Int J Machine Tool Des Res 1981; 21: nos 3/4
12 Hegland DE. Out in front with CAD/CAM at Lockhead-Georgia. Prod Eng 1981
13. Descotte Y, Latombe J. GARI: A problem solver that plans how to machine mechanical parts. In: Proceedings IJCAI, 1981, pp 766–772
14. Matsushima K, Okada N, Sata T. The integration of CAD and CAM by application of artificial intelligence techniques. CIRP 1982; 31: 329–332
15. Emerson C, Ham I. An automated coding and process planning system using a DEC PDP-10. Comput Indust Eng 1982; 6: 159–168
16. Lesko A, El-Midany TT, Davies BJ. 'MIPLAN' implementation at Union Switch and Signal. Presented at Association for Integrated Manufacturing Technology (Numerical Control Society) 20th annual meeting and technology conference, 1983
17. Waldman H. Process Planning at Sikorsky. CAD/CAM technol 1983; Summer
18. Phillips RH, Zhou X-D, Mouleeswaran CB. An artifical intelligence approach to

integrating CAD and CAM through generative process planning. In: Proceedings, ASME international computers in engineering conference, 1984, pp 459–463

19. Allen D, Smith P. Computer aided process planning. Brigham Young University, Computer Aided Manufacturing Laboratory, Provo, Utah, 1984

20. CAM-I. EXPS-E: an experimental system for process planning. CAM-I Report, August, 1984

21. Barkocy BE, Zoeblick WJ. A knowledge-based system for machining operation planning. Autofact 61, 1984

22. Berenji HR, Khoshnevis B. Use of artificial intelligence in automated process planning. Comput Mech Eng 1986, September: 47–55

23. Kramer TR, Jau-shi Jun. Software for an automated machining workstation. Proceedings Third biannual international machine tool technical conference, Session 12, 1986

24. Brooks SL, Hummel KE, Wolf ML. XCUT: a rule based expert system for the automated process planning of machined parts. Technical report BDX-613-3768, Bendix Kansas City Division, Kansas City, MO, June, 1987

25. Nau DS Automated process planning using hierarchical abstraction. Texas Instruments call for papers on AI for industrial automation, 1987

26. Sussman GJ. A computer model of skill acquisition. American Elsevier Publishing Company, New York, 1975

27. Sacerdoti ED. The nonlinear nature of plans. IJCAI 1975; 4: 206–214

28. Cutkosky MR, Kurokawa E, Wright PK. Programmable conformable clamps. In: AUTOFACT 4, Philadelphia, Pennsylvania, November, 1982, pp 11.51–11.58

29. Markus A, Markusz Z, Farkas J, Filemon J. Fixture design using Prolog: an expert system. Robotics Comput Integr Manuf 1984; 1: 167–172

30. Asada H, By A. Kinematic analysis of workpart fixturing for flexible assembly with automatically reconfigurable fixtures. IEEE J Robotics Automat 1985; RA-1: 86–93

31. Ferreira PM, Kochar B, Liu CR, Chandru V. AIFIX: an expert system approach to fixture design. In: Computer/aided intelligent process planning ASME winter annual meeting, 1985, pp 73–82

32. Slocum AH, Peris J, Donmez A. Development of a flexible automated fixturing system. Progress report, US National Bureau of Standards, Automated Production Technology Division, 1985

33. Thompson BS, Gandhi MV. Phase change fixturing for flexible manufacturing systems. J Manuf Syst 1985; 4: 29–39

34. Voelcker H, Requicha AAG. Geometric modeling of mechanical parts and processes. Computer 1977; 10: no. 12

35. Phillips RH, Arunthavanathan V, Zhou X-D. Symbolic representation of CAD data for artificial intelligence based process planning. Comput aided intell process plan 1985; 19

36. Wright PK, Bourne DA. Manufacturing intelligence, Addison-Wesley, Reading, MA, 1988

37. Hayes CC, Wright PK. Automated planning in the machining domain. In: Proceedings ASME meeting on knowledge based expert systems for manufacturing, PED, 1986, vol. 24, pp 221–232

38. Hazen FB, Wright PK. Autonomous fixture loading by a machine tool. In: ASME manufacturing '88, Atlanta, GA

39. Mason MT. Mechanics of pushing. In: 2nd international symposium on robotics research, Kyoto, Japan, 1984. pp 73–80

40. Englert PJ, Wright PK. Principles for part setup and workholding in automated manufacturing. J Manuf Syst 1988

41. Englert PJ. Principles for part setup and workholding in automated manufacturing. PhD thesis, Department of Mechanical Engineering, Carnegie-Mellon University, 1987

42. Bowden FP, Tabor D. The friction and lubrication of solids. Oxford Press, London, 1950

Section D

Tooling Design

This section contains three chapters. The first chapter, by Pillinger et al., describes an intelligent knowledge-based system for designing metal-forming dies. The system combines rule-based techniques with finite-element simulation and is able to improve its design rules automatically as it gains experience. The second chapter, by Nee and Poo, reviews the state of the art in expert CAD systems for jigs and fixtures. The third chapter, by Pham and de Sam Lazaro, presents two knowledge-based programs for jig and fixture design, one to provide design advice and the other to carry out design operations automatically.

Chapter 13

An Intelligent Knowledge-Based System for the Design of Forging Dies

I. Pillinger, P. Hartley, C.E.N. Sturgess and
T.A. Dean

Introduction

Forging is the process whereby the shape of a workpiece (usually metallic) is changed by pressing or hammering the workpiece between two or more dies, with or without the application of heat. Complex shapes are usually forged from the original stock material in a number of stages. Quite often, the dies for several stages are grouped together in the same forging press or hammer, the workpiece being transferred from die cavity to die cavity (manually or by an automatic mechanism) in between successive blows of the machine.

Forging is capable of high rates of production with low material wastage. With suitable die designs, forged components may require little or no subsequent machining particularly with recently developed "near-nett-shape" forming procedures. Forging can also improve the mechanical properties of the component and with careful preform and die design a specified variation in mechanical properties may be achieved.

The design of dies for the forging industry has traditionally been a highly skilled occupation. An expert die designer is required to produce a sequence of dies that fill correctly, do not break under the forging loads, do not suffer from excessive wear, and produce components without defects and with the appropriate distribution of microstructure, all subject to the loading and geometric limitations of the available forging plant. Such expertise can only be acquired as a result of many years' involvement in forging die design. Not surprisingly, it is increasingly difficult to find suitably qualified die designers.

In addition, even the most skilled die designer may not produce a successful set of dies first time. It is quite usual for forging dies to be redesigned several times before a component can be put into full production. Since lengthy workshop trials have to be performed each time the dies are altered, this is both time consuming and expensive.

The lack of skilled die designers can be remedied, to a large extent, by the development of rule-based die-design programs. These have proved to be very successful, but do not avoid the need for extensive production trials when applied to nonstandard components or materials. Also, design rules are necessarily conservative and may not lead to an optimum die design configuration that makes most efficient use of the forging capacity.

Both of these problems can be addressed by computer simulation of material deformation, using techniques such as finite elements (FE). Again, FE simulation programs have been found to be very useful in validating die designs independently of experimental trials, and in providing detailed information about the forging process. The disadvantage of these techniques is that although approximate FE simulations can be obtained quite quickly, detailed studies require large amounts of computing time, which makes them inappropriate for routine use.

Clearly, what is required is a system that uses a rule-based program to design forging dies, and only uses computer techniques to simulate deformation when unfamiliar components, materials or forging conditions (such as temperature or deformation rate) are encountered. Ideally, it should be possible to improve the die-design rules as more experience is gained, so that the rule-based program produces die designs that are better, and that are "right first time" more frequently.

This chapter describes just such an integrated system, which is being developed in conjunction with Rover Group and which has been funded by the Application of Computers to Manufacturing Engineering (ACME) Directorate of the UK Science and Engineering Research Council. The package takes the form of an intelligent knowledge-based system (IKBS) in which details of die designs previously examined by the system are recorded so that they may be used to assess whether the current designs are likely to be successful, or whether more information needs to be obtained by means of numerical simulation techniques. The accumulated knowledge may be used to improve the die-design rules.

The following sections of this chapter will first of all consider previous work carried out in this area, and will then describe in detail how the forging die-design IKBS works. Some details of the implementation will then be discussed and finally mention will be made of how the system could be extended and improved.

Background and Previous Work

Work has been in progress for many years at the University of Birmingham, directed towards producing computer-based tools to help reduce development lead times and costs in the forging industry. The two principal areas of research have been the development of programs, firstly to design dies and secondly to predict metal flow and properties in forging processes. The die-design programs have been developed for a number of different types of forging [1–3]. They are all based on the rules and guidelines currently being used in industry, which are largely empirical. Although these rules differ for the different categories of component, the programs all work in the same way.

The starting point is the description of the shape of the required component. After making an allowance for machining, the program designs the die cavities of the finished forging by adding various features to the shape, such as draft, webs, corner and fillet radii, flash lands and gutters, in accordance with rules based on standard forging practice. Other rules are used to determine what the shape of the preform should be for this stage of the forging sequence. The obvious constraint here is that the volume of the material must be conserved during the process, but other considerations have to be taken into account, such as the maximum deformation that can occur in one go, limiting aspect ratio of the preform (to prevent buckling for example), the reluctance of material to fill deep cavities in the die and so on. The load for this stage of the forging is also estimated.

Once the shape of the preform has been calculated, this determines the die cavity or tool shapes for the previous stage in the sequence, and so the die-design rules can be applied again. Eventually, a preform shape is obtained that corresponds to the original bar stock.

The die-design programs aim to produce the desired component in as few stages as possible, though this aim may not always be achieved in practice. A typical sequence of stages in the forging of an automobile drive flange is shown in Fig. 13.1.

The programs that have been developed at the University of Birmingham to predict metal flow have been based on the FE method. FE techniques are widely used nowadays as design tools, mainly in linear analysis of elastically deforming systems. The fundamental principles of FE analysis of material deformation are broadly similar to those involved in more familiar examples and will not be discussed in detail here (see reference [4] for an excellent account of FE techniques in general).

Some basic aspects of an FE analysis of deformation are illustrated in Fig. 13.2. Essentially, the technique seeks to obtain a set of simultaneous equations expressing the components of force on all the node or grid points of the workpiece in terms of the displacement at all those nodes. This is represented in matrix form in Fig. 13.2 by equating the nodal

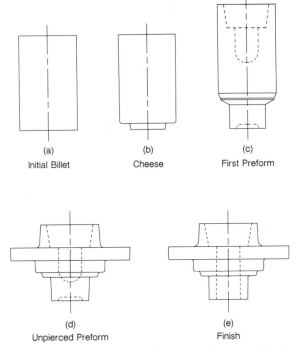

Fig. 13.1 Intermediate stages in the forging of an automobile drive flange.

force vector **f** to the product of the stiffness matrix **K** and the nodal displacement vector **d**. The main differences between FE analysis of forming processes and FE analysis of elastic structures are firstly that the former is highly nonlinear (the stiffness equations in Fig. 13.2 depend upon various quantities that change throughout the deformation, and the boundary conditions also vary) and secondly that care has to be taken to choose the correct physical and mechanical model of the deforming material. In the context of metal deformation, the correct physical law is the Prandtl–Reuss elastic–plastic relationship, which relates strain rate to the current state of stress at a point. The correct mechanical description involves the use of expressions for stress rate and strain rate that are valid for the finite deformation steps employed in metalforming analysis.

In the simulation of metalforming processes, suitable means of specifying the changing boundary conditions represented by moving die surfaces is also very important. Details of all these techniques, and a full description of the FE analysis of metalforming problems, can be found in references [5–7].

Earlier work has centred on integrating FE metalforming analysis and a program that designs dies for upset forging processes [8]. Several die

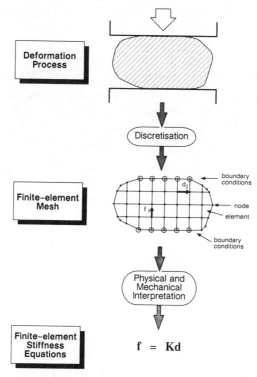

Fig. 13.2 Fundamental principles of the finite-element analysis of deformation processes.

sequences and preform shapes were produced using the die-design program and examined by the FE program to evaluate the system. Other modules were included to allow the overall system to make geometric comparisons of axisymmetric components and also to learn from experience.

Description of the IKBS

Overview

The objectives of the IKBS for the design of forging dies are to:

- Speed up the process of die design by automatically generating a full specification of all the dies and preforms required to produce a finished forging from the description of the component supplied by a CAD system

- Reduce the cost and time involved in obtaining successful forging dies by means of numerical analysis of the forming processes
- Improve and deskill the die-design process and increase likelihood of achieving a successful set of die designs first time, by intelligent use of knowledge relating to all previous dies designed by the system

Figure 13.3 shows an overview of the IKBS illustrating how the different components interact and how information is passed between them. The overall system consists of an interactive IKBS program (in the shaded rectangular frame), a set of computer-based databases and a group of related activities.

The interactive program contains three modules: the sequence design program to design the forging dies; the FE pre-processor to initiate FE simulation of selected forging stages; and a control module to supervise the use of the other two modules and to manage the information stored in the system. There is also an interface allowing communication between the die designer and the computer modules. The three interactive modules

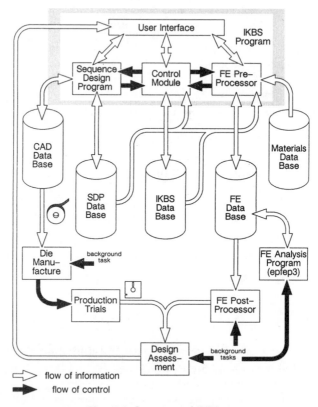

Fig. 13.3 Overview of IKBS.

and the FE analysis will be discussed in greater detail in the following sections.

One important feature of the die-design system as a whole is the ability to generate NC tapes to automate the machining of the required die cavities. Feedback from the production runs is also very important and information such as press stroke/load data is recorded for analysis and assessment of the die designs. Design assessment is partly computer-based, with information passing back into the IKBS database via the IKBS program.

Various databases are used by the system as a means of storing information and of transferring it between modules:

CAD database. A design database managed by the CAD package that is used to produce the original component description.

SDP database. Contains information created and managed by the SDP to describe the die sequences, tool layouts etc.

IKBS database. Contains information about all components previously examined by the system, including details of whether they were successful, whether the designs had to be modified, etc.

FE database. Contains FE datafiles to be run by the FE program and the output from these programs.

Materials database. Contains physical properties (elastic and plastic moduli, thermal coefficients etc.) of the metals considered by the system and also information about lubricants used.

Sequence Design Program (SDP)

The SDP has been developed from the die-design program MODCON [2] and uses a similar method of describing shapes. This resembles that used in industry when volumes have to be calculated for cost estimation. The complete forging is divided into elementary forms or primitives, such as cones or toroids, which may represent either solid material or cavities. The primitives can be described by a few geometric parameters and so the input data are easily obtained. The program can either import the data directly from a solid model created on the CAD system, or can convert a conventional wire-frame model into the required form.

To simplify the design rules, five categories of shape are recognized by the SDP – stub axles, drive flanges, gear blanks, bevel pinions and non-axisymmetric components (such as connecting rods). Typical examples of the different categories of component are shown in Fig. 13.4.

- *Stub axles.* These consist of a flange with a tapering shaft on one side and a flat surface on the other. The flange may be either circular or square in section. The shaft consists of a set of cylinders and/or tapering cones. The flat surface may be surmounted by a ring, the inside of which may be deeper than the surface of the flange.

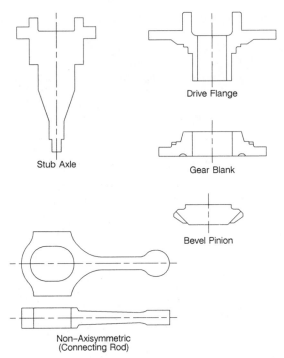

Fig. 13.4 Categories of component recognized by the IKBS.

- *Drive flanges.* These consist of a flange with a stepped boss on one side and possibly a ring on the other. The whole component is pierced by an axial hole.
- *Gear blanks.* These are similar to drive flanges but tend not to be so tall, and there are teeth around the base of the boss.
- *Bevel pinions.* These consist of a conical shape with a short boss on the broader end. There are helical teeth on the sloping face of the cone which are usually machined after forging.
- *Non-axisymmetric components.* The shape of these will vary, but typically, a connecting rod consists of a shaft which has a "big end" at one extremity and a "little end" at the other. The shaft may taper and sometimes will have a web along its middle. The big end consists of an ovoid ring (two semi-circular ends with short straight sections connecting them) with external flanges along the longer sides which is attached to the shaft, possibly with a blended taper, at the mid-point on one of the ends. The little end is in the form of a boss.

Finite-Element Program

Since detailed FE calculations of complex metalforming process can be very time-consuming, these are performed as background tasks. The interactive part of the IKBS contains the FE pre-processor, a program to set up the datafile for the FE analysis using information about workpiece and die shapes stored in the SDP database along with material property data stored in the material database.

The FE program used by the IKBS to model the deformation of the workpiece during various stages of the forging process is called epfep3† (elastic–plastic finite-element program for 3 dimensions). This is a fully three-dimensional program specifically developed to study large deformation processes, which takes into account both elastic and plastic components of strain. It can accommodate a variety of material types by means of the specification of different physical properties such as thermal capacity and conductivity, elastic (Young's) modulus, Poisson's ratio and yield stress. The last of these quantities may vary from point to point within the workpiece as a function of strain, strain rate and temperature. Complex rigid die boundaries may be specified, with an associated level of friction, thermal transfer coefficient and temperature. Die surfaces may be stationary, or they may move progressively during the deformation. The movement may be translational, rotational or a combination of both, and the amount of movement may either be proportional to process time, or vary sinusoidally with it. The latter relationship is useful when mechanical presses are being used to produce the component.

The analysis of the deformation process is divided into a number of steps or increments. At each step, epfep3 generates information about the current shape of the workpiece grid and velocity of the grid points, the distribution of components of strain, stress, and values of strain rate and temperature, as well as the deforming load. This information can be processed by the post-processing module for display in numerical or graphical form. Figure 13.5 shows some typical results, in this case the distribution of temperature within half of a transverse cross-section through a connecting rod during the last stage of a warm forging operation.

Control Module

The control module is the part of the IKBS program that integrates the die-design and deformation simulation modules of the system and performs the functions required to assess whether FE analysis is required in particular instances.

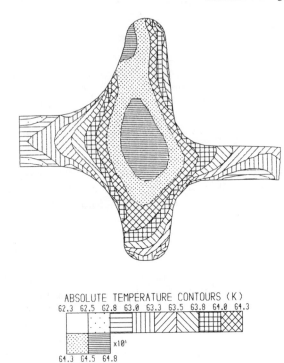

ABSOLUTE TEMPERATURE CONTOURS (K)

Fig. 13.5 Typical finite-element results: temperature distributions in a transverse cross-section of a connecting rod during forging.

Figure 13.6 illustrates the main actions of the control module. In the following description, the letters refer to the labels shown in Fig. 13.6.

a. The examination of a new component begins with the identification of its category (stub axle, drive flange, gear blank, bevel pinion or non-axisymmetric) and the determination of the shape of the finished (as forged) or machined component from the information stored in the CAD database. The material type and processing conditions (temperature and lubrication) are also required.

b. In the first instance, it is required to design the complete sequence of dies up to the finished forging stage.

c. The SDP is used to determine the shape of each of the preforms required to produce the target stage of the forging. To start with, this is the finished component, but on subsequent occasions, it may be an earlier stage in the forging sequence (see "o" below).

The proposed die design sequence will be assessed stage by stage, so for each of the preforms, from the initial bar or billet up to the target stage, a sub-entry is created in the IKBS database entry for the current component containing a modified description of the shape of

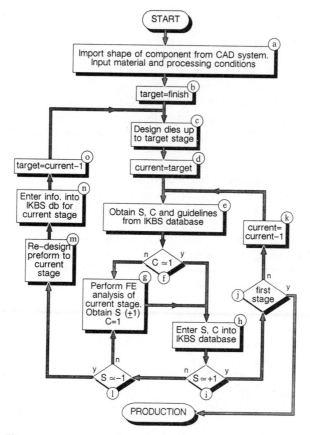

Fig. 13.6 Use of the IKBS in designing and validating forging dies.

this preform, together with an indicator of the success of the die/preform design. For example, for a drive flange, a sub-entry is created for each of: billet, cheese (compressed billet), first preform, unpierced preform and finished forging (see Fig. 13.1). Initially, these fields are null. The success indicator of the last stage is not actually used (since it is not a preform to a forging stage, only a machining operation) and is only included for completeness.

d. The assessment of the die-design sequence begins with the stage that produces the target preform.

e. The current forging stage is compared with entries in the IKBS database for the same stage of the same category of component. This is described in detail later. Two numbers are evaluated:

An indicator (S) of whether the proposed die/preform design is satisfactory, with S varying between -1 (unsatisfactory) and $+1$ (satisfactory).

An indicator (C) of how strongly the evidence in the IKBS database supports the proposed success indicator S, with C varying between 0 and 1. Small values of C would indicate that there is no strong evidence to support the value of S; if C equals 1, then there is no uncertainty in the value of S. This would be the case if S equals +1 or −1. This would occur when the current forging stage is identical to one previously examined by the system.

A text string giving recommendations of any changes that need to be made to the die design may also be returned where appropriate.

f. If the confidence indicator C is not close to 1, then there is not enough evidence to draw any firm conclusions, and FE simulation of all stages up to target one needs to be performed.

g. The description of the shape of the proposed die and preform for the current stage is used to create an FE datafile. The FE analysis is performed as a background task, and the results examined to see whether the FE analysis confirms that the required shape can be forged from the given preform ($S = +1$) or not ($S = -1$). In both cases, the result is known with complete confidence ($C = 1$).

h. At this point, both the success indicator S and the confidence indicator C are known and C will be close or equal to 1. The values are known either as a result of FE analysis or as a result of studying previous examples known to the system. They may therefore be inserted into the results field in the sub-entry for the current stage of the forging.

i. If the success indicator S is close to +1, it is very likely that the proposed die/preform design for the current stage is satisfactory. Previous stages, if there are any, can now be examined.

j. If there are no previous stages, then the whole sequence of die designs has been examined and has been judged to be satisfactory, and manufacture of the dies and production trials can proceed.

k. Otherwise, the previous stage in the forging sequence is examined.

l. If S is close to −1, it is very likely that the proposed die/preform design for the current stage is unsatisfactory, so the preform to the current stage must be redesigned ("m"). If S is neither close to +1 nor close to −1, then more information is needed, so an FE simulation of the current stage of the forging process must be performed ("g").

m. FE analysis or comparison with previous examples has indicated that the die/preform design for the current stage is unsatisfactory. The die shape is determined by the subsequent stage of the forging, so the preform must be redesigned outside the IKBS using the CAD system. If the decision to redesign was made on the basis of comparison with previous examples, guidelines will have been retrieved suggesting how the preform should be changed, otherwise this information must be deduced from the FE results.

n. Whatever changes have been made to the design of the preform are recorded in the results field of the sub-entry for the current stage of forging in the IKBS database.

o. Since the preform to the current stage has been altered, it is necessary to redesign the dies up to the stage before the current one.

Details of the Implementation of the IKBS

General Remarks

The interactive IKBS program has been developed to run on a Unix-based engineering workstation. Since the actual FE analyses are performed independently of the main program, they need not be carried out on the same machine if a more suitable platform for intensive numerical computation is available.

The Sequence Design Program is written in Fortran, but Common Lisp was chosen for the other components of the interactive program because of the ease with which it can manipulate complex and dynamic data structures. The dialect of Common Lisp used allows Fortran and other procedural code to be invoked as Lisp functions (with bi-directional parameter passing) so the interaction between the SDP and the control module is fairly simple.

The interactive program is entirely menu-driven and there is extensive on-line help available. This may be either context related or in the form of system documentation. The package that has been developed at the University of Birmingham utilizes the native window management system of the workstation being used to devote different regions of the screen to menu selection, display of help information, display of warning and error messages and for textual input.

The various activities illustrated in Fig. 13.6 are not all performed within the interactive program, and in any case it is not envisaged that the complete design process would be accomplished in one session. The main menu of the IKBS program therefore provides a range of options relating to the main tasks that need to be performed:

- Design dies (using SDP) for a new component.
- Redesign dies (using SDP) up to a specified stage.
- Assess design of die/preform for a specified stage.
- Initiate (set up data file for) FE analysis of a specified stage.
- Input results of FE analysis.
- Proceed to die manufacture.

Structure and Use of IKBS Database

Since the sequence design program design rules are different for each of the five categories of component, the IKBS maintains a separate partition of the database for each of them.

Each partition of the IKBS database is in the form of a Lisp list. At the highest level, this is a list of entries for each of the components of a particular type examined by the system so far. At a deeper level, the structure is more complicated because a given stage of the forging sequence may have more than one instance, each having a different die or preform shape or possibly different forging temperature or lubrication. This may arise when, for example, the FE results indicate that a particular preform will not produce the desired shape at the end of the current forging stage and so has to be redesigned. Since this can potentially occur at any or all of the stages, the list structure must accommodate repeated branching.

The following semi-formal description of the database for a particular type of component may be given:

db-for-category	(\{component-entry\}*)
component-entry	((component-number [component-information]) preform-sub-entry)
preform-sub-entry	(preform-name (\{instance-of-preform\} +))
instance-of-preform	((\{description-parameters\} +) [(success-indicator confidence-indicator redesign-guidelines)] [preform-sub-entry])
component-number	fixnum uniquely defining component
component-information	simple-string containing information not directly used by IKBS (e.g. date entered, related components, origin of design etc.)
preform-name	symbol labelling the preform of the forging, e.g. FINISH, UNPIERCED etc.
description-parameter	single-float that is either the normalized value of one of the geometric parameters defining the shape of the particular instance of a preform, or the value of a process parameter (forging temperature, friction factor)

success-indicator	single-float between -1 and $+1$ inclusive, indicating whether the die/preform combination satisfactory, i.e. it is the success indicator of the stage that uses the current preform as its starting point
confidence-indicator	single-float between 0 and 1 inclusive indicating how confidently success indicator is known
redesign-guidelines	simple-string of information input by the die designer when the shape of the current preform has to be changed explaining what modifications have to be made and why they are necessary

In the above description, { }* represents any number (including zero) occurrences of the enclosed item, { }⁺ represents one or more occurrences of the enclosed item and [ ] represents zero or one occurrence of the enclosed item.

It should be noted that the sub-entries for the preforms of a forging are nested with earlier preforms inside later ones. Thus the outermost "preform-sub-entry" in a component entry refers to the finish shape, and in "instance-of-preform" the enclosed "preform-sub-entry" refers to the previous preform in the forging.

The database structure chosen is very flexible and can easily accommodate any number of redesigns of a preform. Its recursive definition also simplifies the writing of access and update functions.

Calculation of Design Assessment Values

The values of the design assessment indicators S and C are determined by identifying forging stages in the IKBS database that are similar to the one under consideration, and establishing whether the corresponding die/preform designs were satisfactory or not.

In practice, the success indicator S is calculated as the weighted average of the S values stored in the IKBS. The weighting factor for each stored S value will be a measure of how similar the current forging stage is to the one that produced that result value.

Forging stages are compared for similarity of both geometric and process parameters. The current example is compared only with the same stage of previous components of the same type (stub axle, bevel pinion, gear blank, drive flange and non-axisymmetric) and of the same material, though the forging conditions (temperature, lubrication) may be different. The shape of each preform may be characterized by a set of primitive solid shapes, as in the SDP. Each of these primitives may be uniquely defined by a number of geometric parameters. As explained in the previous section, for each stage of each component examined by the system so far, the IKBS database will contain one or more sets of numbers corresponding to the values of the geometric parameters for the primitive shapes defining one of the proposed forging preforms, in addition to a number of process parameters. The values of the geometric parameters will be the same as those used in the SDP, except that in the IKBS database, they will be normalized with respect to the overall component length, so that the evaluation of a die design sequence will not be affected by actual component size. Suppose there are $m-1$ previous instances of the current stage of the forging of a particular type of component in the IKBS database for which the success indicator is not null. Denote the geometric and process parameters associated with the preform and product of instance i by P_{ij}, where j varies between 1 and n. For each such example define:

$$\delta_i = \left[\sum_{j=1}^{n} w_j (P_{ij} - P_{mj})^2 \right]^{\frac{1}{2}} \tag{1}$$

where P_{mj} is the set of parameters associated with the current (mth) example of the forging stage under consideration, and w_j is a weighting factor for parameter number j indicating the significance of that parameter in determining whether or not a die/preform design is satisfactory. The weighting factor will therefore be a means of ignoring unimportant details of the shape and fine-tuning the design assessment procedure as examples are accumulated in the IKBS database.

Equation (1) can be thought of as expressing a metric in multi-dimensional component parameter space, a space in which the points represent the mappings of individual instances of the current forging stage (Fig. 13.7). In this context, δ_i is the distance, in this space, between the mapping of the example under consideration, and the mapping of previous example i. Note that the position in this space depends upon the initial *and* final shapes of the workpiece for this stage.

A suitable weighting factor, therefore, for the calculation of the average of the stored S values will be the inverse of the measure defined in equation (1), multiplied by the corresponding confidence indicator, with the condition that when δ_i equals zero, the inverse equals some very large, but not infinite, value. Thus:

$$S_m = \frac{\displaystyle\sum_{i=1}^{m-1} \frac{C_i}{\delta_i} S_i}{\displaystyle\sum_{i=1}^{m-1} \frac{C_i}{\delta_i}} \tag{2}$$

where S_i $(-1 \leqslant S_i \leqslant +1)$ is the success indicator for instance i.

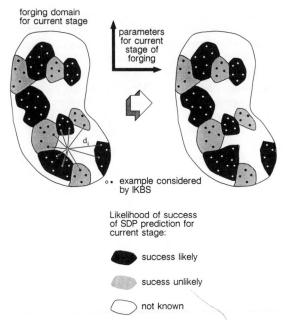

Fig. 13.7 Assessment of the likelihood of success of a proposed die/preform design by comparison with previous examples, and the growth in knowledge in the forging domain.

Equation (2) will always produce a value of S_m between -1 and $+1$, whether the example under consideration is actually similar to any previous instances or not. It will therefore be essential to have an indication of how reliable the value of S_m actually is. What is required is a value based on the distance, in parameter space, between the mapping of the current example and that of the previous example *most* similar to it. Suppose the previous example k is the nearest to the current one in parameter space. Then:

$$C_m = \frac{C_k}{1 + \delta_k} \tag{3}$$

The confidence indicator C_m will therefore be the same as the confidence indicator for example k when the minimum distance in parameter space is zero (i.e. example k is identical to the one under consideration) and will tend to zero as the smallest value of distance tends to infinity.

As a special case, both S_m and C_m will be defined to be zero when the IKBS database contains no previous examples of a particular type.

The Common Lisp functions listed in the Appendix show how the design assessment values can be calculated. The main function "assess-stage" takes four arguments: the type of component under consideration, the forging stage to be assessed, a list of the parameters defining the product of this stage and a list of the parameters defining the preform of this stage. It returns three values: the success indicator for the die/preform design, the corresponding confidence indicator, and a text string which may contain advice about how the preform needs to be modified, where this information is appropriate.

Two global variables are used. *ikbs-db* is an association list of the databases for the separate categories of component. Each of these databases is a list, as described in the previous section. *parameter-weights* is also an association list containing a sublist for each category of component, but in this case the sublists are themselves association lists with entries for each of the forging stages. The entries contain lists of the weighting coefficients w_j. For convenience, there is one list for the product shape parameters and another for the preform shape and processing parameters.

Conclusions

This chapter has described how an intelligent knowledge-based system can be used to combine the best features of rule-based design and numerical simulation programs. In its present form, the design rules embedded in the sequence design program are static. As a consequence, when they prove to be inadequate and certain preforms in the forging sequence have to be redesigned, the SDP will continue to propose unsatisfactory die designs for similar components. The IKBS will, of course, readily determine that they are unsatisfactory, and recommend how the preforms ought to be modified but it would speed up the development time if this step could be circumvented.

In the short term, the SDP design rules can be modified by hand on the basis of information stored in the IKBS database. To facilitate this, database analysis tools could be written to determine which of the SDP rules need to be modified and in what way.

In the longer term, this process could be automated, with the SDP design rules being constantly updated as more and more components are examined by the system. As the design rules are improved, it is to be hoped that the more lengthy FE analyses will need to be performed only very occasionally, though they could still be used to enhance the system knowledge base by means of parametric studies on important features or conditions.

Acknowledgements. The authors would like to thank the ACME Directorate of the SERC for funding the work discussed in this chapter and are grateful to their colleagues at Rover Group for their assistance during the course of the project.

Appendix

```
(defun assess-stage  (type
                      stage
                      product-params
                      preform-params)
  ;; obtain weighting factors for specified stage
  ;; and type of component
  (let*  ((weights
            (cdr
              (assoc
                stage
                (cdr
                  (assoc type *parameter-weights*)))))
          (product-weights (car weights))
          (preform-weights (cadr weights))
          (succ-ind 0)
          (conf-ind 0)
          (nearest-dist most-positive-single-float)
          (nearest-conf 0)
          (nearest-guide " "))
    (declare (single-float succ-ind
                           conf-ind
                           nearest-dist
                           nearest-conf)
             (simple-string nearest-guide))
    ;; Examine each of the entries for components
    ;; of the specified type
    (dolist (comp-entry (cdr (assoc type *ikbs-db*)))
            (multiple-value-setq
                    (succ-ind
                     conf-ind
                     nearest-dist
                     nearest-conf
                     nearest-guide)
                    (examine-component-entry
                     (cadr comp-entry)
                     stage
                     product-params
                     preform-params
                     product-weights
                     preform-weights
                     succ-ind
                     conf-ind
                     nearest-dist
```

```
                        nearest-conf
                        nearest-guide)))
        (unless (zerop conf-ind)
              (setf succ-ind (/ succ-ind conf-ind))
              (setf conf-ind
                    (/ nearest-conf (1+ nearest-dist))))))
        (values succ-ind conf-ind nearest-guide)))


(defun examine-component-entry  (comp-entry
                                 stage
                                 product-params
                                 preform-params
                                 product-weights
                                 preform-weights
                                 succ-ind
                                 conf-ind
                                 nearest-dist
                                 nearest-conf
                                 nearest-guide)
        (declare (single-float succ-ind
                               conf-ind
                               nearest-dist
                               nearest-conf)
               (simple-string nearest-guide))
        ;; examine each of the examples of the product
        ;; defined for this stage of forging
        (dolist (product-entry (cadr comp-entry))
                ;; search for specified stage and examine
                ;; parameters
                (multiple-value-setq
                    (succ-ind
                     conf-ind
                     nearest-dist
                     nearest-conf
                     nearest-guide)
                  (if (eq stage (car comp-entry))
                      (examine-product-entry
                       product-entry
                       product-params
                       preform-params
                       product-weights
                       preform-weights
                       succ-ind
                       conf-ind
                       nearest-dist
                       nearest-conf
                       nearest-guide)
```

```
                        (examine-component-entry
                        (caddr product-entry)
                        stage
                        product-params
                        preform-params
                        product-weights
                        preform-weights
                        succ-ind
                        conf-ind
                        nearest-dist
                        nearest-conf
                        nearest-guide))))
              (values succ-ind
              conf-ind
              nearest-dist
              nearest-conf
              nearest-guide))


(defun examine-product-entry    (product-entry
                                product-params
                                preform-params
                                product-weights
                                preform-weights
                                succ-ind
                                conf-ind
                                nearest-dist
                                nearest-conf
                                nearest-guide)
              (declare (single-float succ-ind
                                conf-ind
                                nearest-dist
                                nearest-conf)
                    (simple-string nearest-guide))
      (let    ((sum 0))
              (declare (single-float sum))
              ;; start to calculate distance
              (dolist (prod (mapcar #'wt-square-dif
                                (car product-entry)
                                product-params
                                product-weights))
                    (declare (single-float prod))
                    ;; product-weights ensures that processing
                    ;; parameters for product stage are ignored
                    (setf sum (+ sum prod)))
              ;; examine each of the examples of the preform
              ;; defined for this stage of forging
              (dolist (preform-entry (car (cdaddr
                                product-entry)))
```

```
;; ignore entries if success indicator
;; is null
(when (caadr preform-entry)
      (let ((dist sum) (ratio 0))
           (declare (single-float
                     sum
                     ratio))
      ;; finish calculation
      ;; of distance – include
      ;; processing parameters
      ;; this time
      (dolist
        (prod2
          (mapcar
            #'wt-square-dif
            (car preform-entry)
            preform-params
            preform-weights))
        (setf dist (+ dist prod2)))
      (setf dist (sqrt dist))
      (setf
       ratio
       (if (zerop dist)
           (* (cadadr preform-entry)
              most-positive-single-float)
           (/ (cadadr preform-entry)
              dist)))
      (setf
       succ-ind
       (+ succ-ind
          (*
            ratio
            (caadr preform-entry))))
      (setf
       conf-ind
       (+ conf-ind ratio))
      (when (< dist nearest-dist)
            (setf nearest-dist dist)
            (setf
             nearest-conf
             (cadadr preform-entry))
            (setf
             nearest-guide
             (car
              (cddadr
               preform-entry)))))))
```

```
(values succ-ind
       conf-ind
       nearest-dist
       nearest-conf
       nearest-guide))

(defun wt-square-dif (p1 p2 w)
      (declare (single-float p1 p2 w))
  (let ((dif (- p1 p2)))
    (declare (single-float dif))
    (* w dif dif)))
```

References

1. Yu GB, Dean TA. A CAD/CAM package for hammer forging dies. In: Proceedings 25th Machine Tool Design Research Conference, Birmingham, April 1985. Macmillan, London, 1985, pp 459–464
2. Choi SH, Sims P, Dean TA. A complete CAD/CAM package for hammer forging dies. Proceedings 25th Machine Tool Design Research Conference, Birmingham, April 1985. Macmillan, London, 1985, pp 451–458
3. Gokler MI, Dean TA, Knight WA. Computer aided die design for upset forging machines. Proceedings 11th NAMRC Conference, 1983, pp 217–223
4. Rao SS. The finite element method in engineering, Pergamon, 1982
5. Pillinger I. The prediction of metal flow and properties in three-dimensional forging using the finite-element method. PhD thesis, University of Birmingham, 1984
6. Pillinger I, Hartley P, Sturgess CEN, Rowe GW. Finite-element modelling of metal-flow in three-dimensional forming. Int J Num Meth Eng, 1988; 25: 87–97
7. Rowe GW, Sturgess CEN, Hartley P, Pillinger I. Finite-element plasticity and metalforming analysis. Cambridge University Press, 1990
8. Hartley P, Sturgess CEN, Dean TA, Rowe GW, Eames AJ. Development of a forging expert system. In: Pham DT (ed) Expert systems in engineering. IFS, Bedford, 1988, pp 425–443

Chapter 14

Expert CAD Systems for Jigs and Fixtures

A.Y.C. Nee and A.N. Poo

Introduction to Automated Fixture Design

The design of jigs and fixtures is a highly complex and intuitive process. There are many mathematical and scientific formulae which can be used to calculate cutting forces, deflection of structural members, tolerance analysis of locating datum, etc. However, many of the good design features of jigs and fixtures such as correct proportioning, ease of loading, safety considerations, ingenuity in securing workpieces, etc. come from the experience and skill of the designer. As a result of the demand for both engineering analysis and craftmanship, the automation of fixture design has not been considered possible in the past. Recent developments in artificial intelligence techniques and, in particular, knowledge representation and inference procedures have opened up great opportunities in automating this field. Computer-aided jigs and fixtures research started in mid-1970s and towards the early 1980s, several prominent research institutions began work in this area.

In automated manufacturing systems such as flexible manufacturing systems (FMS) and computer-integrated manufacturing systems (CIMS), the subsystems consist of automated workstations for fabrication, assembly and other secondary operations linked by robots and material handling equipment. The single most important operating criterion in such systems is the ability to cope with changing products, batch size and mixed orders with maximum utilization of the equipment, or in other words the systems must be truly flexible. The difficulty of fixturing systems in being flexible is often considered a major impediment to the development of computer-integrated manufacturing systems.

In the hierarchy of computer-aided design and manufacture (CAD and CAM) jigs and fixtures design represents an important interface between design and process planning (Fig. 14.1). Process planning has in turn

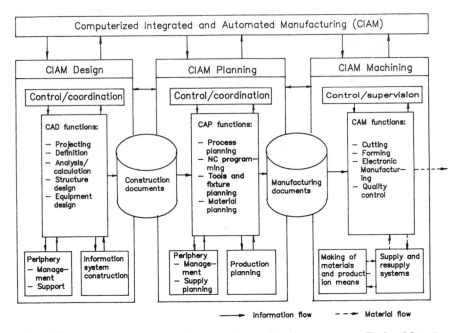

Fig. 14.1 Components of an integrated and automated production system. Tool and fixturing is central to the computer-aided process planning function [1]. (Reprinted with permission from Springer Verlag.)

been considered as the missing link between CAD and CAM. In an effort to automate the process planning procedure without fully automating the fixture design process, it will be difficult to avoid human intervention along the line and errors due to manual interpretation will inevitably result. Conventional fixture design practice utilizing standard fixture components produced by specialized fixture element manufacturers presents certain problems in automating the fixture design process. These standard components, although often belonging to a family, are rather numerous and there is great difficulty in representing them in databases. The decision to use certain elements, for example, which type of clamps, whether toggle, hook, strap or cam operated, is also hard to arrive at without human intervention. The concept of flexible fixturing is difficult to realize using conventional fixture elements.

Modular fixturing, or the concept of using a limited number of building blocks such as V-blocks, base plates, angle supports and clamping straps to arrive at the fixture configuration using the "Lego" approach, has proved to be popular in automated fixture design.

Flexible Fixture Design – Concepts and Techniques

Introduction

As mentioned previously, fixturing is fundamental to many manufacturing operations such as machining, fabrication, assembly and inspection. As the manufacturing scenario is moving towards greater flexibility due to shorter product life cycle and greater product variety, the fixturing process must be equally flexible.

Flexible fixturing refers to fixtures which are capable of accommodating a family of parts of different geometries and sizes. This concept is quite different from conventional fixturing methods where a dedicated fixture is often developed for a specific operation. The fixture is often useless for other products and is also seldom of use when the elements are stripped apart. Most companies, however, would tend to store the fixture for future use or repeat orders rather than taking them apart. This would mean that a large number of fixtures need to be stored and would require unproductive storage space together with some retrieval system to identify them for further use.

Gandhi and Thompson [2] pointed out that the advantages to be gained from employing flexible fixtures would be three-fold, viz., reduced lead time and effort required to design special purpose fixtures, reduced overhead of storing a multiplicity of fixtures and simplified programming requirements (in an FMS). Typical case studies have shown that flexible fixturing systems can reduce jigs and fixtures manufacturing time by 90%, lead time by 85%, fixture manufacturing cost by 80%, fixture material cost by 95% and fixture storage areas by a similar order.

Flexible Fixtures

Several different methodologies have been proposed for flexible fixturing (Fig. 14.2) [3]. Basically they are: modular fixture building blocks, programmable conformable clamps and phase change fixtures.

Modular Fixturing Systems (MFS)

An MFS is a kind of modular universal fixture, which is assembled directly from a set of ready-made, reusable, standardized elements and combined units (Fig. 14.3) [4]. An example of an assembly is shown in Fig. 14.4.

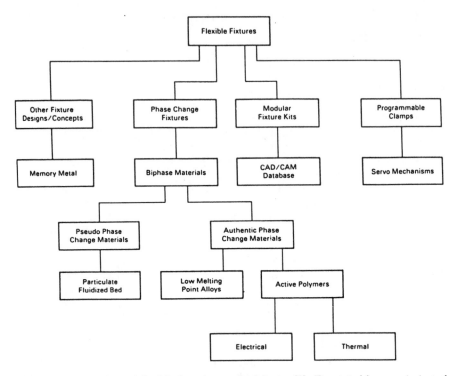

Fig. 14.2 An overview of flexible fixturing methodologies [3]. (Reprinted by permission of the authors, MV Gandhi and BS Thompson, and the Society of Manufacturing Engineers. © 1985 *J Manuf Syst* vol.4/no.1.)

The typical components consist of (1) base elements with holes and T-slots, (2) supporting elements such as V-blocks, angle plates, rectangular supports, (3) positioning elements – location pins, centres, discs, and plates used to position the component accurately in the fixture to the datum, (4) guiding elements such as drill bush guides – generally not used in CNC machining work, (5) fastening and clamping elements such as T-bolts, clamping screws and a wide range of clamps, (6) subassemblies or combined units – such as indexing units, sine bars, angle tables, parallel and wedge clamps, hinged elements and (7) miscellaneous elements such as adjustable button for castings, swivel and connecting plates, clamping handles, compression springs, etc.

The elements resemble Lego and Meccano toy building blocks. A large variety of combinations can be built with a limited number of building elements. These elements and combined units interchange well and provide high accuracy and long service life.

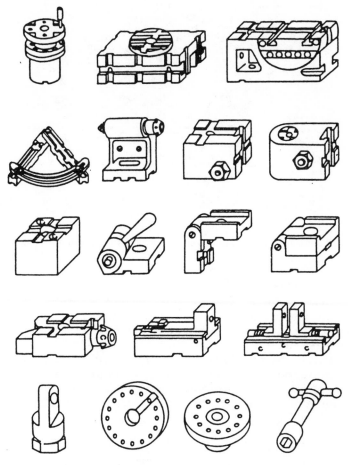

Fig. 14.3 Combined units of CATIC modular fixturing system (MFS) [4]. (Reprinted with permission from IFS Publications.)

The concept of modular fixturing was invented by Wharton and Wilcocks (now WDS Wharton) in the early 1950s. Presently there are a number of suppliers who make modular fixture components. One of the largest manufacturers is CATIC (China National Aerotechnology Import and Export Corporation) of PRC. According to CATIC [4], it typically takes between 20 minutes and 4 hours to complete the assembly of a fixture.

The elements that make up a fixture can be broken down and rebuilt quite easily and this represents substantial savings over conventional methods.

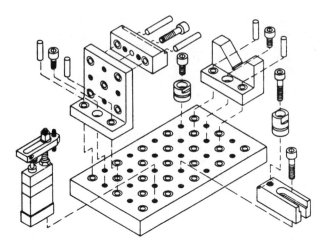

Fig. 14.4 Modular fixture incorporating a base plate with plate and tapped holes. (Courtesy of Fritz Werner Machine Tool Corp.)

Phase Change Fixturing Systems

Flexible fixturing based on the concept of material phase changing [3] makes use of a material which can be changed easily from fluid to solid and vice versa. In essence, a container is filled with this rapid phase change material and a workpiece placed in a desirable orientation in the container surrounded by the material in its liquid form. The liquid is then changed into solid which now firmly holds the part in position for the machining operations. The part is removed by changing the material back to its liquid phase.

Phase change fixturing appears very simple in concept and economical as the material is reusable over a large number of cycles. There are a number of problems which need to be considered, one of which is the accurate positioning of the part. This may require metrological equipment or a robot to hold the part in the correct orientation. If a part is bulky and heavy, this will further aggravate the problem. Selecting a suitable medium which is capable of reversing its state easily is a second major problem.

Programmable/Transformable Clamps

Cutkosky et al. [5] designed a clamping device capable of accommodating a variety of turbine blades produced in an automated forging cell. This clamping device can cater for several hundred blade styles of varied geometry. It is possible to design fixtures with adaptable shapes to suit a family of products. In this way, the need to produce individual fixtures is largely eliminated.

Computer-Aided Fixture Design

Introduction

Although CAD/CAM has been an efficient productivity tool for around 15 years, its application in the area of tool design and in particular, jigs and fixtures, is of recent origin. This is partly due to the highly complex and intuitive process of tool design and the difficulty in embedding intelligent problem-solving capabilities in computer algorithms. While the major stumbling block appears to be that of knowledge representation, promising solutions in overcoming this obstacle are available due to the advent of techniques based on artificial intelligence and rule-based expert systems. By combining the knowledge of manufacturing methods, machine information and special expert system shells, it is now possible to tackle a problem that has been left largely to the skilled tool designer.

Classical CAD of Fixtures

Creating Library Symbols

Early applications of CAD techniques in designing fixtures merely made use of the draughting capabilities of a CAD package in producing drawings. Fixture elements such as pins, clamps, springs, and supports are typically stored as library symbols which can be called and placed anywhere as desired. These library symbols are of fixed shapes. Although they could be scaled, mirrored, rotated, etc., the proportion of the features remain unchanged. This was soon found to be unsatisfactory as a large number of library symbols were needed and the effort required to generate them was formidable. In addition, only a particular supplier's components would usually be built into a library and it would be more difficult to accommodate several catalogues.

Parametric Design

Parametric design refers to the creation of a design – a total construction or only part of it – by means of a specific program. A given basic construction can be used to create a family of parts or subassemblies by merely varying the design parameters. The variations are constructed according to the same principles as the basic construction. The varying parameters may be the main dimensions of the part, while the other dimensions are functions of these prime quantities.

By using parametric design programs, repetitive design work can be avoided and the product manufacturing becomes faster. The designer is able to use his time more effectively when the routine drawings and calculations are made by computers.

A common way of providing a parametric capability in a CAD system is to provide some form of parametrised macro facility. Such systems are often quite general in their capabilities, but they suffer the drawback that users need to acquire the skills of a programmer in order to become an effective macro writer. The majority of users have a draughting or engineering background and it is unreasonable to expect them to learn this new skill.

The normal way to prepare and change macros is by means of a text editor, which is often regarded as an unacceptable method of altering pictures. The main difficulty is that the correspondence between lines in a textual macro and lines or arcs in a drawing is not obvious. Further, simple changes in a dimensioning scheme can lead to radical alterations to the sequence of operations in a macro.

Parts families may also be created by means of a combination of variations in dimensions as well as shapes; this is illustrated in Fig. 14.5.

By a combination of both dimensional as well as geometric parametrised programs, a large number of components can be created easily.

Application of Artificial Intelligence Techniques and Languages in Fixture Design

Review of Pioneering CAD of Fixtures

Imhof and Grahl [6] were probably the first researchers to give a comprehensive treatise on CAD of jigs and fixtures. They considered in great depth the choice of workpiece datum based on dimensional relationship, geometrical form of the workpiece and the condition of the workpiece surface. The arrangement of the fixture elements is based on the computer transformation of the description of the elements in the fixture coordinate system (Fig. 14.6).

Ingrand and Latombe [7] developed an expert system for automatic fixture design using MacLisp. The system applies expert rules to select the resting surfaces and points which eliminate the six degrees of freedom considering factors such as surface quality, dimensional and geometrical relationship of the surfaces.

Eversheim et al. [8] reported a computer-aided fixture design project at Aachen. They planned to construct the fixture element in three dimensions and employ artificial intelligence techniques in establishing some of the fixture design rules.

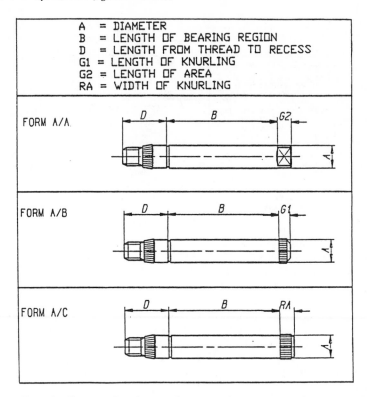

Fig. 14.5 Parts family created with a combination of variations in dimensions as well as shapes.

Markus et al. [9] also agreed that fixture design appears to be an appropriate subject for investigation using artificial intelligence tools and they reported the design of fixtures using Prolog. They proposed a prototype fixture design approach for a family of box-type workpieces from a fixture kit. The input data consisted of the description of the workpiece shape, the machining operations required and the coordinates of the supports and clamps to be used. The algorithm produces a sequence of draught fixture arrangements.

Interactive CAD Fixture Design Approach

Technique by Miller and Hannam. Miller and Hannam [10] reported the development of the interactive fixture design software package JIGS (Jigs by Interactive Graphics). The main elements of their design procedures are shown in Fig. 14.7.

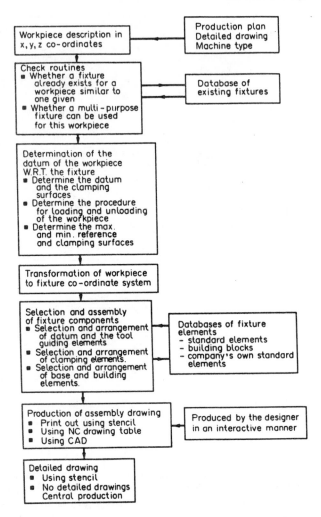

Fig. 14.6 CAD fixture design approach used by Imhof and Grahl. Original diagram in German [6].

In their approach, the user first either defines the three orthogonal views of the workpiece and machine interface or retrieves information from the files previously created. After this, the user interactively selects a datum position on the workpiece in three views. A classification code is used to retrieve any existing designs. As a large list of methods of achieving the various operating functions is possible, the code also functions as a filtering routine for selecting the possible methods. The components are drawn automatically and the user interactively assembles them together. The stages of drawing a fixture are shown in Fig. 14.8.

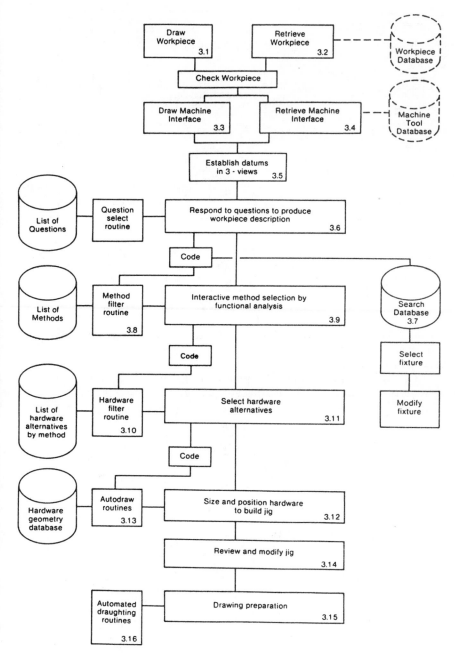

Fig. 14.7 Block diagram representation of CAD fixture design technique used by Miller and Hannam [10]. (Reprinted by permission of the Council of the Institution of Mechanical Engineers from *Proc I Mech E* vol. 199, no. B4, 1985.)

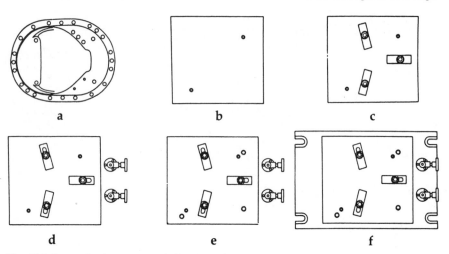

Fig. 14.8 Stages in the automated drawing of a jig (one view): **a** workpiece, **b** location, **c** clamping, **d** support added, **e** loading added, **f** body added [10]. (Reprinted by permission of the Council of the Institution of Mechanical Engineers from *Proc I Mech E* vol. 199, no. B4, 1985.)

The final stage of the design is a review stage where modifications and improvements can be incorporated. Figure 14.9 shows the completed design after editing.

During the initial trial period, a productivity benefit of 5:1 was achieved compared with drawing board-based designs.

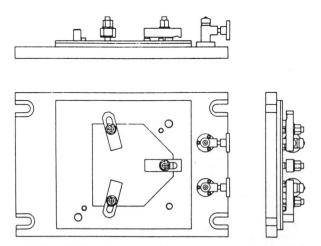

Fig. 14.9 Three views of the completed jig constructed as illustrated for one view as shown in Fig. 14.8 [10]. (Reprinted by permission of the Council of the Institution of Mechanical Engineers from *Proc I Mech E* vol. 199, no. B4, 1985.)

Technique by Nee, Bhattacharya and Poo. In the approach by Nee, Bhattacharya and Poo [11–12], the essential fixture elements are grouped into three main categories, viz.,

• Clamping elements
• Positioning and guiding elements
• Supporting and base elements

The three modules are then integrated to give a complete fixture description as depicted in Fig. 14.10.

Structurally the program is divided into two parts:

• Data processing
• Graphics interface

Three different types of databases are used in linking the above two parts for data retrieval:

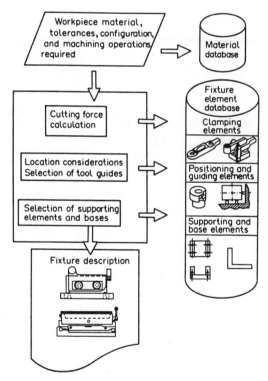

Fig. 14.10 Interactive CAD fixture design approach used by Nee, Bhattacharya and Poo [12]. (Reprinted with permission from *Robotics Comput Integ Manuf* vol. 3, no. 2. © 1987, Pergamon Press.)

- Logical databases
- Linking databases
- Parametric databases

The program structure described above is shown in Fig. 14.11. As the program is being executed, a series of questions will first be directed to the designer requesting workpiece information such as the base area and the height of the probable clamping surface from the base. At this stage the program is completely interactive. The program then displays a menu asking if the designer intends to design any particular element such as clamps, locators, supporting or base elements, or to construct the fixture as a whole. A list of common material is then displayed from which the designer selects the nearest workpiece material. The program automatically retrieves the machinability rating from the logical database. Based on the required machining operation and the user input feed, depth of cut and cutting speed, the major cutting force components will be estimated. The program then searches the clamp database where a list of suitable and available clamps will be displayed. The designer can interactively select the most suitable clamp from the recommended list. The parameters associated with the selected clamp are passed to a draughting routine and the exact proportion of the clamp will be drawn in two views and stored away in a drawing file.

A suitable locating datum will next be determined. An attempt has been made to select a suitable datum for a workpiece having parallel machined or unmachined planes. Generally the factors governing the selection are:

- Premachined surfaces and their relation to the surface to be machined
- Geometrical form of the workpiece and other premachined features such as holes, cylinders and edges
- General condition of the workpiece surface

A least tolerance technique [13–14] is used to select the correct locating surface should there be more than one parallel premachined surfaces. In a premachined workpiece, several machined planes may already exist. Machining of other surfaces may therefore be located from any one of

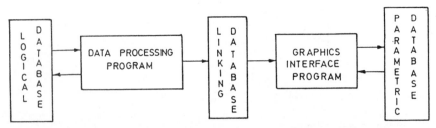

Fig. 14.11 Program structure of CAD of fixtures by Nee, Bhattacharya and Poo [13].

the premachined planes. The choice of the datum plane depends on the tolerance specification and suitability of the surface for location purposes. In the present approach the least tolerance method is used to analyse and select the correct datum plane. The problem of meeting the required tolerance specification is aggravated by the fact that the design drawings of engineering components are often dimensioned from plane to plane. Transferring tolerances to and from a particular datum is tedious and error-prone.

In the least tolerance method, the least or the most stringent tolerance amongst all the other tolerances is first considered. In order to identify the best datum plane, a test plane is progressively transferred from the first plane (Fig. 14.12). The least tolerance method is applied to each case and the transferred tolerances are computed and stored. For a workpiece having N conventionally dimensioned planes, the number of planes to be considered will be $N+1$. The number of transferred tolerances using all the $N+1$ test planes will be $N(N+1)$ and this is stored in the form of a matrix. A subprogram is written to link to the main program for the selection of the correct datum plane based on the least tolerance method.

This subprogram caters for bilaterial, unilateral and mixed tolerancing systems. Figure 14.13 shows the flow chart depicting the subprogram. If the plane selected based on the least tolerance is unsuitable due to reasons such as difficulty in accessing the plane, or the plane being too small and unstable for locating purposes, etc., the next higher tolerance and the associated plane will be recommended as the datum plane. This process is repeated until a suitable datum is selected.

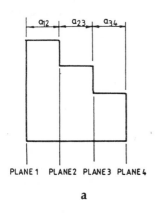

a

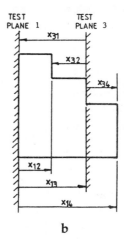

b

Fig. 14.12 Transferring tolerances with respect to test planes. **a** workpiece with parallel premachined planes dimensioned from plane to plane. Original tolerances are a_{12}, a_{23} and a_{34}. **b** tolerances transferred to test planes. Transferred tolerances are x_{12} x_{13} and x_{14} with respect to test plane 1 and x_{31} x_{32} and x_{34} with respect to test plane 3.

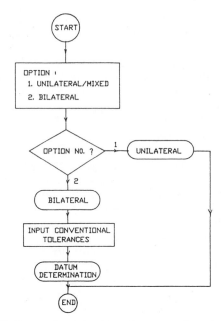

Fig. 14.13 Datum determination subprogram flow chart [14].

In summary, the design rules for selecting the datum plane are:

1. The tolerance specification from the datum plane to the surface where machining operations are required is the least stringent.
2. The datum plane is large enough to support the workpiece, i.e. the centre of gravity of the workpiece must be well within the datum plane.
3. The datum plane can be readily accessed.

The locating elements are selected based on the geometry of the workpiece. The decision made by the program is rather qualitative in nature. It asks for workpiece information such as whether the workpiece is forged or cast, any premachined planes, the basic shape of the workpiece, etc. From the answers provided the program displays a list of locators and their sizes. The designer then selects a locator from the list interactively. The selected element is immediately codified by the program with its accessory specifications for subassembly and the information is stored into the linking database. Drill and ream bushes will be retrieved from the standard tool guiding database. In the absence of a standard-dimension bush, the recommended dimensions of a "customized" drill bush will be displayed.

In selecting the supporting elements and bases, consideration is given to the size of the workpiece, the estimated space allocated to the clamping

and locating elements and the type of machine tools to be used. The structure of the supporting element must be strong enough to absorb cutting, clamping and other inertia, gravitational forces which may be associated with the operation.

Each of the elements mentioned above is constructed parametrically and stored in separate drawing files. The final assembly of the fixture components is manually performed although some of the subassemblies such as clamping elements are performed automatically.

To illustrate the automatic construction of the clamp assembly, the following example is given. From the WDS International Catalogue, a strap clamp assembly consists of the following items:

- Plain clamp strap
- Nut and washer
- Heel pin
- Stud
- Quick return spring

Each of the above elements has its own associated parameters. Typical elements taken from the WDS Catalogue are illustrated in Fig. 14.14.

The governing parameters, in this instance, are the stud diameter and the workpiece height. The stud diameter depends on the clamping force

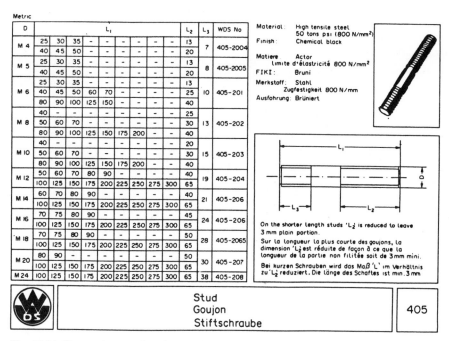

Metric

D	L_1									L_2	L_3	WDS No
M 4	25	30	35	-	-	-	-	-	-	13	7	405-2004
	40	45	50	-	-	-	-	-	-	20		
M 5	25	30	35	-	-	-	-	-	-	13	8	405-2005
	40	45	50	-	-	-	-	-	-	20		
M 6	25	30	35	-	-	-	-	-	-	13	10	405-201
	40	45	50	60	70	-	-	-	-	25		
	80	90	100	125	150	-	-	-	-	40		
M 8	40	-	-	-	-	-	-	-	-	25	13	405-202
	50	60	70	-	-	-	-	-	-	30		
	80	90	100	125	150	175	200	-	-	40		
M 10	40	-	-	-	-	-	-	-	-	20	15	405-203
	50	60	70	-	-	-	-	-	-	30		
	80	90	100	125	150	175	200	-	-	40		
M 12	50	60	70	80	90	-	-	-	-	40	19	405-204
	100	125	150	175	200	225	250	275	300	65		
M 14	60	70	80	90	-	-	-	-	-	40	21	405-206
	100	125	150	175	200	225	250	275	300	65		
M 16	70	75	80	90	-	-	-	-	-	45	24	405-206
	100	125	150	175	200	225	250	275	300	65		
M 18	70	75	80	90	-	-	-	-	-	50	28	405-2065
	100	125	150	175	200	225	250	275	300	65		
M 20	80	90	-	-	-	-	-	-	-	50	30	405-207
	100	125	150	175	200	225	250	275	300	65		
M 24	100	125	150	175	200	225	250	275	300	65	38	405-208

Material: High tensile steel 50 tons psi (800 N/mm²)
Finish: Chemical black
Matière: Actor limite d'élasticité 800 N/mm²
FIKI: Bruni
Merkstoff: Stahl Zugfestigkeit 800 N/mm
Ausfuhrung: Brüniert

On the shorter length studs 'L₃' is reduced to leave 3 mm plain portion.
Sur la longueur la plus courte des goujons, la dimension 'L₃' est réduite de façon à ce que la longueur de la partie non filitée soit de 3mm mini.
Bei kurzen Schrauben wird das Maß 'L' im Verhältnis zu 'L₃' reduziert. Die Länge des Schaftes ist min.3 mm

Stud
Goujon
Stiftschraube   405

Fig. 14.14 Clamp element data from WDS International. (Courtesy of WDS Tooling, UK.)

required while the workpiece height is given by the distance between the datum and the clamping surface. When the above two parameters are varied, other element sizes are also affected. For example, a larger stud size needs a larger nut, washer and clamp strap; a taller workpiece requires a longer heel pin and stud length, etc. It is therefore important to consider all these factors in the program designed to select the correct clamp assembly. Since the various elements have well-defined relative positions with respect to one another, it is possible to incorporate the coordinates in the subassembly program to ensure their correct positions. Several parametrically constructed clamp assemblies are shown in Fig. 14.15.

The draughting package used in the present research is CADAM from IBM. CADAM geometry interface modules allow an application programmer to edit or access the CADAM database system. In this study, CADCD is used as the key module to generate and retrieve drawings created by CADAM. For the linking of a CADCD routine with the CADAM software, a program written in Fortran 66 is used to access to the various CADCD subroutines available from the CADCD library routines (Fig. 14.16).

Figure 14.17 shows the details of a workpiece requiring a milling operation to generate a slot of 40 mm width. The program described above is used interactively to create the necessary fixture elements for this workpiece. Figure 14.18 shows all the elements selected after the

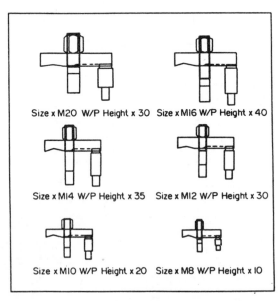

Size x M20 W/P Height x 30   Size x M16 W/P Height x 40

Size x M14 W/P Height x 35   Size x M12 W/P Height x 30

Size x M10 W/P Height x 20   Size x M8 W/P Height x 10

Fig. 14.15 A family of strap clamps constructed parametrically [12]. (Reprinted with permission from *Robotics Comput Integ Manuf* vol. 3, no. 2. © 1987, Pergamon Press.)

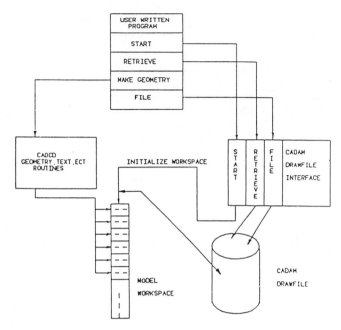

Fig. 14.16 Accessing CADAM database with CADCD modules. (Courtesy of IBM.)

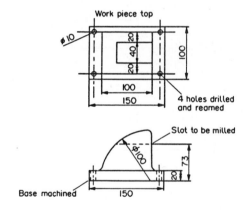

Fig. 14.17 Workpiece to be considered in the program [12]. (Reprinted with permission from *Robotics Comput Integ Manuf* vol. 3, no. 2. © 1987, Pergamon Press.)

program run. Figure 14.19 shows the completely assembled fixture drawing. The final drawing is manually assembled with standard commands such as MOVE, TURN, etc., from the CADAM package. Automatic assembly of the complete fixture is difficult as the connectivity

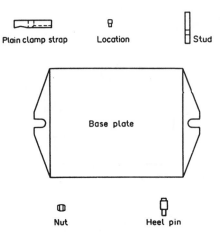

Plain clamp strap   Location   Stud

Base plate

Nut   Heel pin

Fig. 14.18 Fixture elements selected by the program [12]. (Reprinted with permission from *Robotics Comput Integ Manuf* vol. 3, no. 2. © 1987, Pergamon Press.)

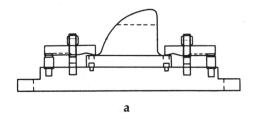

a

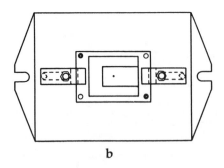

b

Fig. 14.19 Front (**a**) and plan (**b**) view of the assembled fixture [12]. (Reprinted with permission from *Robotics Comput Integ Manuf* vol. 3, no. 2. © 1987, Pergamon Press.)

relation between each element and the workpiece has to be defined. This will make the program too complex and therefore is not attempted here.

Expert System CAD Fixture Design Approach

Expert Machinist. The overall flow chart of the Expert Machinist described by Englert and Wright [15] is shown in Fig. 14.20. These authors used the approach proposed by Hayes-Roth, Waterman and Lenat [16] where the development of an expert system takes five major stages: identification, conceptualization, formalization, implementation and testing.

In identification, the problem boundaries and key assumptions are properly defined. In conceptualization, the team videotaped a number of sessions involving the knowledge engineer and the expert. Good machining practices and heuristic rules are extracted and used in the control strategy of the expert system.

In the formalization stage, the concepts derived are represented in a typical expert system development language which allows for ease of addition or deletion of rules in a knowledge base. The language used by them is cell management language (CML). The data and rule construction based on CML is implemented in the form of tables consisting of columns (fields) and rows (entries) that form unit blocks of information (items). Individual items may contain numerical information, characters, or strings

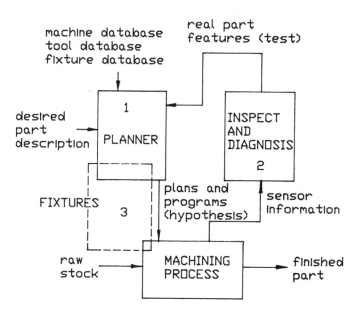

Fig. 14.20 Expert Machinist overall flow chart [15]. (Reprinted with permission from PJ Englert and PK Wright. © 1986 IEEE Int Conf on Robotics and Automation.)

and a combination of items may form an equation or call to another table.

In implementation, the strategies and heuristics used by the machinists are transformed into rules and control structures used by the expert system.

A control strategy for applying the rules must be developed. Figure 14.21 shows the various components of the fixture setup planner. The controller (inference engine) is responsible for deciding which rule is to be used considering tradeoffs between alternative decisions. The controller channels information flow between the static and dynamic databases with the rulebase. The static database contains information pertaining to the fixture elements and tools which do not change. The dynamic database contains updated knowledge after each search. The dynamic database contains information such as the new locating surface after a cut has been made and what will be the new tolerance limits, etc.

One of the techniques used to reach an optimal setup is by placing cost factors and quality ratings along the paths. The controller gives an overall quality rating with cost of the past and future paths to the current state, and this assists in making the optimal decision.

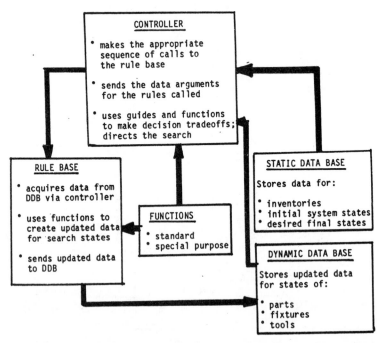

Fig. 14.21 Setup Planner flow chart [15]. (Reprinted with permission from PJ Englert and PK Wright. © 1986 IEEE Int Conf on Robotics and Automation.)

Holdex. A rule-based CAD fixture design system was also reported by Lim and Knight [17] in their HOLDEX (holding device expert system) system. HOLDEX is implemented using PADL-1 and LISP. PADL-1 is used to provide a three-dimensional geometric data structure of the fixture elements while Lisp is used to develop the expert system. The overall structure of HOLDEX is shown in Fig. 14.22.

HOLDEX allows the use of modular fixturing systems as well as conventional fixture elements as it is more economical to use conventional elements for low-quantity production.

Four major tasks are outlined in the computerization of the fixture design process. The first task involves the transformation of geometrical parameters associated with the workpiece into nongeometrical entities.

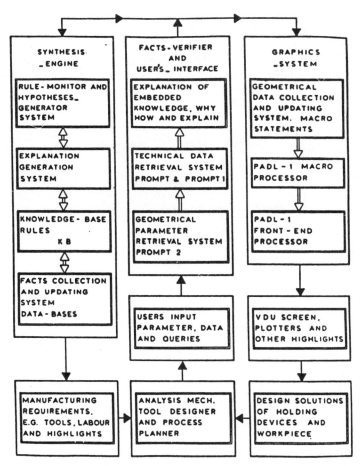

Fig. 14.22 An overview of HOLDEX: the relationships between the user's interface, graphics system and synthesis engine [17]. (Courtesy of BS Lim.)

This makes the program much easier to make inferences as design rules are now expressed in terms of machined features such as dowel holes instead of a finished hole with a certain tolerance and surface finish.

The second task involves the extraction of other nongeometrical entities such as batch size, initial state of material whether it is forged, cast, etc.

The third task involves the synthesis of the nongeometrical entities to produce geometrical parameters based on the rules in the knowledge base.

The final task involves the decomposition of the deduced geometrical parameters of the fixture into two-dimensional drawings which can be used for production and assembly purposes.

HOLDEX has also incorporated in its system a help device which has some limited conversational capability and could record the conversation to enable the user to recognize any deficiencies in the knowledge base. HOLDEX produces both textual and graphical information such as the type and size of tools required. Figure 23 shows a modular fixture constructed by the system.

HOLDEX is also capable of informing the user as to why certain information is needed and how certain conclusions have been established or deduced.

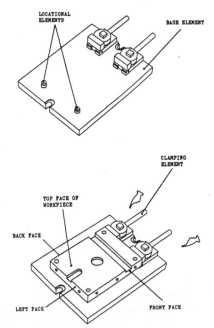

Fig. 14.23 A modular fixture which allows features located on the top, front, back and left face to be machined in one setting. Fixture constructed by HOLDEX [17]. (Courtesy of BS Lim.)

Automated Fixture Design Project at Industrial Technology Institute, Ann Arbor, Michigan. Gandhi and Thompson [2] proposed a methodology for the automated design and assembly of modular fixtures (Fig. 24). Central to their proposed approach is an expert system which captures the expertise of the shop floor personnel.

In addition to the usual features of an expert fixture design package such as ability to explain the reasoning process and respond to queries, the system should also be able to learn and assimilate new knowledge to enhance its decision-making capability. The knowledge base contains information based on subjective information from experts in jig and fixture design as well as formal fixturing rules and fundamental theories based on mathematics and physics.

A postprocessing phase to include finite element models to analyse structural deformation and thermal effects is also suggested. This is the

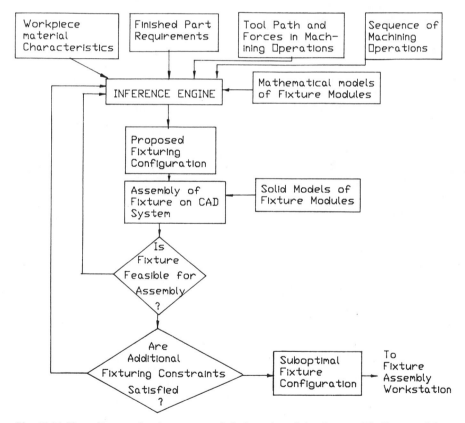

Fig. 14.24 Flow diagram for the automated design of modular fixtures [2]. (Reprinted by permission of the authors, MV Gandhi and BS Thompson, and the Society of Manufacturing Engineers. © 1986, *J Manuf Syst* vol. 5, no. 4.)

first report that proposes to include rigorous mathematical analysis in fixture design.

The authors postulated that the automated fixture design process comprises two distinct stages: mathematical modelling and task specification. The former involves the representation of real objects by their geometries and abstractions. Task specification represents the sequence of operations that need to be specified involving the relative spatial position of the various fixture elements.

Conclusions and Future Developments

Fixture design, originally a manual process demanding expert knowledge in tool engineering, ergonomics and process planning, is now capable of being automated with CAD and AI tools. This is possible through the extensive development in knowledge representation and database management systems linked to CAD modelling packages, and more recently with the use of the object-oriented programming approach.

As with automated process planning, the fixture design process can also be approached using the variant and generative techniques. This will necessitate the formulation of a fixture classification and coding system. In addition to the information provided in the general classification systems designed for workpieces, it is necessary to include in the fixture classification and coding system information on cutting tools, machine tools, positioning and dimensioning tolerances with respect to location and datum surfaces, method of securing the workpieces and their loading sequences, assembly sequences, coolant and chip considerations in the case of machining fixtures, etc. The workpiece coding scheme will become a subsystem of the fixture coding scheme and if the schemes are designed in modular forms, the workpiece scheme can be used independently for other purposes of retrieval and identification.

Figure 14.25 shows the proposed outline of the variant and generative fixture design schemes. Central to the system is a three-dimensional CAD model which provides detailed geometric information together with other textual information. A workpiece coding scheme will be used to retrieve from the workpiece datafile any similar designs and this information will be passed to the fixture coding scheme to retrieve related fixtures, if any. The variant technique will be used to edit the fixture and create new part lists and assembly drawings.

Failing to retrieve any similar designs, a feature extractor will be used to extract relevant geometric and the associated textual information to be passed to the expert generative fixture design system consisting of machining physics (formulae for evaluating cutting forces, stability, strength analysis, etc., using finite element tools) and expert heuristics

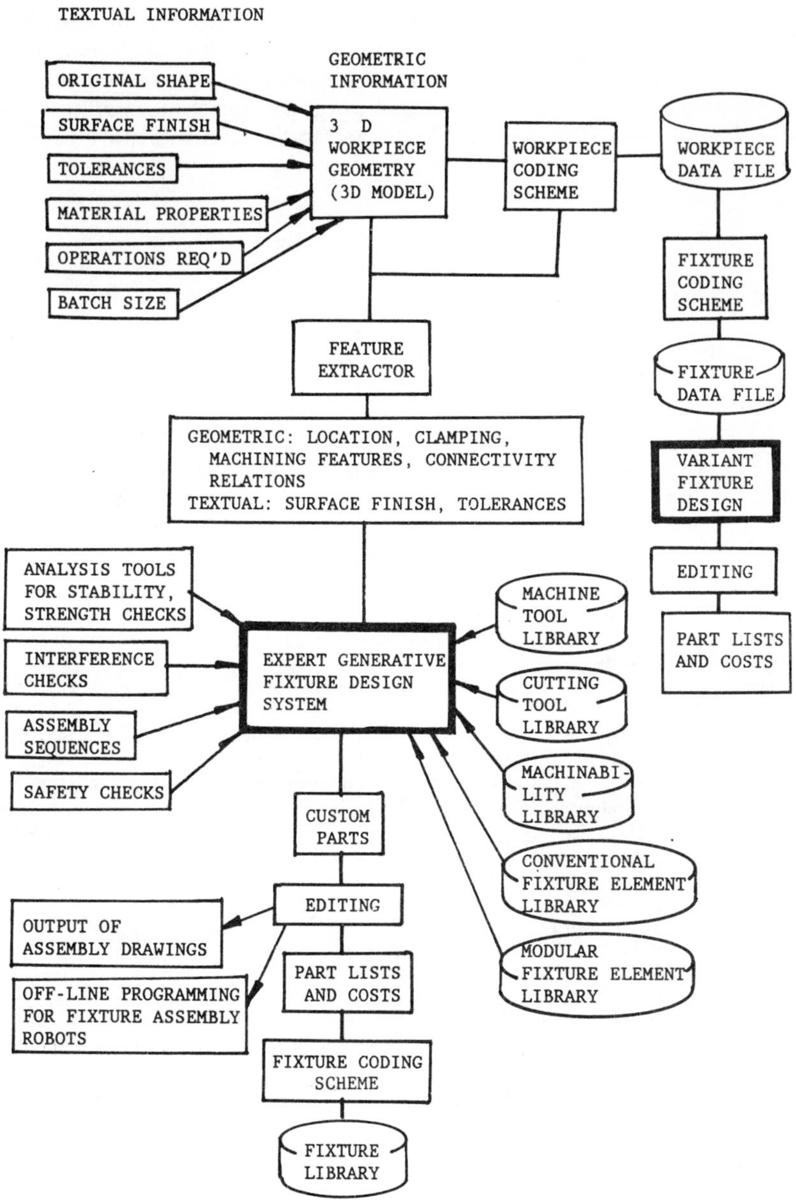

TEXTUAL INFORMATION

Fig. 14.25 Proposed outline of the variant and generative fixture design schemes.

(rule of thumb, good design proportions, ergonomics, etc.). Several libraries will be accessed which include machine tool library, cutting tool library, machinability library and libraries containing conventional and modular fixture elements. Depending on the batch size and other requirements, an economic comparison will be made to decide on the optimized selection of the components. Nonstandard and custom parts required will be advised. The final output consists of detailed parts and assembly drawings, off-line programming for robot-assisted assembly and the newly coded fixture will be added to the fixture library for future references.

References

1. Kochan D. CAM – developments in computer-integrated manufacturing. Springer-Verlag, Berlin, Heidelberg, New York, 1986
2. Gandhi MV, Thompson BS. Automated design of modular fixtures for flexible manufacturing systems. J Manuf Syst 1986; 5: 243–252
3. Gandhi, MV, Thompson BS. Phase-change fixturing for flexible manufacturing systems. J Manuf Syst 1985; 4: 29–39
4. Xu Y et al. A modular fixturing system (MFS) for flexible manufacturing. FMS Mag 1983; October: 292–296
5. Cutkosky MR, Kurokawa E, Wright PK. Programmable conformable clamps. AUTOFACT 4 conference proceedings, Society of Manufacturing Engineers, 1982, pp 11.51–11.58
6. Imhof G, Grahl W. Die Auswahl von Vorrichtungselementen als Bestandteil der Rechnergestutzten projektierenden Konstruktion von Vorrichtungen. Wiss Z Techn Hochsc Karl-Marx-Stadt 1977; 19: 49–61
7. Ingrand F, Latombe J. Functional reasoning for automatic fixture design. Lifia (IMAG) France, 1980
8. Eversheim W, Buchholz G, Knauf A. Rechnerunterstutzte Konstruktion von Baukasten-vorrichtungen. Indust Anz 1985; 107: 13–15
9. Markus A, Markusz Z, Farkas J, Filemon J. Fixture design using Prolog: an expert system. Robotics Comput Integr Manuf 1984; 1: 167–172
10. Miller AS, Hannam RG. Computer aided design using a knowledge base approach and its application to the design of jigs and fixtures. Proc I Mech E 1985; 199: no. B4
11. Nee AYC, Bhattacharya N, Poo AN. A knowledge-based CAD of jigs and fixtures. Technical paper TE85-902, Society of Manufacturing Engineers, 1985
12. Nee AYC, Bhattacharya N, Poo AN. Applying AI in jigs and fixtures design. Robotics Comput Integr Manuf 1987; 3: 195–201
13. Bhattacharya N. A knowledge-based computer aided design of jigs and fixtures. MEng thesis, National University of Singapore, 1986
14. Nee AYC, Bhattacharya N, Poo AN. Computer-aided datum determination using the least tolerance method. Technical paper MS87-905, Society of Manufacturing Engineers, 1987
15. Englert PJ, Wright PK. Applications of artificial intelligence and the design of fixtures for automated manufacturing. IEEE International conference on robotics and automation, 1986, pp 345–351
16. Hayes-Roth F, Lenat DB, Waterman DA (eds). Building expert systems. Addison-Wesley, Reading, MA, 1983
17. Lim BS, Knight JAG. Holdex – holding device expert system. Proceedings 1st international conference on applications of artificial intelligence in engineering problems, 1986, pp 483–493

## Chapter 15

Knowledge-Based Design of Jigs and Fixtures

D.T. Pham and A. de Sam Lazaro

Introduction

Jigs and fixtures are devices used in manufacturing processes such as machining, welding, bonding and assembly. They are employed to locate and hold the workpiece firmly in position and to ensure that it is in a state of stable equilibrium (geometric control) and that dimensional accuracy is maintained during the manufacturing operation (dimensional control). Jigs and fixtures are generally referred to as "fixturing systems" or simply "fixtures".

Fixturing systems have traditionally been designed by experienced designers using empirical rules and formulae. Several factors have motivated researchers to work towards the mechanization of the design process for fixturing systems. Among them are the decline in the number of experienced fixture designers and the long lead time for fixture design which is unacceptable in a flexible manufacturing environment. Progress in the fields of computer-aided design (CAD), modular tooling and knowledge-based (expert) systems has significantly facilitated development of automated fixture design programs by providing the requisite tools and environment for such development. Several systems with various degrees of automation have been developed in the last decade (see Chapter 14). Two such complementary prototype systems have been produced by the authors' group. The first is a knowledge-based, interactive design aid ("Jig and Fixture Designer's Assistant") and the second, a fully automated system ("AutoFix"). Both systems will be described in this chapter.

Jig and Fixture Designer's Assistant

Jig and Fixture Designer's Assistant [1] is implemented using XI Plus, a commercially available expert system shell. The system comprises four separate modules, one for fixture planning, and one each for designing locations, clamps and supports. The elements of the system are shown in Fig. 15.1.

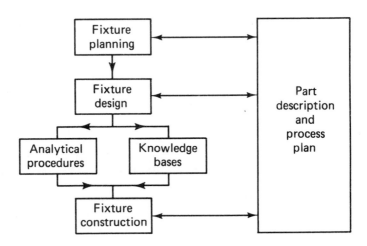

Fig. 15.1 Elements of Jig and Fixture Designer's Assistant.

Before the process of fixture design can commence, it is assumed that process planning has been completed and that the sequence of operations, tolerance, surface finish etc. have been decided. These data, together with the part drawings, are stored in files which interact with the fixture design modules. Information contained in these files is accessible and modifiable during the execution of the program.

Fixture Planning

In fixture planning, it is first decided whether any special fixturing is required for the component in a specific process. Many factors are considered. These are, for example, batch size, tolerance, types of machines, ancillary equipment and cost, the existence of facilities such as computer numerical controllers (CNC), integrated probing systems and palletized manufacturing cells. If special fixturing is needed, the fixturing configuration is determined. This is performed automatically based on the component specifications, and the machines and facilities available. The principles governing fixture planning are structured in a decision tree which forms the skeleton of the fixture planning knowledge base.

Location, Clamp and Support Design

In these design modules, the required locating, clamping and support positions are determined. The modules are run interactively, during which guidelines are provided for the user to implement the design. The principles of design are organized into three separate decision trees.

At the beginning of the location design module, efforts are made to place common locations throughout as many processes as possible. The clamping design module strives to use only one clamp to push the component against all the locators. There is less ambiguity when loading a single clamp fixture than one with multiple clamps.

A separate module is used to add supports. Fixed supports are more economical. However, if accuracy of machining is critical, adjustable supports are recommended. Supports are designed so as not to interfere with the positioning of the locators.

An abbreviated run of the system is given in the appendix. Jig and Fixture Designer's Assistant is an advisory program. Design decisions have to be taken by qualified designers. In AutoFix, described next, the requirement for a designer has been eliminated.

AutoFix

AutoFix is an automated fixture design system using the rule-based language OPS5 and the 'I-Deas' CAD package by Structural Dynamics Research Corporation (SDRC) USA. In the following sections, the main features of OPS5 and I-Deas will be highlighted first. Details of AutoFix will then be examined and finally an example of a typical design procedure utilizing AutoFix will be presented.

Salient Features of OPS5

The rule-based language, OPS5[2], uses data elements which are either vectors or objects with associated attribute-value pairs. For example, to represent the assertion that location of a workpiece in a fixture should be undertaken prior to clamping, OPS5 would use the vector:

(task-order geometric-location clamping)

Similarly, attributes are represented as statements with the distinguishing operator "^". There is no restriction on the number of attributes that may be assigned to a particular object and matching of one or more attributes alone may be required for identification of the object. For example,

(material ^steel 1018 ^hardened yes)

would represent hardened steel with ANSI No. 1018.

Rules would have the format: $\langle Ai_1 \rangle \langle Ai_2 \rangle \ldots \langle Ai_k \rangle \rightarrow \langle Ci_1 \rangle \langle Ci_2 \rangle \ldots \langle Ci_j \rangle \ldots \langle Ci_n \rangle$ where Ai_k is an antecedent of the rule and Ci_j is a consequent which would turn true, should all antecedents be matched with some data/knowledge from the knowledge base. A consequent can also be matched to other antecedents and thus trigger other rules or programs/routines. Typical actions which could be ordered by a consequent are MAKE (a new element in the working memory) REMOVE/MODIFY (an existing element) or WRITE (a line of text).

Variables are denoted by symbols enclosed in angle brackets, e.g. $\langle MAT \rangle$. They are said to be bound to a data element if they match.

The inference engine (interpreter) performs forward chaining, applying the following strategies in case more than one rule is satisfied:

1. Rules satisfied with the most recently created working memory elements are triggered first (recency strategy).
2. Rules which are more specific are triggered in preference to more general ones (specificity strategy).
3. Once a rule has been triggered it is prevented from looping on itself (refraction strategy).

One of the strengths of OPS5 is its generality, making it easy to tailor the design of a knowledge-based system to fit the characteristics of its domain. It also has an effective pattern recognition mechanism, a feature which at times averts lengthy database searches. The principal weaknesses reside in the lack of a front-end facility for expanding or modifying the database and for the introduction of explanation clauses. Also, most numerical computations have to be carried out by external scientific languages like Fortran and Pascal. Communication with other programs can be achieved either by conversion of symbolic atoms, which are OPS5 entities, into appropriate types or by using buffer data files. In AutoFix, the latter option was adopted. The program can be halted at any stage and the contents of the buffer files examined. This provides the user with a way to monitor data and to check the validity of parameters during the execution of a program.

I-Deas CAD Package

I-Deas is an interactive CAD package which has solid modelling and engineering analysis facilities. Artifacts or objects can be generated either in two or three dimensions, in wire frame or solid models. For AutoFix, the three-dimensional solid model option was selected as it enabled the system to detect interference between components during assembly. Objects are described interactively and their attributes stored in "keystroke" files, which are records of the operator's inputs from the keyboard. These files are later "played back" to generate the object. It is possible to run them in batch mode with all dimensions and attributes assigned to variables, the values of which are governed by a control program. Figure 15.2 is an example of one such keystroke file for a cylindrical primitive. Several such files may be run in tandem to produce a workpiece or a fixture. Shapes, both fixture elements and workpiece primitives, are stored as shells in a database. Dimensions, as decided by AutoFix, are then supplied by a batch control file.

A finite element analysis program is also available with the CAD package. The solid object model is prepared for engineering analysis in the preprocessing stage. Here the finite element mesh consisting of elements and nodes is generated manually by the operator. The load set (consisting of external force vectors) and the restraining set (containing the reactive forces at the support points) are combined with the set of elements and nodes to form a "case set". The analysis of the case set is done in another package, model solution, and the results – displacement, stress, strain etc. – are sent to the postprocessing package. The output (deformed geometry, stress/strain curves etc.) is presented graphically and also stored in files. The latter capability has been utilized by AutoFix to automate the process of finite element analysis, as will be explained later.

Because I-Deas has been designed to be run interactively and is highly compartmentalized, communication is necessary to permit the process of design to advance automatically from one stage to another. A communication interface was written to enable the system to execute commands and transfer data between modules.

Details of AutoFix

Based on the program structure, AutoFix can be divided into four sections: the input stage, the knowledge base, satellite programs and the output stage. A block diagram of the system is given in Fig. 15.3.

Input stage

The input to the system is served by two distinct programs. The part geometry is extracted from the CAD system and converted into information

```
K : /
K : mf
K : pr
K : #color=4        ; 'K :' indicates that the command
K : #n=1            ; on the line indicated is a keystroke.
K : #d1=20          ; The '#' sign is to identify variables
K : #d2=0           ; and their values.
K : #d3=10
K : #xcoord=10
K : #ycoord=-0.5
K : #zcoord=5
K : #orientx=0
K : #orienty=0
K : #orientz=0
K : r               ; run
K : cylinder        ; the file name 'cylinder'
K : /               ;
K : c               ;
K : sp              ; This section is a generic
K : d1              ; description of a file to
K : 16              ; create a cylinder, accessing
K : color           ; values from those declared above.
K : or
K : tr
K : 0 d2 0
K : "/"
K : "STO"
K : 550
K : :
K : c
K : o
K : 0 d2 0
K : 0 0 0
K : d1
K : no
K :
K : "B"
K : j
K :
K : 550
K : or
K : r
K :
K : orientx, orienty, orientz
K : tr
K : xcoord, ycoord, zcoord
K : /
K : ma
K : de
K : 550
K : y
```

Fig. 15.2 Keystroke file for a cylindrical primitive.

which is readable by the knowledge-based system. This portion of the
input stage also determines the overall dimensions of the workpiece, the
thickest and thinnest points and the availability of flat surfaces. The
second program ascertains from the operator information which cannot
be extracted from the CAD package. Data such as batch size, workpiece
finish, machine configuration, type(s) of operation and so on are facts
known only to the operator. This information is global in nature and
would be required for any type of machining operation. Additional

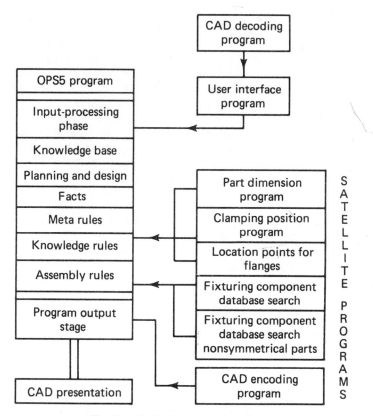

Fig. 15.3 Block diagram of AutoFix.

information would be called for as the knowledge-based system solves specific design problems. Questions such as "Do you propose to carry out gang drilling?" or "Are multiple milling cutters being used for this operation?" are asked in order to define the particular problem more clearly. Very rudimentary knowledge is expected of the operator and the input stage is mainly menu-driven.

The Knowledge Base

The most important section of the program is the knowledge base. It consists of facts, meta-rules, knowledge rules for design and assembly and organizational rules. In addition there are rules introduced mainly for communication and file handling. Facts are either built into the system or are acquired from the operator during the input phase. This information does not change and may be used anywhere in the program.

A built-in fact, for example, is that fixtures can be classified into fixed, shuttle, swivel, tilt, indexing and universal. This would appear in the program as: (fixture ˆtype ⟨⟨ fixed shuttle swivel tilt indexing universal⟩⟩) where "fixture" would represent the class of object and "type", its attribute, the value of which is one of the "atoms" from within the double angle brackets. Facts ascertained from the operator would generally be process and machine data and these would be "bound" to the attributes of certain objects. They would generally have only one value.

Meta-rules are production rules governing the groups of rules which will be used under a given set of conditions. In practice, they generate working memory elements which trigger groups of rules required for a particular configuration of facts. If, for example, the size of the workpiece is small, the batch size very large and a CNC machine with an indexable table is available (all facts ascertained during the input phase), then the rules for multiple fixtures on a cube should be examined. This would appear as the production rule shown in Fig. 15.4.

```
(p cube_fixture::
(workpiece ˆsize small)
(production_level ˆvalue very_large)
(machine ˆtype cnc ˆindexable yes)

-->

(make cube-fitted_fixture))
```

Fig. 15.4 A meta-rule.

The working memory element "cube_fitted_fixture" will be used to trigger another group of rules from among the knowledge rules.

The bulk of the rules in the knowledge-based system are knowledge rules. Three rules, with comments, are shown in Fig. 15.5. It will be noticed that all these rules require the presence of the element "cube_fitted_fixture" in the working memory to trigger them. Depending on various other conditions, such as type of material, size and shape of workpiece, type of chips being produced and so on, the final configuration of the fixture is worked out. These rules, and those used to derive information necessary to arrive at a decision/fixture configuration, form the bulk of the knowledge rules for design. Another group of knowledge rules are used for assembly. These generally concentrate on factors such as optimum clamping positions, type and size of baseplate, clamps and other structures, and sequence of assembly of components. If components

```
;The next rule caters for a small rectangular workpiece
;which could be accommodated four on a cube face.

        (p cube-fitted_fixture1::
        {(cube-fitted_fixture) <check>}
        (workpiece ^shape rectangular ^length < 100 )
        (chips ^type discontinuous)(process ^type milling)
             (machine ^cutters single)    etc...
        :            -->
        (make part ^partname location_pin
               ^dimension (compute 0.1 * <d3>)
           (make part ^partname cube_for_CNC_workholding
               ^dimension (compute 0.85 * <table_width>)
           (remove <check>))

;The following rule is for a cube-mounted cylindrical .
;workpiece. Several such workpieces may be mounted on a
;single cube.

        (p cube-fitted_fixture2::
           {(cube-fitted_fixture) <check>}
           (workpiece ^shape cylindrical ^length < 100)
           (machine ^cutter single) etc....
           -->
           (make part ^partname three_jaw_chuck
                ^dimension (compute 2.2 * <d3> )
                ^x <x   ^y <y>  ^z <z> )
              (make part ^partname cube_for_CNC_workholding
                ^dimension (compute 0.85 * <table_width>)
              (remove <check>)

; In the next rule a mounted fixture is suggested with a
;four-jaw chuck since continuous chips are expected.

        (p cube-fitted_fixture3::
        {(cube-fitted_fixture) <check>}
        (workpiece ^shape rectangular ^length < 100 }
        (chips ^type continuous)(process ^type milling)
        (machine ^cutter multiple) .... etc
             -->
        (make part ^partname four-jaw_chuck
               ^dimension (compute 0.65 <d3>
               ^x <x>    ^y <y> ^z <z> )
           (make part ^partname cube_for_CNC_workholding
               ^dimension (compute 0.85 * <table_width> )
           (remove <check> ))

;Note:The variables x,y and z are the coordinates of the
;centre of the device and each chuck will have one set of
;such coordinates.
```

Fig. 15.5 Examples of knowledge rules.

need to be designed specifically for the workpiece and operation under
consideration, then this is achieved by a set of design rules. The
positioning of the components is implemented using the assembly rules.

Organizational or control rules are necessary to guide the course of the
program according to design requirements. These rules ensure that all
input parameters are entered first, all input and output files are open at
the appropriate times, data are made available for design and assembly
is started after the entire design process has been completed. An
illustration of how organizational rules start and partition the program
is given in Fig. 15.6.

```
(startup
(watch 2) ;Sets the level of information reported back.
(disable halt) ;Enables use of a command file.
(openfile infile1 opsin.txt in) ; Opens input file.
(make start1); Makes the first working memory element
                    ; which starts the first section of the
                    ; program.
(make input ^status active); activates the input phase.
    (run)) ;

;There now follows a series of productions for input of
;information and computing the part geometry.
                    :
                    :
                    :
;The next rule activates the design stage of the program.


    (p checkpoint1::
        (input ^status terminated) - (stop1)
        -->
        (make design ^status active) (make stop1))
```

Fig. 15.6 Organizational rules.

Satellite Programs

As mentioned earlier, symbolic languages like OPS5 are not designed to carry out computation of any complexity greater than simple arithmetic. In order to solve more complex mathematical problems, it is necessary to use a high level language. In this case Fortran has been adopted. The external programs developed with this system have the features of being autonomous and at the same time are able to communicate with the main program. This gives them the status of satellite programs. Nine such satellites communicate with the main program via buffer data files. This serves two useful purposes:

1. It enables the user to examine the data fed into the system in order to ascertain validity of the information. It is also possible to halt the progress of the design at any stage and to examine, and perhaps alter, certain parameters to suit a particular design.
2. Alterations can be made to the satellite programs without changes to the main program.

This structure makes parallel processing possible, thereby potentially reducing the time for design. At present the system runs on a single processor but with simple modifications it is capable of being executed in a multi-processing environment.

The Output Stage

The output stage is the final stage in the process of design – namely the presentation of the results. The fixture components selected by the system are now brought up on the screen of the CAD workstation using a driver program. This program creates a keystroke file to call various units from the CAD database, dimension them and position them relative to the global system of axes. It uses the output information from the OPS5 program to create this file and format it as shown in Fig. 15.2. Since the amount of data that could be output from the OPS5 program would vary from one unit (in case of a chuck) to over 15 (in a more complicated fixture), the output program had to be made as flexible as possible. Also the nature of the data varies, with some components having only two or three parameters to define their geometry and others requiring ten. The output program allows for up to 15 parameters and automatically regulates the size of each block of data depending on the number of parameters.

Execution of AutoFix

The workpiece blank which is to be fixtured is drawn on the CAD package in terms of primitives. The prototype of AutoFix permits up to six primitives, but this number could be extended. The operator then types in a single command (namely, "fix") and then answers a series of questions from a menu. These relate to characteristics of the machine and workpiece and other information which is not available from I-Deas. AutoFix proceeds to design the fixture with no assistance from the operator. In the course of program execution, certain other questions may arise and these generally require yes/no answers. There are four stages in the design process – planning, preliminary design, final design and construction/assembly. Over 500 design rules are examined (coded in approximately 1750 productions). AutoFix accesses a database of modular components and attempts to design the fixture using the components from the database. If no appropriate component is available, the system will design one to suit the workpiece. Once the fixture has been designed, AutoFix generates a batch file which contains the parameters for the workpiece and the fixture. This is run on I-Deas and the design is presented on the CAD workstation screen.

Finite Element Analysis

In the course of the design, AutoFix generally positions one or two supports near the centroid of the workpiece. This is to reduce deflections due to clamping and cutting forces. In the case of thin workpieces or if the cutting forces are high, these deflections could be appreciable, leading to deviations from dimensions or even damage to the component. In either case, this would translate into rejected parts. AutoFix, therefore, first carries out a finite element analysis to determine the point of maximum deflection. The results of this analysis (deflections at each node) are stored in a file as mentioned earlier. This file is then examined by AutoFix and the node at which the maximum deflection occurs is ascertained. Finally, the coordinates of this node are obtained from I-Deas and AutoFix repositions the support at this point. This process has been almost completely automated; only mesh generation and force positioning remain to be carried out by the operator.

Conclusion

A well-engineered fixture requires a great deal of skill and effort to design. Until recently, that task has had to be performed laboriously by experienced designers. The advent of expert system technology has enabled their knowledge to be encapsulated in computer programs and the design process to be mechanized. The wide-ranging degrees of mechanization have been demonstrated in the two expert systems presented here. The first system, based on a low cost shell, acts only as a guide to jig and fixture design. Its use enables the design task to be deskilled to a large extent. The second, employing much more computing power and an artificial intelligence language, completely removes the necessity for skilled intervention from the operator.

Two examples of fixtures generated by AutoFix for cylindrical and rectangular workpieces are shown in Figs 15.7 and 15.8.

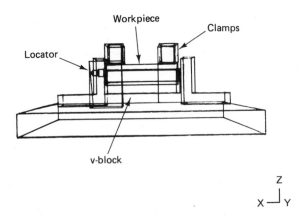

Fig. 15.7 Example of a fixture for a cylindrical workpiece.

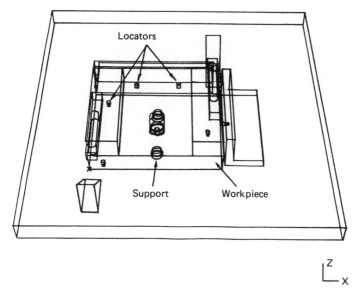

Fig. 15.8 An example of a fixture for a workpiece consisting of four primitives.

Appendix. Sample Run of Jig and Fixture Designer's Assistant

Note: responses selected by the user are shown in **bold** type.

FIXTURE PLANNING

1. The number of parts to be manufactured = ?
   ⟩ **5000**
2. The tolerance (in mm.) specified for the operation = ?
   ⟩ **0.01**
3. Please confirm availability of:
   indexing table
   palletized system
   integrated probing system
   none
4. Is probing an acceptable time loss:
   acceptable
   not acceptable
   (Help: Probing may be acceptable when the batch size is small as it is not economically viable to build a special fixture.)
5. Please indicate the type of operation:

turning
milling
drilling
6. Please indicate whether the machine is:
manual
automatic
numerically controlled

REPORT: THE FIXTURE RECOMMENDED IS A SINGLE PART HOLDER WITH AUTO CLAMP.

LOCATION SYSTEM DESIGN

1. Process Sequence Number = ?
   ⟩ **1**
2. It is always advisable to use common locations when performing multiple operations. Are there any common location points on the workpiece?
   yes
   no
3. By surveying the process plan, check whether it is possible to introduce common location points for all the processes.
   possible
   not possible
4. Have the locations already been specified in the part drawing?
   already specified
   not specified
5. The component is:
   prismatic
   rotational
6. The datum is:
   flat
   axis
7. Do you need assistance with determining the optimum placement of locators? Note that if you do the clamping positions would also be suggested to you.
   yes
   no
Place *THREE* locators on the datum plane such that:
(a) The position of the locators should be as far apart as possible.
(b) If required, the locators may be replaced by a planar locator, in which case the locating area should be kept as small as possible.
(c) Rest locators should embrace the centre of gravity.
8. Number of rest locators still needed = ?
   ⟩ **0**
9. Choose one of the following types for rest locators:
   plane type
   button/pin type

(Help: Three button/pin locators on a plane can be replaced by a planar locator. If the latter is used, the locating area should be kept as small as possible. Pin locators are recommended on rough surfaces.)

10. Choose one of the following types for side locators:
       circular type
       button/pin type
11. Choose one of the following types for the stop locator:
       circular type
       button/pin type

Place *TWO* side locators on a surface normal to the surface of the rest locators such that:

(a)   The locators should be as far apart as possible
(b)   The locators are preferably on a machined surface.

Place *ONE* stop locator on a surface normal to the other two locating surfaces, preferably one that has been previously machined.

CLAMPING SYSTEM DESIGN

1. Deflection could be minimized if the locators are positioned directly opposite to the direction of the tool forces. Has this been achieved in this case?:
       yes
       no
2. Does the tool force push the component against the locators? (Lighter clamps may be employed if this is so.):
       yes
       no
3. Identify the clamping surface(s). Check whether one clamp is enough:
       enough
       not enough
   (Help: (a) Clamping surfaces should be as close to the tool force as possible.
          (b) Clamping forces should not directly oppose tool forces.
          (c) It is preferable to have only one clamping force, holding the component firmly against all locators.)
4. Check whether the tool force can substitute for any clamping force. The tool force can:
       substitute for clamping forces
       not substitute for clamping forces
   (Help: In order to keep the number of clamps to a minimum, the tool forces in some light work, may be utilized to retain the workpiece against the locators.)
5. Check whether two clamps are enough:
       enough
       not enough
6. Critical finished or semi-finished surfaces may be damaged by the clamps. Check whether the surfaces selected are:
       critical
       not critical

REPORT: THE CLAMPING SYSTEM INCLUDES THREE OR MORE CLAMPING POSITIONS.

SUPPORT SYSTEM DESIGN

1. Is the deflection of the workpiece under its own weight:
       considerable
       negligible
   (Help: When a workpiece is heavy and the rest locators are positioned far apart, then it may deflect under its own weight leading to machining inaccuracies and even distortion.)
2. Is the deflection of the workpiece under the clamping force:
       considerable
       negligible
3. Is the deflection of the workpiece under the tool force:
       considerable
       negligible

REPORT: NO SUPPORT IS REQUIRED.

Acknowledgements. The authors would like to thank the Minnesota Supercomputer Institute for financing part of this work. The Designer's Assistant was developed in collaboration with their colleague, Dr M.J. Nategh, whose help they are pleased to acknowledge.

References

1. Pham DT, Nategh MJ, de Sam Lazaro A. Jig and fixture designer's assistant. Int J Adv Manuf Technol 1989; 4: 26–45
2. Brownston L, Farrel R, Kant E, Martin N. Programming expert systems in OPS5: an introduction to rule-based programming. Addison-Wesley, Reading MA, 1986

Section E

Systems Design

This section contains four chapters. The first chapter, by Li et al., describes part of an intelligent software package for designing hydraulic systems. The second chapter, by McGuire and Wee, discusses the design of control systems and the use of knowledge-base technology in a tool for generating control programs for programmable logic controllers. The third chapter, by Kusiak and Heragu, presents a knowledge-based system for selecting equipment for a manufacturing plant. The system, which also peforms optimization, is to be used in the design of the plant. The fourth and final chapter, by Parthasarathy and Kim, considers fundamental issues in designing intelligent manufacturing systems and discusses the representation and utilization of design knowledge by formal decision rules and performance measures.

Chapter 16

An Expert System for Designing Hydraulic Schemes

Li Congxin, Huang Shuhuai and Wang Yungan

Introduction

At present there are few well-considered systematic theories and determinate methods for designing hydraulics. Engineers in various fields mainly rely on their own practical experience and use the manual method of design. The quality of designed projects largely depends on the ability of designers. However, experts are limited in number, so it is necessary to research the design theory and practice of hydraulics, sum up the experience of experts and use the computer to simulate experts' design procedures.

Applying graphics, informatics and the techniques of decomposition and synthesis of information, the authors have carried out research on the theory and method of designing hydraulics. They have proposed computer-aided design (CAD) integration of hydraulic schemes and a new method, the counter-march method, applicable to computer design of hydraulics. Based on the above, an expert system, EDH, has been built and has given satisfactory results in practical application of designing hydraulic schemes. In addition, the authors have advanced a semantic tree representation of knowledge and a corresponding control strategy.

Research on Theory and Method of Designing Hydraulics

The combination method of basic circuits is used in traditional design of hydraulics. It is an approximate method which only considers the overall picture between the design requirements and basic circuits. The chosen circuits may interact with one another and some hydraulic components may be superfluous for complicated hydraulic systems [1, 2].

Carefully analysing the design procedure of complicated hydraulics and basic circuits, the authors have obtained the following conclusions:

1. The pressure at and flow rate through actuator ports are directly important to the choice of hydraulic components.
2. Generally, users' requirements cannot be directly applied to the design of hydraulics and some of them need to be decomposed into information regarding actuator ports.
3. Experts adopt the techniques of information synthesis in designing hydraulics.

So it is necessary to analyse basic circuits and collect the experience to choose hydraulic components.

For example, in Fig. 16.1, the ports of actuator C1 are A and B. If the pressure of port A needs to be maintained for a long time at a process, pilot check valve V1 ought to be chosen and connected with port A in all processes to prevent leaking of the spool valve that drives C1. Pilot valve V1 is controlled for closing or opening by the state of port V13.

Thus the information about port A is further decomposed and passed to port V11 and V13 of pilot valve V1 [3].

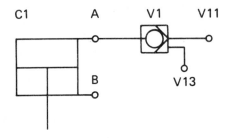

Fig. 16.1

Graphs are used for describing the connection between hydraulic components in various processes. Connections in the processes of downfall, press forward, maintaining pressure and return are shown in Fig. 16.2 (a–d) where vertices indicate the ports of hydraulic components and edges denote the components themselves (C1 is the cylinder; V1 is the pilot check valve; V3 is a sequence valve; edges marked by a number are pipes).

If the process number is n, the connection G in all processes is expressed by:

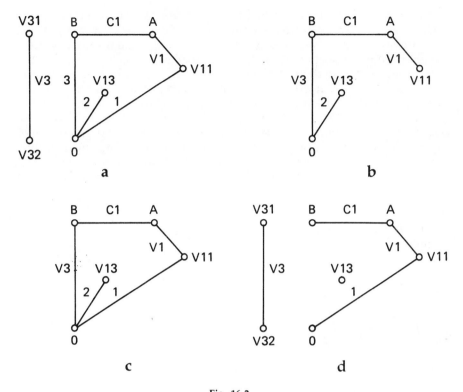

Fig. 16.2

$$G = G1 \cup G2 \cup \ldots \cup Gn \tag{1}$$

$$Gi = (Vi, Ei) \qquad (i = 1, 2, \ldots, n) \tag{2}$$

where Vi and Ei are the sets of vertices and edges at process i; Gi is the connection graph during process i. Gi is the union of many subgraphs consequently. For convenience the subgraph containing actuator ports is defined as the main subgraph. For Fig. 16.2, Vi and Ei are expressed as follows [4]:

$$
\begin{aligned}
V1 &= (A,V11,V13,B,0,V31,V32)\\
V2 &= (A,V11,V13,B,0)\\
V3 &= (A,V11,V13,B,0)\\
V4 &= (A,V11,V13,B,0,V31,V32)\\
E1 &= \{(A,B),(A,V11),(V11,0),(V13,0),(B,0),(V31,V32)\}\\
E2 &= \{(A,B),(A,V11),(V13,0),(B,0)\}\\
E3 &= \{(A,B),(A,V11),(V11,0),(V13,0),(B,0)\}\\
E4 &= \{(A,B),(A,V11),(V11,0),(V31,V32)\}
\end{aligned}
\tag{3}
$$

The design of hydraulic schemes can be treated as an operation to produce new edges and vertices continuously from the ports of actuators. The pressure and flow valves being determined, a group of the graphs describing the connection of components in all processes can be obtained.

In order to choose the hydraulic source, it is necessary to consider the nodes which are last produced in main subgraphs and ought to be connected with hydraulic source. Only port $V11$ in Fig. 16.2(b), port B and port $V13$ in Fig. 16.2(d) need to be connected with the hydraulic source. For this simple example, fixed displacement pump $P1$ and relief valve $V4$ may be chosen as the hydraulic source. Fig. 16.2 is thus developed into Fig. 16.3 where $V41$, $V42$ and $V43$ are the inlet, outlet and control ports of relief valve $V4$; $P11$ and $P12$ are inlet and outlet of pump $P1$.

In nonmain subgraphs, the vertices may be:

1. Connected with a tank
2. Stopped up

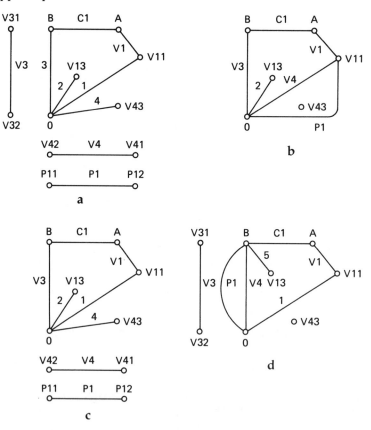

Fig. 16.3

3. Connected with the hydraulic source

4. Connected with a vertex in main subgraphs

The most appropriate treatment should be decided according to the connection of the vertices in other main-subgraphs and the character of hydraulic components which the vertices belong to. For instance, Fig. 16 3(a) can be changed into Fig. 16.4 according to the design experience: "If the maximum pressure of a cylinder port is less than the pre-regulated pressure of the sequence valve connected with the port during a process, then the sequence valve can be connected between the port and tank during the process." By doing this, not only the number of selectors will be reduced in the next logic design, but also the pressure surge caused by the transition of selectors may be reduced.

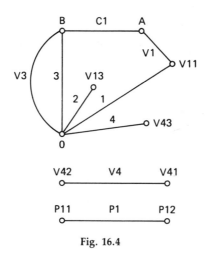

Fig. 16.4

In fact, some vertices in nonmain subgraphs may have several connections. Vertex $V11$ in Fig. 16.2(c) would be connected with either the tank or pump. Therefore, it is important to apply design experience, since using the method of permutation and combination to find good connections will result in combinatorial explosion.

The method above is called the counter-march method as the direction of producing vertices in main subgraphs is opposite to the direction of fluid flow. The method is most suitable for computerized design of hydraulics, because it is difficult to manually deal with all the vertices of complicated hydraulics.

CAD Integration of Hydraulic Schemes

There are currently a few software packages used for building mathematical models, simulation, calculation, and plotting in CAD of hydraulics [5–8]. However, using them is difficult for people who are unfamiliar with hydraulics and computers. The authors propose the CAD integration of hydraulic schemes as shown in Fig. 16.5. The integration is such a large task that it has to be divided into four main parts:

1. Design of hydraulic circuits
2. Logic design of hydraulics
3. Drawing hydraulic schemes
4. Simulation of hydraulics

This chapter puts emphasis on the first part used for choosing working pressure, parameters of actuators, pressure and flow control components and hydraulic sources.

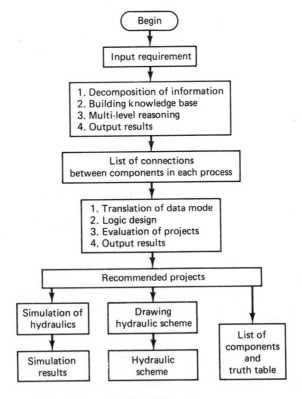

Fig. 16.5

After integration, users only need to input the requirements of hydraulics and make a few choices. They can then get corresponding hydraulic schemes, lists of components, truth tables and the results of simulation.

The Expert System for Designing Hydraulic Schemes

Because the mathematical model for designing hydraulics could not be found and there are a great deal of nonnumerical experience and laws in design of hydraulic circuits, expert system techniques are used. Figure 16.6 is a schematic diagram of the expert system for designing hydraulic schemes [9].

In Fig. 16.6 the knowledge base consists of a rule base and a fact base. The contents of the fact base come from users' requirements and nonexperiential information in the design knowledge of experts, while

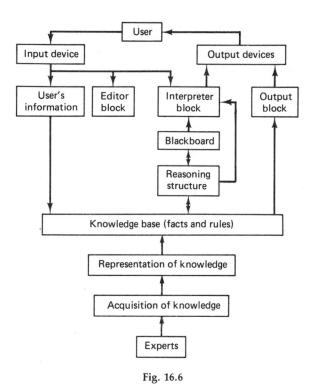

Fig. 16.6

the contents of the rule base come from the design experience of experts.

Facts in the fact base are easily acquired and represented with simple predicate expressions. For example, the mechanical efficiency of actuators and the elastic modulus of steel can be expressed as:

(mechanical_efficiency 0.9)
(e 2.1E11)

The facts "actuator 1 is a cylinder" and "the velocity of actuator 1 at process 2 is 0.1 m/s" can be expressed as:

(mode (actuator 1) cylinder)
(velocity 2 (actuator 1) .1)

However, the representation of design experience is more complicated. There is both procedural and nonprocedural knowledge in design experience. For this reason, the authors propose a synthetic representation of semantic tree and frame based on predicate calculus.

Predicate calculus is the earliest representation of knowledge in artificial intelligence. A predicate expression represents a relation in the domain of definition. It consists of predicates and arguments. In the facts above "mode" and "velocity" are predicate symbols and "actuator", "cylinder" and number are arguments.

The rule "If the pressure of the port of an actuator is greater than or equal to zero and the flow rate is less than zero (i.e. the flow is out from the port), then the port ought to be connected with a tank" cannot be directly represented by the simple predicate expression above and has to be first expressed as the complex expression

$$(\exists \ PORT)(\forall \ PRESSURE) \ (\forall \ FLOW)[[(GE \ PRESSURE \ 0)\wedge$$
$$(LT \ FLOW \ 0)]$$
$$\rightarrow(connect \ PORT \ tank)] \tag{4}$$

Then using de Morgan, equivalent, distributive, negative laws and the following steps:

1. Temporarily remove the implication symbol.
2. Reduce the jurisdiction of negative symbol.
3. Cut out the existential quantifiers.
4. Move all universal quantifiers in front of the expressions and then remove them.
5. Restore the implication symbol and transmit it into the set of linear expressions.

Expression (4) can be transformed as follows:

$$\sim(\exists \ PORT)(\forall \ PRESSURE)(\forall \ FLOW)[[(GE \ PRESSURE \ 0)\wedge(LT$$
$$FLOW \ 0)]\vee(connect \ PORT \ tank)]$$
$$(\forall \ PORT)(\exists \ PRESSURE)(\exists \ FLOW)[[(GE \ PRESSURE \ 0)(LT \ FLOW$$
$$0)]\vee(connect \ PORT \ tank)]$$

(∀ PORT)[~[(GE (PRESSURE-P PORT) 0)∧(LT (FLOW-P PORT) 0)]∨(connect PORT tank)]

~[(GE (PRESSURE-P PORT) 0)∧(LT (FLOW-P PORT) 0)]∨(connect PORT tank)]

(GE (PRESSURE-P PORT) 0)∧(LT (FLOW-P PORT) 0)]→(connect PORT tank)]

where "GE" and "LT" are the essential functions in Lisp; "PORT" is a variable; "PRESSURE-P" and "FLOW-P" are Skolem functions, their return values are the pressure and flow rate of the PORT.

In the set of linear expressions the Skolem functions are introduced, which may result in a large number of new Skolem functions and greatly increase the amount of programming [10]. Besides, the readability of rules will worsen.

Because in a sentence there is only a predicate expressing the main idea of the sentence, the authors employ a tree to express a sentence, i.e. to take the predicate as a root of the tree and define the predicate as a centre predicate. Thus the sentence "teacher gave students the books that his brother wrote in 1987" can be expressed as a semantic tree shown in Fig. 16.7 where predicate "gave" is the root of the semantic tree; constants "teacher", "students" and "in_1987" are leaves; variable "BOOKS" is introduced to link predicates "gave" and "wrote"; variable "BROTHER" is introduced to link predicates "wrote" and "of".

The method of producing a semantic tree is:

1. Change a compound sentence into several simple sentences and express them as predicate expressions.

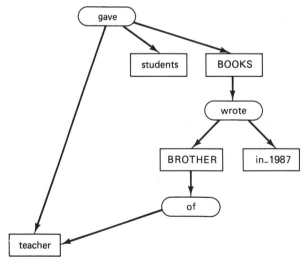

Fig. 16.7

2. Consider the centre predicate as the root.
3. If the term after the predicate is a constant, then regard the constant as a leaf.
4. If there is no term after the predicate, then take the predicate as a leaf.
5. If the term after the predicate is a variable which is only relative to the predicate, then also take the variable as a leaf.
6. If the term after the predicate is a variable which is also a term of other predicates, then treat the variable as a new node and repeat the procedure from 3.
7. If the term of the predicate is an expression, then treat it as a new node and repeat the procedure from 3.

So it is possible to express any complex events as corresponding trees, put them together and compose a large semantic tree. The root of the last produced tree is called the centre predicate of the tree. The centre predicate is a relative concept. A semantic tree is indeed a special semantic network, but it places more emphasis on the direction of production.

For example, the rule "If the pressure of the port of an actuator is greater than or equal to zero and the flow rate is less than zero, then the port ought to be connected with a tank" can be described as the semantic tree shown in Fig. 16.8 where "GE", "LT" and "connect" are the centre predicates of the events respectively; variables "ACTUATOR", "PROCESS" and "PORT" are common arguments of predicates "pressure", "flow" and "connect"; the centre predicate of the semantic tree is "rule".

Figure 16.8 expresses a rule which consists of three events. However, there are more rules in the knowledge base of an expert system, so the frame technique is used for managing and searching the rules [11]. A frame is a practical realization of a semantic tree, and Fig. 16.8 can be expressed as a frame:

```
(rule (if (pressure    PROCESS (actuator ACTUATOR PORT) P)
          (GE P 0)
          (flow         PROCESS (actuator ACTUATOR PORT) F)
          (LT F 0))
      (then (connect PROCESS (actuator ACTUATOR PORT) tank)))
```

where "rule" is the name of the frame; "if" and "then" are the names of slots; "pressure" and "GE" are the names of attributes; (actuator ACTUATOR PORT), "Process" and "0" are values; (actuator ACTUATOR PORT) further consists of the name of attribute "actuator" and values "ACTUATOR" and "PORT".

The rulebase can be further expressed as the following greater frame:

```
(rules (rules1 (rule1 (if ...) (then ...))
               (rule2 (if ...) (then ...))
               . . . . . .)
```

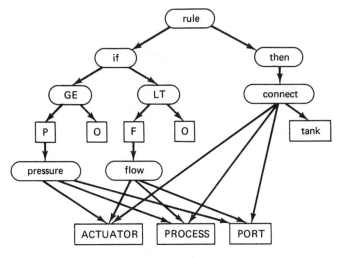

Fig. 16.8

(rules2 (rule1 (if ...) (then ...))
           . . . . . .)
       . . . . . .))

In the expert system for designing hydraulic schemes forward reasoning
is applied. However, traditional reasoning is so slow that it cannot be
used in practice, since the number of facts in the factbase increases as
reasoning goes on [12, 13]. Therefore, multi-level reasoning and data
streams are used for linking the static rulebase and factbase as shown
in Fig. 16.9 where dotted and full lines express data streams and control
streams respectively.

In Fig. 16.9 the reasoning machine does not directly act on the static
rulebase and factbase. The control structure selects a subrulebase from
the static rulebase in order, puts it in a dynamic rulebase and fires the
reasoning machine which only deals with the dynamic rulebase and
factbase. The control structure automatically collects the facts from the
static factbase which are necessary for reasoning of the subrulebase and
puts the facts into dynamic factbase. When the reasoning is terminated,
the control structure puts back the facts of the dynamic factbase into the
static factbase and repeats the above procedure.

Multi-level reasoning can greatly promote reasoning, conveniently
realize the deep reasoning and the translation of reasoning levels with
the rules in subrulebase.

Cutting and backtracking techniques are used in reasoning. Cutting is
useful for interrupting unnecessary reasoning. When the user finds an
input error, backtracking can be used to stop reasoning immediately and
make the program run backwards to the error level.

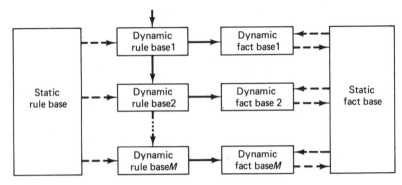

Fig. 16.9

In forward reasoning it is essential to test every rule. The test involves the matching between the premises of rules and facts of the dynamic factbase. When all the premises of a rule are satisfied, the conclusions of the rule are added to the dynamic factbase.

The blackboard technique is used for limiting the domain of variables. Before every rule is tested, the blackboard must be cleaned. If a variable is not on the blackboard, it is recorded on to it and defined as a value matching with the variable. If the variable has already been on the blackboard, then the variable and its value will be matched. After a rule has been tested, the variable on blackboard must be both erased and unbound.

In the reasoning structure there are the domain blackboard as mentioned above and control blackboards which are used in the dynamic rulebase and factbase.

A Practical Example

The expert system EDH can be run in an IBM-PC/XT or compatible computers. It is written in Macro Lisp. EDH consists of a reasoning block, control block, special function blocks and 18 subrulebases in which there are over 300 rules. Appendix 1 shows the rules of subrulebase R-VC for choosing the types of velocity regulation and their practical expressions in the computer (see also Fig. 16.10).

Using EDH, the design of several existing hydraulic systems has been tested. The hydraulic system of press YD32-315 is one of them. After reasoning with subrulebase R-INP, the user's requirements are automatically expressed as the following file:

```
INPUT_DATA:
(surge 1.5)
(load_min 7 (actuator 2) 120000.0)
(load_min 2 (actuator 1) 1.0E6)
(position 0 (actuator 1) 0)
(velocity 0 (actuator 1) 0)
(load 0 (actuator 1) 0)
(position 1 (actuator 1) .5)
(velocity 1 (actuator 1) .08)
(load 1 (actuator 1) 0)
(position 2 (actuator 1) .8)
(velocity 2 (actuator 1) .008)
(load 2 (actuator 1) −3150000)
(position 3 (actuator 1) .8)
(velocity 3 (actuator 1) 0)
(time 3 (actuator 1) 5)
(load 3 (actuator 1) −3150000)
(position 4 (actuator 1) 0)
(velocity 4 (actuator 1) −.042)
(load 4 (actuator 1) 600000)
(position 5 (actuator 1) 0)
(velocity 5 (actuator 1) 0)
(load 5 (actuator 1) 0)
(position 6 (actuator 1) 0)
(velocity 6 (actuator 1) 0)
(load 6 (actuator 1) 0)
(position 7 (actuator 1) 0)
(velocity 7 (actuator 1) 0)
(load 7 (actuator 1) 0)
(position 8 (actuator 1) 0)
(velocity 8 (actuator 1) 0)
(load 8 (actuator 1) 0)
(field metal_forming)
(support (actuator 2))
(support (actuator 1))
(fixed_type (actuator 2) 4)
(direction (actuator 2) 90)
(mass (actuator 2) 300)
(fixed_type (actuator 1) 4)
(direction (actuator 1) 270)
(mass (actuator 1) 3000)
(rod (actuator 2) single)
(rod (actuator 1) single)
(acting (actuator 2) double)
(acting (actuator 1) double)
(mode (actuator 2) cylinder)
(mode (actuator 1) cylinder)
```

(actuator 2)
(process 8)
(time 0 (actuator 2) 10)
(time 0 (actuator 1) 10)
(load 8 (actuator 2) 250000)
(velocity 8 (actuator 2) −.095)
(position 8 (actuator 2) 0)
(load 7 (actuator 2) −400000)
(velocity 7 (actuator 2) −.008)
(position 7 (actuator 2) .1)
(load 6 (actuator 2) 0)
(time 6 (actuator 2) 5)
(velocity 6 (actuator 2) 0)
(position 6 (actuator 2) .25)
(load 5 (actuator 2) −350000)
(velocity 5 (actuator 2) .065)
(position 5 (actuator 2) .25)
(load 4 (actuator 2) 0)
(velocity 4 (actuator 2) 0)
(position 4 (actuator 2) 0)
(load 3 (actuator 2) 0)
(velocity 3 (actuator 2) 0)
(position 3 (actuator 2) 0)
(load 2 (actuator 2) 0)
(velocity 2 (actuator 2) 0)
(position 2 (actuator 2) 0)
(load 1 (actuator 2) 0)
(velocity 1 (actuator 2) 0)
(position 1 (actuator 2) 0)
(load 0 (actuator 2) 0)
(velocity 0 (actuator 2) 0)
(position 0 (actuator 2) 0)

The design results of EDH are the connection lists of hydraulic components in distinct processes and the parameters of chosen components as shown in Appendix 2. The inlet of a component is numbered 1, the outlet 2, and the control port 3. Predicate "connect*" denotes that the components are connected in all processes. For instance, (connect* (cartridge_value 1 1) pump 1 2)) denotes that port 1 of cartridge valve 1 is connected with port 2 of pump 1 in all processes; (connect 2 (shuttle_valve 1 1) tank) symbolizes that port 1 of shuttle valve 1 is connected with tank in process 2. (flow_of_valve (check_valve 1) 0.0000736939) expresses that the flow rate of check valve 1 is 0.0000736939 (m³/s) according to which standard the check valve will be chosen. Appendix 2 represents the connection of hydraulic components in eight distinct processes.

Based on the results in Appendix 2, the design shown in Fig. 16.10 is obtained by using the logic design program which is the second part of

CAD integration of the hydraulic schemes. The design of Fig. 16.10 is better than that shown in Fig. 16.11, which is practically adopted in Hanyang Forging Machine Factory at present. There are fewer components in Fig. 16.10 than in Fig. 16.11. EDH may thus be used for testing the rationality of existing hydraulics.

In designing, it is possible that two choices do not have obvious distinction. In Fig. 16.10 it is possible either to choose cartridge valve CPV3, check valve CH1 and relief valve RV3 or to replace them with a pilot check valve. For this reason a software switch is set in EDH. If users put it on "T", then EDH will consult users as in the above case. Otherwise, EDH does not ask users and makes selections in terms of the experience of experts.

Owing to the variety and complexity of hydraulics, the rules in EDH are still imperfect. By means of an editor block, users can add, change and delete the design rules according to the special knowledge of their domain.

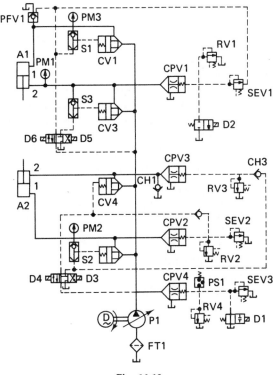

Fig. 16.10

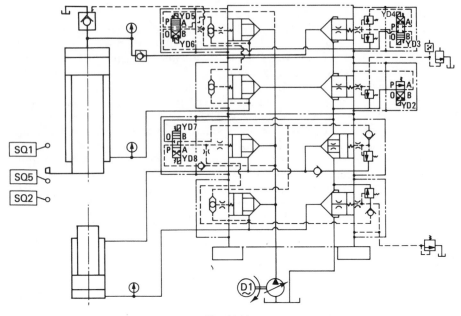

Fig. 16.11

Concluding Remarks

1. There is no well-considered systematic theory and determinate method of designing hydraulics at present. This chapter proposes a countermarch method applicable to computerised design of hydraulic schemes.
2. The CAD integration of hydraulic schemes is based on developing key software.
3. It is rational to use an expert system for designing hydraulic schemes. The authors have built the expert system EDH which has given satisfactory results in practical application.
4. EDH may be used for designing new hydraulics, testing the rationality of existing hydraulics and making suggestions for improvement.
5. When an expert system is built, it is necessary to use the appropriate representation of knowledge and control strategy. This chapter advances a semantic tree representation and corresponding control strategy.

Appendix 1

Rule 1

If variable displacement pump is used,
then reasoning is cut.

Rule 2

If an actuator is single acting cylinder, the actuator needs velocity regulation in a process, the power of the actuator in the process is less than or equal to 5000 watt,
then meter-in system is used for the actuator in the process.

Rule 3

If an actuator is single acting, the actuator needs velocity regulation in a process, the power of the actuator in the process is greater than 5000 watt,
then bleed-off system is used for the actuator in the process.

Rule 4

If an actuator is double acting, load varying of the actuator in a process is larger, the mass of actuator is greater,
then meter-in and meter-out systems are used for the actuator in the process.

Rule 5

If an actuator is double acting, load varying of the actuator in a process is larger, the mass of the actuator is smaller, the actuator is not supported,
then meter-in and meter-out systems are used for the actuator in the process.

Rule 6

If an actuator is double acting, load varying of the actuator in a process is larger, the mass of actuator is smaller, the actuator is supported, then meter-in system and back pressure are used for the actuator in the process.

Rule 7

If an actuator is double acting, load varying of the actuator in the process is larger, the power of the actuator in the process is less than or equal to 5000 watt, the direction of load of the actuator in the process is opposite to the direction of velocity,
then meter-in system is used for the actuator in the process.

Rule 8

If an actuator is double acting, load varying of the actuator in the process is larger, the power of the actuator in the process is greater than 5000 watt, the direction of load of the actuator in the process is opposite to the direction of velocity,
then bleed-off system is used for the actuator in the process.

Rule 9

If an actuator is double acting, load varying of the actuator in the process is larger, the power of the actuator in the process is greater than 5000 watt, the direction of load of the actuator in the process is the same as the direction of velocity,
then meter-out system is used for the actuator in the process.

```
R-VC :
(rule_vc1 (IF (pump 1 variable_displacement_pump))
          (THEN (Cut)))
(rule_vc2 IF (acting (actuator ACTUATOR) single)
             (velocity_min PROCESS (actuator ACTUATOR) VR)
             (power_actuator PROCESS (actuator ACTUATOR) P)
             (LE P 5000))
          (THEN (meter_in PROCESS (actuator ACTUATOR))))
(rule_vc3 IF (acting (actuator ACTUATOR) single)
             (velocity_min PROCESS (actuator ACTUATOR) VR)
             (power_actuator PROCESS (actuator ACTUATOR) P)
             (GT P 5000))
          (THEN (bleed_off PROCESS (actuator ACTUATOR))))
(rule_vc4 (IF (acting (actuator ACTUATOR) double)
              (load_varying PROCESS (actuator ACTUATOR) large)
              (mass (actuator ACTUATOR) M)
              (GT M 10000))
```

```
                         (THEN (meter_in PROCESS (actuator ACTUATOR))
                               (meter_out PROCESS (actuator ACTUATOR))))
        (rule_vc5 (IF (acting (actuator ACTUATOR) double)
                      (No (support (actuator ACTUATOR)))
                      (load_varying PROCESS (actuator ACTUATOR) large)
                      (mass (actuator ACTUATOR) M)
                      (LE M 10000))
                  (THEN (meter_in PROCESS (actuator ACTUATOR))
                        (back_pressure PROCESS (actuator ACTUATOR))))
        (rule_vc6 (IF (acting (actuator ACTUATOR) double)
                      (support (actuator ACTUATOR))
                      (load_varying PROCESS (actuator ACTUATOR) large)
                      (mass (actuator ACTUATOR) M)
                      (LE M 10000))
                  (THEN (meter_in PROCESS (actuator ACTUATOR))
                        (meter_out PROCESS (actuator ACTUATOR))))
        (rule_vc7 (IF (acting (actuator ACTUATOR) double)
                      (load_varying PROCESS (actuator ACTUATOR) small)
                      (power_actuator PROCESS (actuator ACTUATOR)
                      POWER)
                      (LE POWER 5000))
                      (velocity PROCESS (actuator ACTUATOR) V)
                      (load_mass PROCESS (actuator ACTUATOR) F)
                      (LE (* V F) 0))
                  (THEN (meter_in PROCESS (actuator ACTUATOR))))
        (rule_vc8 (IF (acting (actuator ACTUATOR) double)
                      (load_varying PROCESS (actuator ACTUATOR) small)
                      (power_actuator PROCESS (actuator ACTUATOR)
                      POWER)
                      (GT POER 5000))
                      (velocity PROCESS (actuator ACTUATOR) V)
                      (load_mass PROCESS (actuator ACTUATOR) F)
                      (LE (* V F) 0))
                  (THEN (bleed_off PROCESS (actuator ACTUATOR))))
        (rule_vc9 (IF (acting (actuator ACTUATOR) double)
                      (load_varying PROCESS (actuator ACTUATOR) small))
                      (velocity PROCESS (actuator ACTUATOR) V)
                      (load_mass PROCESS (actuator ACTUATOR) F)
                      (GT (* V F) 0))
                  (THEN (meter_out PROCESS (actuator ACTUATOR))))
```

Appendix 2

```
(connect* (cartridge_valve 1 1) (pump 1 2))
(connect* (cartridge_valve 3 1) (pump 1 2))
```

(connect* (cartridge_valve 2 1) (pump 1 2))
(connect* (cartridge_valve 4 1) (pump 1 2))
(connect* (check_valve 1 1) tank)
(connect* (actuator 2 2) (check_valve 1 2))
(connect* (cartridge_pressure_valve 4 1) (pump 1 2))
(connect* (cartridge_pressure_valve 4 2) tank)
(connect* (cartridge_pressure_valve 4 3) (sequence_valve 3 1))
(connect* (sequence_valve 3 2) tank)
(connect* (relief_valve 4 2) tank)
(connect* (cartridge_pressure_valve 4 3) (relief_valve 4 1))
(connect* (actuator 1 2) (pressure_meter 1))
(connect* (actuator 2 1) (pressure_meter 2))
(connect* (actuator 1 1) (pressure_meter 3))
(connect* (pump 1 1) (filter 1 2))
(connect* (filter 1 1) tank)
(connect* (actuator 1 1) (cartridge_valve 1 2))
(connect* (cartridge_valve 1 2) (shuttle_valve 1 2))
(connect* (cartridge_valve 1 3) (shuttle_valve 1 3))
(connect* (actuator 1 2) (cartridge_valve 3 2))
(connect* (cartridge_valve 3 2) (shuttle_valve 3 2))
(connect* (cartridge_valve 3 3) (shuttle_valve 3 3))
(connect* (actuator 2 1) (cartridge_valve 2 2))
(connect* (cartridge_valve 2 2) (shuttle_valve 2 2))
(connect* (cartridge_valve 2 3) (shuttle_valve 2 3))
(connect* (actuator 2 2) (cartridge_valve 4 2))
(connect* (actuator 1 1) (prefill_valve 1 2))
(connect* (prefill_valve 1 1) tank)
(connect* (prefill_valve 1 3) (shuttle_valve 1 1))
(connect* (relief_valve 2 2) tank)
(connect* (cartridge_pressure_valve 2 3) (relief_valve 2 1))
(connect* (cartridge_pressure_valve 2 2) tank)
(connect* (actuator 2 1) (cartridge_pressure_valve 2 1))
(connect* (relief_valve 3 2) tank)
(connect* (cartridge_pressure_valve 3 3) (relief_valve 3 1))
(connect* (cartridge_pressure_valve 3 2) tank)
(connect* (actuator 2 2) (cartridge_pressure_valve 3 1))
(connect* (relief_valve 1 2) tank)
(connect* (cartridge_pressure_valve 1 3) (relief_valve 1 1))
(connect* (cartridge_pressure_valve 1 2) tank)
(connect* (actuator 1 2) (cartridge_pressure valve 1 1))
(connect* (cartridge_pressure_valve 1 3) (sequence_valve 1 1))
(connect* (cartridge_pressure_valve 2 3) (sequence_valve 2 1))
(connect* (sequence_valve 2 2) tank)
(connect* (sequence_valve 1 2) (shuttle_valve 1 1))
(connect* (cartridge_pressure_valve 2 3) (check_valve 2 1))
(connect* (check_valve 2 2) (cartridge_valve 4 3))
(connect* (cartridge_pressure_valve 3 3) (check_valve 3 1))

(connect* (check_valve 3 2) (shuttle_valve 2 1))
(connect* (cartridge_pressure_valve 4 3) (pressure_switch 1))
(connect 0 (shuttle_valve 1 1) (pump 1 2))
(connect 0 (shuttle_valve 3 1) (pump 1 2))
(connect 0 (shuttle_valve 2 1) (pump 1 2))
(connect 0 (cartridge_valve 4 3) (pump 1 2)
(connect 0 (cartridge_pressure_valve 1 3) stop_up)
(connect 0 (cartridge_pressure_valve 4 3) tank)
(connect 1 (shuttle_valve 1 1) tank)
(connect 1 (shuttle_valve 3 1) (pump 1 2)
(connect 1 (shuttle_valve 2 1) (pump 1 2)
(connect 1 (cartridge_valve 4 3) (pump 1 2))
(connect 1 (cartridge_pressure_valve 1 3) tank)
(connect 1 (cartridge_pressure_valve 4 3) stop_up)
(connect 2 (shuttle_valve 1 1) tank)
(connect 2 (shuttle_valve 3 1) (pump 1 2))
(connect 2 (shuttle_valve 2 1) (pump 1 2))
(connect 2 (cartridge_valve 4 3) (pump 1 2))
(connect 2 (cartridge_pressure_valve 1 3) stop_up)
(connect 2 (cartridge_pressure_valve 4 3) stop_up)
(connect 3 (shuttle_valve 1 1) (pump 1 2))
(connect 3 (shuttle_valve 3 1) (pump 1 2))
(connect 3 (shuttle_valve 2 1) (pump 1 2))
(connect 3 (cartridge_valve 4 3 (pump 1 2))
(connect 3 (cartridge_pressure_valve 1 3) stop_up)
(connect 3 (cartridge_pressure_valve 4 3) tank)
(connect 4 (shuttle_valve 1 1) (pump 1 2))
(connect 4 (shuttle_valve 3 1) tank)
(connect 4 (shuttle_valve 2 1) (pump 1 2))
(connect 4 (cartridge_valve 4 3) (pump 1 2))
(connect 4 (cartridge_pressure_valve 1 3) stop_up)
(connect 4 (cartridge_pressure_valve 4 3 (pump 1 2))
(connect 5 (shuttle_valve 1 1) (pump 1 2))
(connect 5 (shuttle_valve 3 1) (pump 1 2))
(connect 5 (shuttle_valve 2 1) tank)
(connect 5 (cartridge_valve 4 3) (pump 1 2))
(connect 5 (cartridge_pressure_valve 1 3) stop_up)
(connect 5 (cartridge_pressure_valve 4 3) stop_up)
(connect 6 (shuttle-valve 1 1) (pump 1 2))
(connect 6 (shuttle_valve 3 1) (pump 1 2))
(connect 6 (shuttle_valve 2 1) (pump 1 2))
(connect 6 (cartridge_valve 4 3) (pump 1 2))
(connect 6 (cartridge_pressure_valve 1 3) stop_up)
(connect 6 (cartridge_pressure_valve 4 3) tank)
(connect 7 (shuttle_valve 1 1) tank)
(connect 7 (shuttle_valve 3 1) (pump 1 2))
(connect 7 (shuttle_valve 2 1) (pump 1 2))

(connect 7 (cartridge_valve 4 3) (pump 1 2))
(connect 7 (cartridge_pressure_valve 1 3) stop_up)
(connect 7 (cartridge_pressure_valve 4 3) stop_up)
(connect 8 (shuttle_valve 1 1) (pump 1 2))
(connect 8 (shuttle_valve 3 1) (pump 1 2))
(connect 8 (shuttle_valve 2 1) (pump 1 2))
(connect 8 (cartridge_valve 4 3) tank)
(connect 8 (cartridge_pressure_valve 1 3) stop_up)
(connect 8 (cartridge_pressure_valve 4 3) stop_up)
(process 8)
(flow_of_pump 1 .00167)
(actuator 2)
(mode (actuator 1) cylinder)
(mode (actuator 2) cylinder
(power_of_motor 1 22000.0)
(pump 1 limited_pressure_variable_
displacement_pump)
(working_pressure 3.2E7)
(flow_of_valve (filter 1) .00102531)
(flow_of_valve (check_valve 1) .0000736939)
(flow_of_valve (check_valve 2) .0001)
(flow_of_valve (check_valve 3) .0001)
(check_valve 3)
(pressure_meter 3)
(pressure_switch 1)
(flow_of_valve (cartridge_value 4) 0.000875115)
(flow_of_valve (cartridge_valve 3) 0.00102274)
(flow_of_valve (cartridge_valve 2) .0010205)
(flow_of_valve (cartridge_valve 1) .00102531)
(flow_of_valve (cartridge_pressure_valve 1) 0.00194808)
(flow_of_valve (cartridge_pressure_valve 2) 0.0014915)
(flow_of_valve (cartridge_pressure_valve 3) 0.000598763)
(flow_of_valve (cartridge_pressure_valve 4) 0.00102531)
(prefill_valve 1)
(flow_of_valve (prefill_valve 1) .0102531)
(flow_of_valve (sequence_valve 1) .0001)
(flow_of_valve (sequence_valve 2) .0001)
(flow_of_valve (sequence_valve 3) .0001)
(acting (actuator 2) double)
(acting (actuator 1) double)
(sequence_valve 3)
(pressure_of_valve (sequence_valve 3) 2.67034E7)
(pressure_of_valve (sequence_valve 2) 475484.0)
(pressure_of_valve (sequence_valve 1) 1.43141E6)
(cartridge_pressure_valve 4)
(cartridge_valve 4)
(fixed_type (actuator 2) 4)

(fixed_type (actuator 1) 4)
(rod (actuator 1) single)
(rod (actuator 2) single)
(diameter (actuator 1) .4)
(diameter_rod (actuator 1) .36)
(diameter (actuator 2) .14)
(diameter_rod (actuator 2) .09)

References

1. Yeaple F. Fluid power design handbook. Marcel Dekker, New York, 1986
2. Warring RH. Hydraulic handbook. Gulf Publishing Company, Houston, 1983
3. Maddison RN. Information system methodologics. Wiley Heyden, London, 1983
4. Bondy JA and Murty USR. Graph theory with applications. Macmillan, 1976
5. Backe W. Component design by DSH program. In: Computer aided design in high pressure hydraulic systems, Mechanical Engineering Publications, London, 1983
6. Hull SR. The development of an automatic procedure for the digital simulation of hydraulic systems, Mechanical Engineering Publications, London, 1983
7. Lamb WS. A micro-computer based aid for hydraulic system designers. In: Computer aided design in high pressure hydraulic systems. Mechanical Engineering Publications, London, 1983
8. Bowns DE. Applications of a hydraulic automatic simulation package to the design of fluid power system. Proceedings Beijing international fluid power symposium, China, 1984
9. Weiss SM, Kulikowski CA. A practical guide to designing expert systems, Rowman and Allanleld Publishers, 1984
10. Nilsson NJ. Principles of artificial intelligence. Tioga Publishing Co., 1980
11. Harman P, King D. Expert systems. Artificial intelligence in business. John Wiley, New York, 1985
12. Alty JL, Coombs MJ. Expert systems concepts and examples. The National Computing Centre, 1984
13. Hayes-Roth F. Building expert systems. Addison-Wesley, London, 1983

Chapter 17

Rule-Based Programming for Industrial Automation
B.R. McGuire and W.G. Wee

Introduction and Summary

The design of control systems for industrial automation requires knowledge of both mechanical configuration and control logic. A design tool, named rule-based controls development system (RBCDS), was developed to apply knowledge-based system technology to the engineering discipline of control system development. RBCDS aids developers of automated equipment, such as machine tools, material-handling equipment, and industrial manufacturing processes, to design and test software-based control systems.

The second section introduces programmable logic controllers (PLCs) and the development computing environment. The concepts of frame-based knowledge representation and rule-based knowledge representation are introduced to show applicability to representing engineering design knowledge. A discussion of knowledge-base structure and knowledge-base reasoning, specific to the KEE software environment, introduces terminology which is used to describe RBCDS.

The third section describes the RBCDS. The system design and architecture include general characteristics of the RBCDS system. Facilities available to control system designers, including the creation of machine rules and simulation of machine designs, are explained.

The fourth section covers a sample machine to illustrate how RBCDS may be applied. The sample machine is a conveyor which will be controlled with a PLC. Examples of knowledge-base diagrams, rules, KEE units, generated programs and other program listing fragments are presented in this section.

The conclusion discusses the results of the project, evaluates the development environment, and suggests directions for further work.

Software for PLCs is typically developed by coding with a programming language after the mechanical design, electrical design, and sequence of operation are completed. These prerequisite steps typically result in a set of drawings which must be available to the PLC programmer.

RBCDS offers facilities to design a computer model of the desired system concurrently and in conjunction with the mechanical and electrical design tasks. The elements of knowledge which describe the system, such as the mechanical configuration, characteristics of sensors and actuators, and sequence of operation logic are combined to define the knowledge base which represents a "machine" or a process. The knowledge base serves as a working model for testing by simulation. The operation of the machine is simulated through rule chaining and knowledge-base manipulation to allow verification of the system design.

Finally, with the design verified, executable programs may be generated for runtime operation in PLCs.

The concept for a design tool with the capabilities of RBCDS was originated from a combination of practical need and exposure to the technology of knowledge-based systems and expert systems. The combination of Intellicorp KEE software and a Texas Instruments Explorer workstation was selected as the development computing environment. Next, the structure of the knowledge base was developed to permit the most natural representation for the domain of industrial machine control systems. Finally, a graphical user interface was designed to give RBCDS the look and feel of a computer-aided engineering design tool. A simple conveyor control system was designed with RBCDS and is described here as an example.

The initial phase of this work concluded with a demonstration system which executes on an Explorer workstation. Further work is required to support specific commercial PLCs. Details such as mapping I/O addresses to various types of inputs and outputs and generating software in proprietary program formats are required to support each particular model of PLC.

Survey of Previous Work and Preliminaries

A brief discussion of concepts and terminology is presented here to introduce concepts upon which RBCDS is based. An expert system for engineering design should permit definition of constraints for a design: technical feasibility, functionality between components, and consequences of natural laws. An ability to detect potential failures allows the designer to add robustness to a functionally correct design. In summary, an expert system for design must allow incremental development, a model to contain design constraints, and simulation to test the quality of the design [1].

The attraction of expert systems is the potential capability to package and duplicate human expertise. In some ways, computers compensate for human limitations in limited problem domains. Since people don't seem able to step through very long chains of reasoning, experts explain their knowledge in short cause-and-effect reasoning chains of five or fewer links. A highly knowledgeable expert can construct a large number of these short chains. Reasoning is the process of thought and the application of knowledge. Rule-based expert systems are a useful technology of inference (reasoning) that reveals the knowledge and facts that lead to a conclusion [2]. Control system designers using RBCDS have the role of experts since they possess the knowledge of the structure and specifications for the automated system. Control logic may be defined as a collection of simple rules for machine operation and the simulation of machine operation is analogous to problem solving with an expert system. Inferencing with rule chaining modifies the knowledge base which represents the state transitions of I/O devices leading to a conclusion, the goal state of the machine operation.

Programmable Controllers

PLCs, are employed to control the operation of industrial processes including manufacturing and material handling. PLC software is typically programmed in relay-ladder logic or in procedural programming languages. A general purpose software development system for programmable controllers may be implemented with rule-based programming concepts.

PLCs are interfaced to equipment through I/O devices (electrical inputs and outputs) and are programmed to perform sequences of automatic operations. A PLC input is an electrical circuit which has two-state logic, either active (voltage present) or inactive. An input circuit typically includes a switch which interrupts power flow to the PLC input for control system feedback. A PLC output is an electrical source which is energized with voltage when appropriate conditions exist, according to the execution of program logic. An output circuit is typically wired to an electrical load such as a solenoid, relay, or a lamp.

Development Computing Environment

RBCDS is a rule-based programming system which is implemented with Intellicorp KEE (knowledge engineering environment) on a Texas Instruments Explorer workstation. KEE is a high-end expert system shell for developing knowledge-based systems. It allows the integration of frame-based knowledge representation with a production rule system.

The Explorer workstation is a Lisp machine with a custom Lisp processor. The system software provides a Common Lisp interpreter and compiler, an integrated editor optimized for Lisp programming, virtual-memory management, multi-tasking program execution, and an inter-active user interface with high-resolution graphics, a mouse, and a keyboard for user input. A consequence of the underlying Lisp program-ming language is that programming of custom Lisp functions were required in the course of the project. The symbolic-processing facilities of the Lisp language are particularly well suited to the design of knowledge-based systems. RBCDS, like many other engineering design tools, conceals the programming environment in an effort to use the terminology and symbols which are customary to the designer. An Ethernet local area network interface allows connection to other computer systems.

Frames and Rules for Knowledge Representation

Frames, as in frames of reference, are the primary structure for knowledge representation in KEE. Frames consist of collections of slots that contain attributes which describe an object, a class of objects, a situation, an action, or an event. Frame-based reasoning seeks confirmation of attributes in slots or fills and modifies attributes in slots by inference. Slots provide a structure for data in a knowledge base in a manner similar to the way an expert typically organizes data [3].

A frame provides a structured representation of objects and classes of objects. Each class of objects may be both a subclass (specialization) of a more general class and/or a superclass of less general classes. The relationships among frames create "automatic" inferences through inheritance, which allows slots in class member frames to be influenced by slots in class frames [4].

Rules are analogous to the way people think: large collections of relatively simple relationships [5]. Rules are pieces of knowledge which have the form: IF ⟨circumstances⟩ THEN ⟨do action or make conclusion⟩. Production systems, the simplest type of rule-based systems, have three components: working memory, production memory and a rule interpreter. The working memory is a set of data structures to represent the current state of the system. The production memory is a set of rules that cause changes in working memory, given the existence of specific memory patterns. The rule interpreter applies the production memory (rules) to the working memory [5].

The production rule interpreter repeatedly seeks rules whose condition (left-hand side) matches the working memory. When a rule condition is present, the rule action (right-hand side) modifies the working memory and the rule is said to have "fired". A conflict resolution strategy must determine the order of rule firing if multiple rules may be fired by a pattern in working memory.

KEE Knowledge Bases

A KEE knowledge base [6–7] is the highest level of knowledge representation for a domain. KEE uses a frame-based representation of objects and their attributes. The five basic building blocks are: units, slots, slot values, facets and facet values. Units, which typically represent physical objects, are the "frames" in KEE's frame-based knowledge representation system. A slot contains a characteristic of the object. For example, an object such as an electrical switch is represented as the unit, SWITCH, which has the slot STATE which may have a value of ON or OFF. A facet describes a slot by defining the possible values of the slot.

The KEE Rulesystem uses the frames structure to implement production rules for rule-based reasoning. Each rule is contained in a unit with slots to contain the rule premise, rule action, and other rule characteristics. The integration of frames and production rules into a single unified knowledge representation facility allows the strengths of frame-based reasoning and rule-based reasoning to complement each other [4]. Frame-based reasoning is extended to management of rules to allow the grouping of rules into classes. Rule-based reasoning allows the characteristics of one object to affect the characteristics of other objects through relationships which cannot be represented with reasoning by inheritance.

KEE Knowledge Base Reasoning

The KEE Rulesystem [6–7] provides an inference engine for IF-THEN production rules which allows forward and backward chaining. Rules represent knowledge as relationships between facts and KEE allows the organization of hierarchies of rule classes according to the knowledge they represent.

A QUERY command begins backward chaining by "querying" the knowledge base for a fact which becomes the goal for rule chaining. If the fact is present in the knowledge base, or the fact can be proved (asserted into the knowledge base) through backward chaining, the goal is satisfied. All rules which have the goal fact in the rule action may be considered, one at a time, according to a rule ordering mechanism. When a rule is considered, each of the facts in the rule premise is queried and becomes the subgoal for backward chaining unless the fact is already true. For a rule to fire, the facts in the rule premise must exist in the knowledge base either prior to rule chaining or as a result of chaining. When the rule is fired, the facts in the rule action are asserted into the knowledge base. Once started, backward chaining will continue until the original goal fact is asserted into the knowledge base or all possible backward chaining fails.

Rule-Based Controls Development System (RBCDS)

System Design

RBCDS simplifies the task of programming industrial control systems. RBCDS offers facilities to design and test an automated system in software as a knowledge base (a collection of facts and rules). Once designed and tested, runtime software is generated for commercially-available programmable controllers. The frame-based knowledge representation system is well suited for modelling the structure of mechanical assemblies and industrial processes in terms of input and output devices. Rules represent cause-and-effect relationships between input and output devices and rule chaining determines sequences of operations required to drive an automatic system from an initial state to desired goal states.

System Architecture

Several knowledge bases form the RBCDS system. The RBCDS knowledge base (KB) is the first KB loaded into memory and contains KEE units which describe classes of I/O devices. The interface KB controls the user interface. Finally, each machine is represented by a machine KB. For the purposes of RBCDS, a machine is a collection of I/O devices which are monitored and controlled by a single programmable controller. KEE units within a machine KB represent individual I/O devices, which are linked as class members to class units in the RBCDS KB. Control logic is stored as rules which are organized as rule classes (classes of rule units) in each machine KB. The complete, runtime program for one programmable controller is generated from the I/O structure (units) and control logic (rules) in a machine KB.

Designing Machines with RBCDS

RBCDS permits development and testing with the focus on a single machine though more than one machine KB may be loaded into memory and selected at will. The scope of a "machine" includes all of the I/O subsystem which interface to a single, independent PLC and will be controlled by the program in the PLC. Libraries of machines and subassemblies may be used for reference and as points of departure for new machines. Any existing machine may be chosen with the machine selection menu for continued development or testing. Once the machine I/O has been defined, rules are created to control the machine according to its operating requirements. Each rule defines a combination of I/O devices at specific states which represent a machine or process condition and asserts states to one or more I/O devices. When I/O devices and rules have been defined, the machine design may be tested. A typical machine will be designed incrementally, with I/O devices and rules added incrementally and tested, as the design reaches completion. Once verified, the PLC program is generated from the final machine knowledge base.

Machine Rules

In RBCDS a machine's rules are organized as three rule types: control rules, simulation rules and situation rules. Control rules contain control logic which would typically be coded into a PLC program. An example of a control rule is a rule which inspects the state of a limit switch (input device) and deactivates a solenoid-operated hydraulic valve (output device) to stop the motion of a mechanical assembly.

Simulation rules represent the response of the machine to energized output devices and simulate feedback by modifying the states of input devices. Most simulation rules assign new states to inputs as though a physical machine were operating. An example of a simulation rule is a rule which inspects the state of a solenoid valve (output device) and activates the state of a limit switch (input device). Simulation rules are used primarily for testing and verification of control rules.

Situation rules are designed to define expected situations where the machine's operations must perform to some specification. These rules are changed frequently during development and represent external events over which the control system has no direct control. A set of initial conditions, defined by I/O devices with corresponding initial states, is used to initialize each "situation" where the system is to be tested. Situation rules are designed to cause or react to specific events which may occur during the operation of the machine.

Machine Simulation

The machine control system which has been designed with RBCDS must be tested and verified by simulation before a runtime program is generated. A machine may be designed and simulated incrementally, at various stages of completion, ranging from prototype to final design. Even with RBCDS, it is generally not possible to test a machine design for all possible situations. It is possible to define a set of situations where the machine must operate to specification, as well as situations which represent the most likely failures or worst case failure modes. A situation represents the physical world where the machine must operate with controllability limited to the set of inputs and outputs which are available to the PLC for which a program will be generated. Every machine must have at least one situation which defines the typical operating cycle of the machine so the control logic may be tested.

An RBCDS "situation" includes a goal, initial conditions, and a set of situation rules. The situations created for each machine must be designed to test its various modes of operation thoroughly. A goal is the desired state for an I/O device which is to be achieved as the last step in an operating sequence and is expected to occur while simulating the machine's operation. The initial conditions provide a starting point for a machine simulation though the states of I/O devices may be "unknown" if not specified in the initial conditions. Realistic testing and verification of a new design for a control system depends on the degree to which situation rules simulate the physical world.

Runtime Program Generation

Runtime program generation allows a choice of relay-ladder logic or procedural program formats. A specific delivery system must be selected from the menu of supported PLCs. For this project, only a generic runtime programmable controller is supported for demonstration of the relay-ladder and the procedural program formats. A software driver for a commercial PLC would include configurations and constraints according to manufacturer specifications.

The algorithms for generating relay-ladder PLC programs and pro-cedural programs are different. Relay-ladder logic requires that all logic which controls a particular output be combined into a single "network" of contacts that energizes a relay coil (representing the output device). All of the logic in the control rules (as opposed to the simulation rules and situation rules) will be included as relay-ladder network elements. Procedural programs are formed by tracing rule chaining during simulation for each defined machine situation and creating a report of the sequence of operations. Process steps, which are recorded as procedural program statements, are appended to the sequence report in the reverse of the backward chaining order. Rule chaining proceeds depth first until a rule is fired by initial conditions. The first rules fired represent the earliest steps in the sequence. The last rules fired represent the final steps leading to the goal state. Simulation and situation rules append "WAIT" statements to the sequence report as they are fired during machine simulation and control rules append "OUTPUT" assignment statements to the sequence report. The operating sequence report is the simplest form of a procedural program which may be translated into various procedural programming languages. Finally, a translator/compiler for a specific programmable controller would perform file conversion to a binary or text source format to be down-loaded to a PLC.

Example Machine

To demonstrate an RBCDS, a machine named CONVEYOR-A was created. CONVEYOR-A is a conveyor which serves as a buffer to transport "parts", such as metal shafts or totes containing loose materials, between two manufacturing processes. This "machine" represents one type of control system which may be designed using RBCDS. The conveyor has two modes of operation which must be implemented in the control logic, automatic mode and manual mode. In manual mode, the pushbutton input signals cause the conveyor to stop and start at the appropriate times unless inhibited by safety interlock logic. In automatic mode, the conveyor interfaces with the "input" process and the "output" process to start and stop in conjunction with the rate of parts passing from one process to the other. The input process is the source of incoming workpieces to be loaded on the conveyor. The output process receives the workpieces by unloading the conveyor.

Conveyor I/O Specification (see Table 1)

CONVEYOR-A is a good example because it is a simple machine with few I/O devices and is well within the capacity of a single PLC. External I/O provides logical connections between the conveyor and other automatic equipment to synchronize the control system with other PLCs. Manual control functions are typically required with production machines to permit testing and manual setup in preparation for automatic operation.

Table 1. Conveyor I/O devices

Inputs	
PART-LOAD-POSITION	Proximity switch
CONVEYOR-FULL	Proximity switch
NEAR-FULL	Proximity switch
NEW-PART-READY	Proximity switch
EMERGENCY-STOP	Pushbutton
RUN-CONVEYOR-PB	Pushbutton
INDEX-TIMEOUT	Timer input
Outputs	
RUN-MOTOR	Relay coil
PART-UNLOAD-READY	External output
CONVEYOR-READY	External output
Registers	
COUNT-PARTS	Register
INDEX-TIMER	Register

CONVEYOR-A has a motor, position detection switches, external inputs and outputs, and an operator panel. The motor drives the conveyor and is controlled by a power relay which is, in turn, energized by the output, RUN-MOTOR, from the PLC. When RUN-MOTOR is energized, the motor runs and the conveyor advances to carry parts toward the output process. The conveyor carries parts in part holders which are connected to drive chains. Once running, the conveyor is stopped by de-energizing the output, RUN-MOTOR.

Proximity switches detect the motion of the conveyor by the presence of parts or part holders on the conveyor. The arrival of a part holder at the part-loading position indicates the conveyor has "indexed" the distance between part holders and that a part holder is in the conveyor loading position. A switch at the loading position detects the presence of a part in the part holder at the load position. Switches at the conveyor unload position and a fixed distance ahead of the unload position detect the arrival of parts on the conveyor at those positions.

Flow Diagram for Conveyor Operation

Figure 17.1 shows the flow diagram which specifies the automatic sequence of operation for CONVEYOR-A. As the conveyor transports parts from the load end (part loading position) to the unload end (part-unloading position), inputs indicate the presence or absence of parts (workpieces) at various positions along the length of the conveyor. In general, the conveyor indexes by the distance between parts on the conveyor after each new part is loaded and detected by the switch NEW-PART-READY. The presence of a part detected by the CONVEYOR-FULL switch inhibits the automatic conveyor indexing. The index timer triggers conveyor indexing if a new part fails to be loaded on the conveyor within a predefined time interval to allow a steady flow of parts for unloading if the loading of new parts is interrupted. In effect, CONVEYOR-A has a buffering function in addition to transporting parts between two undefined machines.

RBCDS Knowledge Base Diagram

The graph of the RBCDS main knowledge base, RBCDS KB, in Fig. 17.2 shows the hierarchy of I/O devices. The greatest degree of specialization, to the right-hand side of the graph defines characteristics of common types of I/O devices used with PLCs.

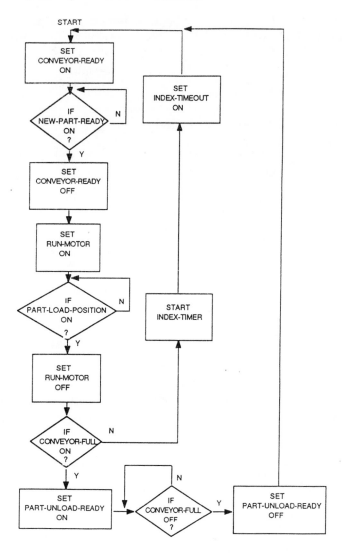

Fig. 17.1 CONVEYOR-A flow diagram.

CONVEYOR-A Knowledge Base Diagrams

Figures 17.3 and 17.4 show graphs of the machine knowledge base, CONVEYOR-A KB, including the hierarchies of machine I/O devices, each of which is a class member of a corresponding device type in the RBCDS KB. The solid lines represent class to subclass relationships. The rules are organized into the rule classes: control rules, simulation rules

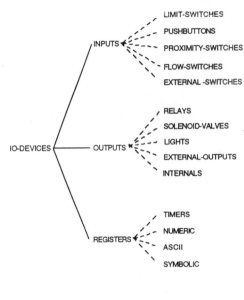

Fig. 17.2 Graph of RBCDS KB.

and situation rules. The structure under situations includes initial conditions and specific rules for each of three situations: EMERGENCY-STOP, MANUAL-CYCLE and STANDARD-CYCLE (automatic operation).

Examples

RBCDS Rules. A sample listing of rules for the machine CONVEYOR-A is shown in Appendices 1–3. Each rule is a simple cause-and-effect relationship.

Example of a KEE Unit for an I/O Device. Figure 17.5 shows how the proximity switch PART-LOAD-POSITION is represented as a unit in KEE [6–7].

Example of a Lisp function to create a machine KB. Figure 17.6 illustrates how a knowledge base for a new machine such as CONVEYOR-A is created using a Lisp function [8].

Examples of outputs from RBCDS. For the CONVEYOR-A problem, examples of relay-ladder networks and procedural programs generated by RBCDS are shown in Figs 17.7 and 17.8.

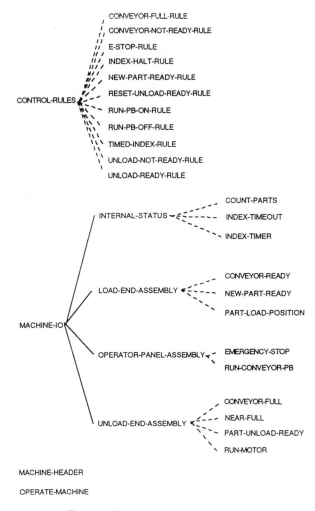

CONVEYOR-FULL-RULE
CONVEYOR-NOT-READY-RULE
E-STOP-RULE
INDEX-HALT-RULE
NEW-PART-READY-RULE
RESET-UNLOAD-READY-RULE
CONTROL-RULES
RUN-PB-ON-RULE
RUN-PB-OFF-RULE
TIMED-INDEX-RULE
UNLOAD-NOT-READY-RULE
UNLOAD-READY-RULE

INTERNAL-STATUS
COUNT-PARTS
INDEX-TIMEOUT
INDEX-TIMER

LOAD-END-ASSEMBLY
CONVEYOR-READY
NEW-PART-READY
PART-LOAD-POSITION

MACHINE-IO

OPERATOR-PANEL-ASSEMBLY
EMERGENCY-STOP
RUN-CONVEYOR-PB

UNLOAD-END-ASSEMBLY
CONVEYOR-FULL
NEAR-FULL
PART-UNLOAD-READY
RUN-MOTOR

MACHINE-HEADER

OPERATE-MACHINE

Fig. 17.3 Graph of CONVEYOR-A KB.

Conclusion

The RBCDS project demonstrates that knowledge-based software technology may be applied to engineering design tools for control system development for industrial automation. RBCDS serves as a prototyping and modelling system which generates executable program code for PLCs. Each I/O device and mechanical assembly which is part of a machine is added to the design and tested in a manner that is analogous to the assembly and testing of the actual machine. The ability to test

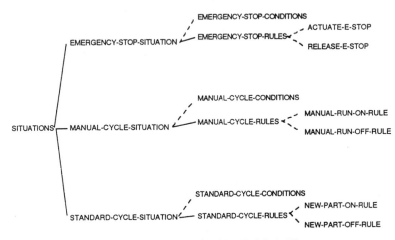

Fig. 17.4 Graph of CONVEYOR-A KB.

control system software before committing it to physical hardware adds an additional opportunity to review the design prior to investment in production equipment.

A significant characteristic of a rule-based system is the difficulty with representing and simulating the repetitive operation required by most PLC programs. RBCDS uses rule chaining to determine sequences of machine operations in terms of input and output state transitions. PLC programs are generated as sequences of relay ladder networks or procedural program statements which compose a program or subprograms. A PLC program is executed repeatedly with each pass through the program occurring in one "scan" or "cycle" which begins with an I/O "exchange" to update the states of I/O devices periodically. The primary impact during testing is that the simulation does not directly resemble the operation of a PLC in continuous operation. The simulation could be improved by creating a PLC emulator which executes a generated PLC program in a manner identical to the PLC operation.

The combination of KEE software and the Explorer workstation was an effective development environment for the RBCDS project. The integration of frame-based knowledge and rule-based knowledge proved to be useful for this application. A number of improvements to RBCDS

```
---------------------------------------------------------
(OUTPUT) The PART-LOAD-POSITION Unit in CONVEYOR-A
---------------------------------------------------------
Unit: PART-LOAD-POSITION in knowledge base CONVEYOR-A
Created by MCGUIRE on 2-5-88 10:22:03
Modified by MCGUIRE on 2-11-88 13:49:22
 Superclasses: NONE
 Member Of: LOAD-END-ASSEMBLY in CONVEYOR-A,
            PROXIMITY-SWITCHES in AIPC
 Members: NONE
---------------------------------------------------------

Own Slot: FORCE-ON from PROXIMITY-SWITCHES
  Inheritance: METHOD
  ValueClass: METHOD
  Values: (LAMBDA (THISUNIT) (PUT.VALUES THISUNIT
          'STATE 'ON))

Own Slot: FORCE-OFF from PROXIMITY-SWITCHES
  Inheritance: METHOD
  ValueClass: METHOD
  Values: (LAMBDA (THISUNIT) (PUT.VALUES THISUNIT
          'STATE 'OFF))

Own Slot: STATE from PROXIMITY-SWITCHES
  Inheritance: OVERRIDE.VALUES
  ValueClass: (ONE.OF ON OFF)
  Values: ON

Own Slot: IO-ADDRESS from PROXIMITY-SWITCHES
  Inheritance: OVERRIDE.VALUES
  ValueClass: NUMBER
  Values: 10010

---------------------------------------------------------
```

Fig. 17.5 Example of KEE unit for an I/O device.

```
Function: CREATE-MACHINE

Purpose: This function creates a knowledge
         base for a new machine.


(DEFUN CREATE-MACHINE (MACHINE-NAME FILE-NAME)
       (CREATE-KB MACHINE-NAME)
       (CREATE-UNIT 'MACHINE-HEADER MACHINE-NAME)
       (CREATE-SLOT 'FILE
                    (LIST 'MACHINE-HEADER MACHINE-NAME)
                    'OWN)
       (PUT-VALUE (LIST 'MACHINE-HEADER MACHINE-NAME)
                  'FILE FILE-NAME)
       (CREATE-SLOT MACHINE-NAME
                    '(MACHINE-HEADERS AIPC)
                    'OWN)
       (PUT-VALUE '(MACHINE-HEADERS AIPC)
                  MACHINE-NAME FILE-NAME)
       (PARENT-LINK
         (LIST 'MACHINE-HEADER MACHINE-NAME)
         (LIST 'MACHINE-HEADERS 'AIPC)
         'MEMBER)
       (CREATE-UNIT 'MACHINE-INDEX MACHINE-NAME)
       (CREATE-UNIT 'SITUATION MACHINE-NAME)
       (CREATE-UNIT 'OPERATE-MACHINE MACHINE-NAME)
       (CREATE-RULE-CLASS 'CONTROL-RULES MACHINE-KB)
       (CREATE-RULE-CLASS 'SIMULATION-RULES MACHINE-KB)
)
```

Fig. 17.6 Example of Lisp function to create machine KBs.

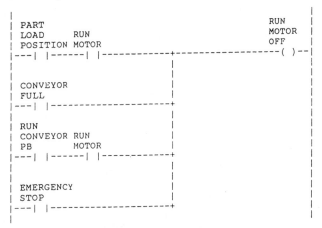

```
For Output: RUN-MOTOR

Primary Network: ENERGIZING
```

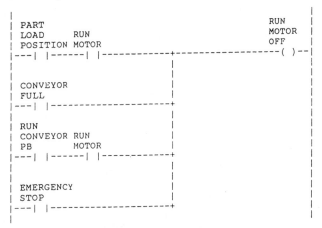

Fig. 17.7 Generated runtime relay ladder network.

are needed to make the system more useful. A facility is needed to print graphical charts to show timing relationships among selected I/O devices for verification of the design. To support a specific PLC, a driver must be written to generate runtime programs for that PLC and a utility is required to download programs from the engineering workstation. Once the underlying design for a machine is entered and tested, runtime programs may be generated for delivery on any supported type of PLC. A utility which converts existing PLC programs into rules may be possible by creating a knowledge base from the program. As an

```
PROCEDURAL LOGIC (SIMPLIFIED)

TASK: STANDARD-CYCLE-TASK

REPEAT
WAIT (NEW-PART-READY == ON)
RUN-MOTOR = ON
WAIT (CONVEYOR-FULL == ON)
RUN-MOTOR = OFF
END


TASK: MANUAL-CYCLE-TASK

REPEAT
WAIT (RUN-CONVEYOR-PB == ON)
RUN-MOTOR = ON
WAIT ((RUN-CONVEYOR-PB == OFF) OR
       (CONVEYOR-FULL == ON))
RUN-MOTOR = OFF
PART-UNLOAD-READY = ON
END


TASK: EMERGENCY-STOP-TASK

REPEAT
WAIT (EMERGENCY-STOP == ON)
RUN-MOTOR = OFF
END
```

Fig. 17.8 Generated procedural logic.

intermediate form of control system software, existing PLC programs could be modified, retested, and subsequently generated for upgraded PLC without major redesign.

References

1. Menges G, Hovelmanns N, Baur E (eds). Expert Systems in Production Engineering. Lecture Notes in Engineering 29, Springer-Verlag, Berlin Heidelberg New York, 1987
2. Tennant H. The redesign of thought. Texas Instr Eng J 1986; 3: no. 1
3. Wolfgram DD, Dear TJ, Galbraith CS. Expert systems for the technical professional. John Wiley, 1987
4. Fikes R, Kehler T. The role of frame-based representation in reasoning. Commun ACM 1985; 28: no. 9
5. Charniak E, McDermott D. Introduction to artificial intelligence. Rule-Based Progr 1985; 7: 437–440
6. Intellicorp. KEE Software development system user's manual, KEE version 3.0, 1986
7. Intellicorp. KEE software development system core reference manual, KEE version 3.0, 1986
8. Winston PH, Horn, BKP. Lisp, 2nd edn. Addison-Wesley, 1984

Appendix 1. Control Rules for CONVEYOR-A

NEW-PART-READY-RULE:
    (IF
        (THE STATE OF NEW-PART-READY IS ON)
        (THE STATE OF PART-LOAD-POSITION IS ON)
        (THE STATE OF CONVEYOR-FULL IS OFF)
    THEN
        (THE STATE OF INDEX-TIMER IS ON)
        (THE STATE OF RUN-MOTOR IS ON)
        (THE STATE OF CONVEYOR-READY IS OFF))

    INDEX-HALT-RULE:
        (IF
            (THE STATE OF PART-LOAD-POSITION IS ON)
            (THE STATE OF RUN-MOTOR IS ON)
        THEN
            (THE STATE OF RUN-MOTOR IS OFF)
            (THE STATE OF CONVEYOR-READY IS ON))

    CONVEYOR-NOT-READY-RULE:
        (IF
            (THE STATE OF CONVEYOR-READY IS ON)
            (THE STATE OF RUN-MOTOR IS ON)
        THEN
            (THE STATE OF CONVEYOR-READY IS OFF))

    RESET-UNLOAD-READY-RULE:
        (IF
            (THE STATE OF PART-UNLOAD-READY IS ON)
            (THE STATE OF RUN-MOTOR IS OFF)
            (THE STATE OF CONVEYOR-FULL IS OFF)
        THEN
            (THE STATE OF PART-UNLOAD-READY IS ON))

    TIMED-INDEX-RULE
        (IF
            (THE STATE OF INDEX-TIMEOUT IS ON)
            (THE STATE OF CONVEYOR-FULL IS OFF)
        THEN
            (THE STATE OF INDEX-TIMER IS OFF)
            (THE STATE OF RUN-MOTOR IS ON))

    CONVEYOR-FULL-RULE:
        (IF
            (THE STATE OF CONVEYOR-FULL IS ON)

THEN
    (THE STATE OF RUN-MOTOR IS OFF))

UNLOAD-READY-RULE:
    (IF
        (THE STATE OF NEAR-FULL IS ON)
        (THE STATE OF RUN-MOTOR IS ON)
    THEN
        (THE STATE OF PART-UNLOAD-READY IS ON))

UNLOAD-NOT-READY-RULE:
    (IF
        (THE STATE OF PART-UNLOAD-READY IS ON)
        (THE STATE OF RUN-MOTOR IS OFF)
        (THE STATE OF CONVEYOR-FULL IS OFF)
    THEN
        (THE STATE OF PART-UNLOAD-READY IS OFF))

RUN-PB-ON-RULE:
    (IF
        (THE STATE OF RUN-CONVEYOR-PB IS ON)
        (THE STATE OF RUN-MOTOR IS OFF)
    THEN
        (THE STATE OF RUN-MOTOR IS ON))

RUN-PB-OFF-RULE:
    (IF
        (THE STATE OF RUN-CONVEYOR-PB IS ON)
        (THE STATE OF RUN-MOTOR IS ON)
    THEN
        (THE STATE OF RUN-MOTOR IS OFF))

E-STOP-CONVEYOR-RULE:
    (IF
        (THE STATE OF EMERGENCY-STOP IS ON)
    THEN
        (THE STATE OF RUN-MOTOR IS OFF))

Appendix 2. Simulation Rules for CONVEYOR-A

CONVEYOR-FULL-RULE:
(IF
    (THE STATE OF COUNT-PARTS IS 2O)
    (THE STATE OF RUN-MOTOR IS ON)
THEN
    (THE STATE OF NEAR-FULL IS ON))

NEAR-FULL-RULE:
(IF
    (THE STATE OF NEAR-FULL IS ON)
    (THE STATE OF RUN-MOTOR IS ON)
THEN
    (THE STATE OF CONVEYOR-FULL IS ON))

ADD-PART-RULE
(IF
    (THE STATE OF PART-LOAD-POSITION IS ON)
    (THE STATE OF RUN-MOTOR IS OFF)
THEN
    (THE STATE OF COUNT-PARTS IS (+ COUNT-PARTS 1)))

REMOVE-PART-RULE:
(IF
    (THE STATE OF PART-UNLOAD-READY IS ON)
    (THE STATE OF CONVEYOR-FULL IS OFF)
THEN
    (THE STATE OF COUNT-PARTS IS (− COUNT-PARTS 1)))

Appendix 3. Situations and Situation Rules for CONVEYOR-A

STANDARD-CYCLE-SITUATION

STANDARD-CYCLE-CONDITIONS:

OUTPUT RUN-MOTOR	OFF
INPUT PART-LOAD-POSITION	ON
INPUT NEW-PART-READY	OFF
INPUT CONVEYOR-FULL	OFF

INPUT RUN-CONVEYOR-PB   OFF

STANDARD-CYCLE-RULES:

NEW-PART-ON-RULE
  (IF
    (THE STATE OF NEW-PART-READY IS OFF)
  THEN
    (THE STATE OF NEW-PART-READY IS ON))

NEW-PART-OFF-RULE
  (IF
    (THE STATE OF NEW-PART-READY IS ON)
    (THE STATE OF RUN-MOTOR IS ON)
  THEN
    (THE STATE OF NEW-PART-READY IS OFF))

GOAL: (THE STATE OF CONVEYOR-FULL IS ON)

MANUAL-CYCLE-SITUATION

MANUAL-CYCLE-CONDITIONS:

OUTPUT RUN-MOTOR   OFF
INPUT NEW-PART-READY   OFF
INPUT CONVEYOR-FULL   OFF
INPUT RUN-CONVEYOR-PB   OFF

MANUAL-CYCLE-RULES:

MANUAL-RUN-ON-RULE:
  (IF
    (THE STATE OF RUN-CONVEYOR-PB IS OFF)
  THEN
    (THE STATE OF RUN-CONVEYOR-PB IS ON))

Chapter 18

Knowledge-Based Programs for Manufacturing System Design
A. Kusiak and S.S. Heragu

Introduction

Manufacturing industry has witnessed significant developments in recent years, as measured by the increase in the number of automated systems in use. The key to the success of these systems is effective exploitation of available resources such as machines, tools, fixtures and material handling systems. Proper use of these resources has increased productivity. At the same time, the design problems related to the modern manufacturing systems have become more complex. Designers and users of automated manufacturing systems have attempted to develop new tools to cope with these complexities. Knowledge-based systems represent a class of modern tools that have been applied to improve the design and management functions in automated manufacturing systems.

Although optimization techniques can solve complex manufacturing problems efficiently, they are not always easy to apply because:

- The data required by the optimization models may not be easily available
- Their scope of availability may be limited
- They require human experts
- The algorithms are often not able to provide optimal solutions for Industrial problems because of their complexity

On the other hand, knowledge-based systems have typically been used to solve problems that are either too complex for mathematical formulation or too difficult to solve using optimization. Some of the difficulties encountered when using optimization techniques may be offset by combining them with knowledge-based systems. Traditionally, human experts have solved problems using the optimization approach; if knowledge-based systems are to be used successfully, they should be applied in such a way that they replace human experts rather than the optimization approach.

In this chapter, knowledge-based systems for manufacturing system design are discussed. The manufacturing system design problem is considered in the next section. Then two types of knowledge-based systems, namely, stand-alone and tandem knowledge-based systems, are described in the third section. Three variants of the tandem knowledge-based system, are also provided with examples. Models which can be used to formulate the manufacturing equipment selection problem are provided in the fourth section. The problem solving approach of KBSES a system for equipment selection is detailed in the fifth section. In the sixth section, the components of KBSES are presented. A numerical example is given in the seventh section. Conclusions are drawn in the last section.

Manufacturing System Design

Manufacturing system design is a complex activity that requires solving a number of problems typically arranged in a hierarchy; one problem in the hierarchy may have to be solved more than once. For example, the solution to the equipment selection problem at a higher level may suggest that an automated guided vehicle (AGV) be used as the material handling system (MHS) for a set of machines T_s. While solving the machine layout problem at a later stage, the constraints (for example, space limitations) may suggest that a robot be used as the MHS. If this is the case, the equipment selection problem with the modified constraints has to be resolved in order to determine the new set of machines T_s.

Two hierarchical approaches to designing manufacturing systems are the *four-level hierarchical approach* and the *two-level hierarchical approach*.

Four-Level Hierarchical Approach

The concept of group technology (GT) is the basis for the four-level hierarchical approach which is illustrated in Fig. 18.1. This approach involves the following problems:

- Equipment selection
- Machine cell formation
- Machine layout
- Cell layout

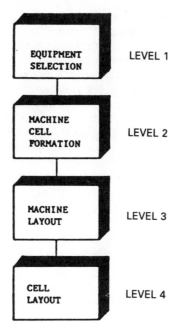

Fig. 18.1 Four-level hierarchical approach to design of automated manufacturing systems.

Equipment Selection

Because the purchase of modern manufacturing equipment may involve very high capital cost, it is considered a critical aspect in the manufacturing system design. By selecting the right number and type of equipment a company can:

- Reduce capital costs
- Reduce maintenance and operating costs as the number of machines purchased is optimal or near-optimal
- Increase machine utilization
- Improve layout of facilities as only the required number of machines are purchased
- Increase efficiency of the production facility due to improved layout and reduced traffic congestion

The equipment selection problem has been analysed and modelled using a number of approaches. Miller and Davis [1] have surveyed, classified and compared a number of deterministic and stochastic approaches. Current approaches make certain assumptions that may not hold in practice. For example, the dynamic programming approach [2] assumes that the machines in a manufacturing system are homogeneous. The mathematical programming formulations do not consider the problem's qualitative aspects. A formulation for the equipment selection problem that would be acceptable in practice is difficult to develop [3]. The artificial intelligence approach, which considers qualitative as well as quantitative factors, is suitable for solving the problem.

The equipment selection problem can be further divided into problems of manufacturing equipment selection and MHS selection. The MHS selection problem involves the type and number of material handling equipment; the manufacturing equipment selection problem involves the equipment, other than MHS, that is needed to produce the given parts.

Three issues are regarded as important in the equipment selection problem. They are:

- Manufacturing equipment selection
- MHS selection
- Budget consideration

Manufacturing Equipment Selection

The type and number of manufacturing equipment required depends on the type of parts to be manufactured and manufacturing processes to be used. It is not necessary to know the exact details regarding the parts in order to solve the problem. The knowledge about representative operations/parts might suffice. A *representative operation (part)* is defined as an operation (part) that represents a large number of other operations (parts) with the same requirements and characteristics as the representative operation (part). For example, if part p_r is a representative of part of n parts $\{p_1, ..., p_n\}$ and if part p_r can be processed using machine T_1 and material handling system H_1, then any of the n parts $\{p_1, ..., p_n\}$ can also be processed using the equipment T_1, H_1. In many instances, manufacturing equipment requirements determined in the initial design stages may have to be modified later.

MHS Selection

The representative operations/parts may indicate the MHS type that is required for material handling. However, this is not the only factor that has an impact on the selection. Machine and machine cell layout must also be considered. For example, using a polar-coordinate robot as a handling device imposes the arrangement of machines in a circular pattern as shown in Fig. 18.2. If an AGV has been selected to tend these machines, it may be necessary to revise the decision since, in a circular layout, a robot handles material more efficiently than an AGV.

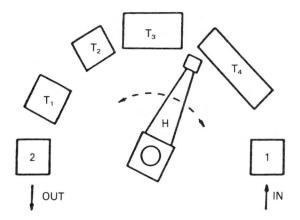

Fig. 18.2 Circular layout of machines. T_1–T_4: machines; 1: pallet with incoming parts; 2: pallet with outgoing parts; H: handling robot.

Budget Consideration

If budget is not a constraint in the design of a manufacturing system, then the MHS selection problem can be postponed until the machine layout is determined. In practice, however, budget always imposes a severe constraint on equipment selection; hence, it is necessary at least to:

1. Determine the type of MHS to be used while solving the equipment selection problem
2. Allocate a percentage of the budget for the MHS that will be selected after the machine layout has been determined

If procedure 1 is adopted, then the equipment selection problem may have to be solved again when the machine layout is determined; the initial MHS selection may be unsuitable for the determined layout. If procedure 2 is adopted, then it calls for a proper allocation of the budget for MHS selection and involves stochastic data.

Machine Cell Formation

Some manufacturing systems require the use of the group technology (GT) concept, where parts are grouped into part families and machines into cells based on the similarity of process plans. The implementation of GT provides:

1. Better planning and control of the manufacturing process – it is easier to schedule a family of parts than each individual part.

2. Reduction in setup cost – scheduling parts with identical machining requirements reduces setup time and hence setup cost considerably.

3. Better space utilization – GT implementation results in machine cells, each being a compact group of machines. As a result, space utilization improves.

4. Higher equipment utilization – when compared with a functional manufacturing facility, a given part in a cellular manufacturing facility spends much less time in travel. As a result, parts are available for processing which in turn reduces machine idle time.

Machine Layout

A frequently used objective for arrangement of machines in a cell is to minimize material handling cost. If there are more than 10 machines to be arranged in a cell, it is difficult to obtain an optimal solution and hence heuristic algorithms are used. While solving the machine layout problem, factors such as space constraints, safety, type of material handling system, etc. must be considered.

Cell Layout

This involves the location of each machine cell on its corresponding site in a way that minimizes plant wide material handling costs. If all the cells are of the same size, then the cell layout problem can be modelled as a quadratic assignment problem (QAP). Heragu and Kusiak [4] have classified existing heuristic algorithms for the facility layout problem into the following four categories:

1. Construction algorithms
2. Improvement algorithms
3. Hybrid algorithms
4. Graph-theoretic algorithms

Two-Level Hierarchical Approach

Although group technology has a number of benefits, it is not always implementable. The main reasons are:

● Parts and machines may not form clusters
● Data collection may be time consuming and expensive

In such cases, the manufacturing system design involves only two steps – equipment selection and layout of machines – as shown in Fig. 18.3.

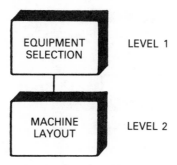

Fig. 18.3 Two-level hierarchical approach to design of manufacturing systems.

This two-level hierarchical approach has been used for the design of classical manufacturing systems.

Classes of Knowledge-Based Systems

Based on their operational mode, two classes of knowledge-based systems can be identified [5]:

- Stand-alone knowledge-based systems
- Tandem knowledge-based systems

Stand-Alone Knowledge-Based Systems

A knowledge-based system in the stand-alone mode uses data and constraints pertinent to the problem and solves it using rather simple procedures. It does not use the optimization approach which involves modelling the given problem and solving the model using an optimal or a "good" heuristic algorithm. Many existing knowledge-based systems fall into the stand-alone class. An example is TOM [6], which is a system used for generating machining sequence for parts.

Tandem Knowledge-Based Systems

The tandem knowledge-based system, on the other hand, combines the optimization approach with the knowledge-based system approach for problem solving. It can be thought of as a knowledge-based system linked to a model and algorithm base (Fig. 18.4). The basic approach

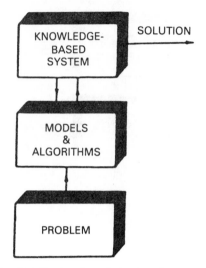

Fig. 18.4 Tandem knowledge-based system.

utilized in a tandem knowledge-based system is as follows. A suitable model is either selected or constructed for the given problem. To solve the model, an optimal or heuristic algorithm (available in the algorithm base) is selected. The solution generated by the algorithm is modified, if necessary, so that qualitative aspects not considered in the model can be incorporated. There are three variants of the tandem knowledge-based system, namely:

- Data-modifying system
- Model-based system
- Model-modifying system

The basic approach in all the three variants of the tandem knowledge-based system is the same; i.e., they use optimization to solve the problem. However, the actual problem-solving approach in each variant is different. The main function in a data-modifying system, is to modify, i.e., generate or reduce data (from among available data) as required by the model which is chosen by the system. In a model-based system, a suitable model and algorithm is selected for the given problem. The model-modifying system attempts to modify a selected model, for example, add or delete constraints depending on the problem at hand, or construct a suitable one. In each of the above mentioned variants of the tandem knowledge-based system, a suitable algorithm for solving the model is selected. The solution produced by the algorithm is modified, if necessary, in order to incorporate qualitative aspects not considered in the model.

Data-Modifying System

In a data-modifying system, first a model is selected for the problem considered. If the model requires additional data to be generated or a subset of data to be extracted from the available set of data, then the knowledge-based system performs the required function, i.e., data generation or data reduction. The resulting data are then used in the model. Next, an appropriate algorithm is selected for solving the model. The algorithm is applied and the solution generated by the algorithm is modified, if necessary, to make it implementable and to account for qualitative aspects of the problem.

The idea is demonstrated using an expert system for the tool path selection problem which occurs in process planning. For details of process planning refer, for example, to Chang and Wysk [7]. Consider the mechanical part shown in Fig. 18.5. The part can be obtained by removing material volumes v_1, v_2, v_3, v_4, v_5, v_6. A tool path p_j is any combination of these volumes. Tool path selection involves determining the set of tool paths with minimum corresponding volume removal costs.

The tool path problem is represented using a 0–1 incidence matrix $[a_{ij}]$. For the part in Fig. 18.5 the incidence matrix (1) is shown below.

$$
[a_{ij}] = \quad
\begin{array}{c}
\\ v_1 \\ v_2 \\ v_3 \\ v_4 \\ v_5 \\ v_5
\end{array}
\begin{array}{cccccccc}
p_1 & p_2 & p_3 & p_4 & p_5 & p_6 & p_7 & p_8 \\
\left[\begin{array}{cccccccc}
1 & & & & & 1 & & \\
& 1 & & & & & 1 & 1 \\
& & 1 & & & & 1 & 1 \\
& & & 1 & & & & \\
& & & & 1 & & & \\
& & & & & 1 & &
\end{array}\right]
\end{array}
\qquad (1)
$$

$$c_j \quad 1.2 \quad 1.4 \quad 1.2 \quad 1.4 \quad 0.9 \quad 1.8 \quad 3.1 \quad 2.9$$

In its simplest form, the tool path selection problem can be modelled as the set covering problem described by (2)–(4). For a detailed discussion of the algorithms for solving the model see Balas and Ho [8].

$$\text{Minimize } \sum_{j=1}^{n} c_j x_j \qquad (2)$$

$$\text{subject to } \sum_{j=1}^{n} a_{ij} x_j \geq 1 \text{ for all } i \qquad (3)$$

$$x_j = 0 \text{ or } 1 \text{ for all } j \qquad (4)$$

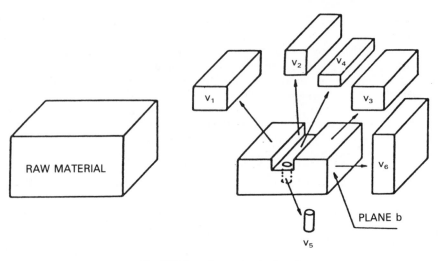

Fig. 18.5 Sample mechanical part.

where:

n number of tool paths (columns in matrix $[a_{ij}]$)

c_j cost of tool path p_j

$$a_{ij} = \begin{cases} 1 \text{ if volume } v_i \text{ is to be removed in tool path } p_j \\ 0 \text{ otherwise} \end{cases}$$

$$x_j = \begin{cases} 1 \text{ tool path } p_j \text{ is selected} \\ 0 \text{ otherwise} \end{cases}$$

The expert system selects the above model ((2)–(4)) for the tool path selection problem. In many applications, the incidence matrix $[a_{ij}]$ required by the model is very large, for example, $500 \times 10\,000$. Hence, the generation of its entries requires considerable computational effort. Either all possible tool paths or only their subset can be generated. The former case would result in a matrix of unmanageable size. Generating only a subset of all possible tool paths produces a smaller incidence matrix and the model becomes easier to solve. An expert system (with the help of rules such as Rule 22 below) can be used to generate the incidence matrix.

Rule 22: IF   mill diameter $d_m > b_r$, where b_r is breadth of raw material

AND   required surface roughness sr $<$ sr$_0$,

THEN generate tool path $p_j = \{v_1, ..., v_i, ..., v_q\}$, where $\{v_1, ..., v_i, ..., v_q\}$ is a set of volumes included in same layer of raw material.

Application of Rule 22 to the part in Fig. 18.5 results in tool path $p_7 = \{v_1, v_2, v_3\}$ in matrix (1). In a similar manner, the other tool paths, p_1, p_2, ..., p_6 and p_8 (in matrix $[a_{ij}]$) are generated. Using the algorithm of Balas and Ho [8], the above set covering the model in (2)–(4) may be solved for the above data $[a_{ij}]$. Qualitative aspects, if any, are then incorporated to the solution. It should be noted that, in general, for the tool path selection problem, solving the model yields an implementable solution. As a result, an implementable solution which incorporates qualitative aspects is obtained.

Model-Based System

The model-based system is suitable for problems which can be formulated using a number of models, each model being suitable for a particular situation. For a given problem, the knowledge-based system first selects an appropriate model and then an algorithm. The problem is solved by the algorithm and the solution produced is evaluated. If the solution is implementable, the knowledge-based system accepts it. For example, in the case of the machine layout problem, the solution (layout) is implementable if space constraints are satisfied and adjacency requirements are met in the layout produced by the knowledge-based system. If the solution is not implementable, then the system may take one of the following actions:

1. Modify certain parameters in the algorithm (if possible) and apply the algorithm again to the problem in order to generate a new solution, check whether it is implementable and repeat the above procedure until an implementable solution is obtained.
2. Modify the solution in order to make it implementable.

Of course, alternative 1 may not be applicable to all algorithms. Even if it is applicable to a particular algorithm, the corresponding parameter can be modified only to a certain extent, beyond which any modification fails to produce acceptable solutions. In such a case, i.e., when the parameter(s) in the algorithm cannot be modified any further, and if the solutions produced thus far are not implementable, the knowledge-based system adopts alternative 2 mentioned above. Note that the system may use alternative 1 to also improve the current solution. KBML [9], which is a model-based system, uses alternative 1 to improve the current solution and alternative 2 to make a solution implementable (if necessary).

The knowledge base in KBML consists of rules for solving the machine layout problem. There are five classes of rules, namely:

- Class 1 rules for determining the type of layout or the type of material handling system (MHS)
- Class 2 rules for selecting an appropriate model and algorithm for the layout problem

- Class 3 rules for making initial assignments based on input data
- Class 4 rules for varying parameters within the algorithm (if applicable)
- Class 5 rules for implementing the layout

To solve the layout problem, the five classes of rules are applied sequentially beginning from Class 1 rules which consist of four meta-rules and 10 first-order rules. A sample meta-rule and first-order rule are shown below. The meta-rules activate the first-order rules. The first-order rules are further categorized into two classes of rules, namely Class 1A and 1B rules. If the type of layout is known and the type of MHS is not, Class 1A rules are activated. If the type of MHS is known and the type of layout structure is not, Class 1B rules are activated.

Meta-rule:

Rule R2:   IF   type of layout is unknown and type of MHS is
                   known
           THEN apply rule R11 of Class 1B.

First-order rule:
Rule R12: IF   type of MHS is gantry robot
           THEN use multi-row layout.

Class 2 rules are capable of selecting an appropriate model and algorithm for solving the given problem. A sample rule which selects the model and algorithm for a given problem is provided below.

Rule R15: IF   number of machines to be assigned is < 8
           THEN select model M1 and solve the model optimally
                   using algorithm A1
           ELSE   select model M1 and solve the model
                   suboptimally using the heuristic algorithm A2.

Note that M1, A1 and A2 are the model and algorithms stored in the database of KBML. The models are represented as lists and the algorithms are defined as functions. The model and algorithm selected by KBML depend upon the nature of the problem, namely whether:

- The machines are of equal or unequal sizes
- The number of machines in the layout problem is < 8 or ≥ 8
- The machines are to be arranged in single-row, double-row or multi-row pattern

Class 3 rules are used to make initial assignments. These may be user-specified or decided by KBML. User-desired assignments have priority over the assigments done by KBML. Class 3 consists of six rules.

It is possible to modify certain parameters in some of the algorithms. Class 4 rules are used for changing these parameters. For every modified value of the parameter, the algorithm often provides a different solution (layout). The solution generated by the algorithm is evaluated for each value of the modified parameter. If the modification of the parameter in the algorithm leads to a better solution, the process is continued; otherwise it is terminated and the layouts obtained are evaluated for implementability by Class 5 rules. A layout is implementable if:

- Adjacency requirements (between pairs of machines) specified by the user are met
- Location restrictions of machines specified by the user are satisfied
- Space constraints are not violated

KBML uses a forward-chaining inference strategy. The inference engine attempts to match the data concerning type of MHS and type of layout with the IF part of the meta-rules in Class 1. If the match with the IF part of a rule is successful, then the rule fires other first-order rules. These first-order rules suggest either the type of layout or the type of MHS to be used depending upon which rule has been fired. The control is then directed to Class 2 rules. Again, the inference engine attempts to match the data provided by the user (number of machines to be assigned) and the data created by the first-order Class 1 rules (type of layout) with the IF part of Class 2 rules. If a successful match is found in any rule, the THEN part indicates the model and algorithm that is to be used to solve the layout problem considered.

Similarly, using the forward-chaining strategy, the inference engine uses Class 3 rules to perform the user-desired assignments and also some assignments based on the domain knowledge stored in the knowledge base. As mentioned before, such knowledge is represented in the form of production rules in KBML. The inference engine applies Class 4 and Class 5 rules in a way similar to the other rules.

The knowledge base in KBML consists of 35 rules. New rules can be easily added to it. The system is coded in Common Lisp and implemented on a Symbolics 3650 machine.

Model-Modifying System

Knowledge-based systems belonging to this class are, in general, more difficult to develop because they require a careful consideration of the knowledge base. The task of modifying a model to suit the problem environment is a difficult one even for human experts, since a number of factors such as data availability, ease of applying algorithms, constraints, etc., have to be examined. It is therefore understandable that it is difficult to develop a knowledge-based system, which like a human expert is capable of modifying a model to suit the problem considered.

The model-modifying approach involves two steps. In the first step, the knowledge-based system either constructs an appropriate model for the given problem or selects one from the available set of models. In the second step, it determines whether there are any algorithms available to solve the model. If no appropriate algorithms are found, then the system either modifies the current model such that a suitable algorithm for solving it is available, or constructs another model for the problem. If the knowledge-based system is unable to do either of these, then it considers modifying the algorithm or parameters within the algorithm, so that the problem can be solved. Of course, modification of the solution generated by the algorithm, in order to make it implementable or to incorporate qualitative factors, is done, if necessary.

A similar approach has been suggested by Kanet and Adelsberger [10]. However, perhaps the first attempt to build a model-modifying system was by Dolk and Konsynski [11]. In the data-modifying and model-based systems, the models are simply selected from the model base and are not manipulated. Hence their representation is not an important issue. But, in a model-modifying system, since models are constructed or modified, their representation should be given careful consideration. Evans et al. [12] have also attempted to develop a model-modifying system that is capable of selecting or constructing models for the production-planning problem. The system developed by Dolk and Konsynski [11] is discussed below. These authors examined the issues involved in knowledge representation for model management systems. They proposed a "model abstraction" approach to represent models in a database. It involves representing a model in three sections, namely: data objects, procedures and abstractions. The data objects section lists the data items of the model. For example, the data items for a linear programming (LP) model are objective function, constraints, decision variables, etc. The procedures section lists (1) the procedures available, i.e., addition of constraints, deletion of constraints, etc., (2) the data objects the model accesses and (3) the data objects it returns. The assertions section lists information about data objects and procedures and also their relationships. For example, the assertions section of an LP model may specify that all the expressions be linear in the decision variables.

Using the above mentioned approach, Dolk and Konsynski show how LP models may be represented in a system. More importantly, they demonstrate how constraints, for example, integrality constraints, may be added to an LP problem to make it an integer-programming problem. The model management system that they developed constructs a model based on the statement of the problem provided by the user and identifies a similar model stored in the database using a pattern-matching technique. It then solves the constructed model using an appropriate algorithm. The above model management system is classified as a model-modifing expert system because it is capable of constructing a model for the given problem.

Modelling the Manufacturing Equipment Selection Problem

Before discussing the structure and problem solving approach of KBSES, it is necessary to provide the model used to formulate the manufacturing equipment selection problem. In the past, a number of approaches have been used to model and solve the manufacturing equipment selection problem. Miller and Davis [1] have surveyed, classified and compared a number of them. The dynamic resource allocation model developed by Miller and Davis [13], the machine procurement model in Murty [14], the production equipment selection model in Kusiak [3], the aggregate production planning and machine requirements planning model in Behnezhad and Khoshnevis [15] are examples of the models which have been used to formulate the manufacturing equipment selection problem. In this section, an integer-programming formulation of the problem is presented. The following notation is used:

- c_{ij} cost of performing operation o_i on machine M_j
- h_{ij} cost of handling part P_i using material-handling carrier H_j
- r_{ij} time required to perform operation o_i on machine M_j
- s_{ij} time required to transport part P_i using material-handling carrier H_j
- T_j time available on machine M_j
- S_j time available on material-handling carrier H_j
- N_i number of operations o_i to be performed
- M_i number of units of part P_i to be manufactured
- U_j cost of machine M_j
- V_j cost of material-handling carrier H_j
- B total budget available
- x_{ij} number of operations o_i to be performed on machine M_j
- y_{ij} number of units of part P_i to be transported on material-handling carrier H_j
- u_j number of units of machine M_j selected
- v_j number of units of material-handling carrier H_j selected

Model M1

The objective function of model M1 minimizes the procurement cost of machines and material-handling carriers as well as the operating and handling cost of the parts to be manufactured.

$$\text{Minimize} \sum_{i=1}^{m} \sum_{j=1}^{n} c_{ij} x_{ij} + \sum_{i=1}^{p} \sum_{j=1}^{q} h_{ij} y_{ij} + \sum_{j=1}^{n} U_j u_j + \sum_{j=1}^{p} V_j v_j \qquad (5)$$

$$\text{subject to} \sum_{j=1}^{n} x_{ij} \geq N_i \qquad\qquad\qquad i=1, \dots, m \quad (6)$$

$$\sum_{i=1}^{m} r_{ij} x_{ij} \leq T_j u_j \qquad\qquad\qquad j=1, \dots, n \quad (7)$$

$$\sum_{j=1}^{q} y_{ij} \geq M_i \qquad\qquad\qquad i=1, \dots, p \quad (8)$$

$$\sum_{i=1}^{p} s_{ij} y_{ij} \leq S_j v_j \qquad\qquad\qquad j=1, \dots, q \quad (9)$$

$$\sum_{j=1}^{n} U_j u_j + \sum_{j=1}^{p} V_j v_j \leq B \qquad\qquad\qquad (10)$$

$$x_{ij} \geq 0, \text{ integer} \qquad\qquad\qquad i=1, \dots, m \quad (11)$$
$$j=1, \dots, n$$

$$y_{ij} \geq 0, \text{ integer} \qquad\qquad\qquad i=1, \dots, p \quad (12)$$
$$j=1, \dots, q$$

$$u_j \geq 0, \text{ integer} \qquad\qquad\qquad j=1, \dots, n \quad (13)$$
$$v_j \geq 0, \text{ integer} \qquad\qquad\qquad j=1, \dots, q \quad (14)$$

Constraints (6) and (8) ensure that the required number of operations are performed and the required number of parts are transported, respectively. Constraints (7) and (9) impose that the time available on each machine and material-handling carrier is not exceeded. Constraint (10) ensures that the available budget is not exceeded. Constraints (11)–(14) impose nonnegativity and integrality.

Note that model M1 can be used for the selection of machines and material-handling carriers. However, if the material-handling carriers have already been selected, then model M1a can be used. Note that model M1a is a simplified form of model M1.

Model M1a

$$\text{Minimize} \sum_{i=1}^{m} \sum_{j=1}^{n} c_{ij} x_{ij} + \sum_{j=1}^{n} U_j u_j \qquad (5a)$$

$$\text{subject to} \sum_{j=1}^{n} U_j u_j \le B \tag{10a}$$

and constraints (6), (7), (11) and (13).

On the other hand, if the required machines have already been selected, then the following variant of model M1 can be used to select the required material-handling carriers.

Model M1b

$$\text{Minimize} \sum_{i=1}^{p} \sum_{j=1}^{q} h_{ij} y_{ij} + \sum_{j=1}^{p} V_j v_j \tag{5b}$$

$$\text{subject to} \sum_{j=1}^{p} V_j v_j \le B \tag{10b}$$

and constraints (8), (9), (12) and (14).

If the available budget is not known, then model M1 without constraint (10) can be used for the manufacturing equipment selection. Thus, model M1 can be modified to suit the various problem scenarios that might be encountered. As will be seen later, this property of model M1 adds modelling flexibility to KBSES.

Problem-Solving Approach in KBSES

KBSES which has the features of data-generating and model-modifying knowledge-based systems is discussed in the rest of this chapter. Its problem solving technique is outlined in the following seven steps:

Step 1. Collect data from the user.

Step 2. Represent declarative knowledge concerning parts, operations and machines, as frames.

Step 3. Construct a model for the problem considered.

Step 4. Eliminate appropriate machines (material handling carriers) from consideration, so that the number of integer variables in the problem and hence its size is reduced.

Step 5. Generate the following matrices as necessary:

Operation-machine cost matrix
Operation-machine time matrix
Part-material-handling carrier cost matrix
Part-material-handling carrier time matrix

Generate the following vectors:

Time available on each machine
Time available on each material-handling carrier
Number of times each operation is to be performed
Number of units of each part to be manufactured
Cost of each machine
Cost of each material-handling carrier

Step 6. Solve the model formulated in step 3 for the data generated in step 5, using the algorithm stored in the model and algorithm base.

Step 7. Evaluate the solution produced by the algorithm, incorporate qualitative factors not considered in the model formulation step, and modify the solution, if necessary.

First, the system obtains the required data from the user and stores them in the database. These data are then used to represent the declarative knowledge concerning parts, machines and operations. Next, the system either selects or constructs a model for the manufacturing equipment selection problem. If all the data required by model M1 are provided, then model M1 is selected; otherwise, a new model which is a variant of model M1 is formulated. Then, it attempts to eliminate some machines (material-handling carriers) from consideration, in order to reduce the problem size. Note that the fewer the number of machines (material-handling carriers), the fewer the integer variables in the model. The system considers procurement cost, machine (material-handling carrier) capabilities, etc., to determine whether a piece of equipment is to be eliminated from consideration. A sample rule which is used to eliminate machines from consideration is shown below.

IF   machine M_1 can perform operation o_1
  AND   machine M_2 can perform operation o_2
  AND   machine M_3 can perform operation o_1 and o_2
  AND   the cost of the machines M_1 and M_2 is greater than that of machine M_3
THEN   eliminate machines M_1 and M_2 from consideration.

Then the system applies knowledge to the data stored in the data base, in order to generate a set of data suitable for the model formulated. The data generated are in the form of matrices and vectors. Using the algorithm available in the model and algorithm base, the model is solved for the data generated by the system. The solution produced is evaluated, qualitative factors not considered in the model are incorporated in the solution and the revised solution is provided to the user.

Structure of the Knowledge-Based System for Equipment Selection

KBSES consists of the following five components (Fig. 18.6):

- Database
- Knowledge base
- Model and algorithm base
- Explanation module
- Inference engine

Database

The database consists of data pertaining to the problem to be solved. Such data are obtained from the user in an interactive mode. The user is required to enter only the information concerning parts and machines. For each part, the system requests the following data:

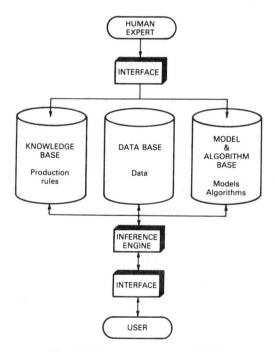

Fig. 18.6 Basic components of KBSES.

- Code
- Number of units to be manufactured
- Code of each operation to be performed
- Number of times each operation is to be performed
- Type of available material-handling carrier(s) that can be used to transport the part, the corresponding operating cost(s) and time(s) involved in transporting the part, and the carrier(s) procurement cost

If the code of the part entered by the user matches with that of a part already existing in the knowledge base, the system requests only the number of units to be manufactured; the remaining data are not requested since they are already available in the knowledge base.

For each machine, the following data are requested:

- Code
- Code of each operation it can perform
- Corresponding operating cost and operating time
- Procurement cost

As before, if the code of a machine matches with that of a machine already existing in the knowledge base, the system does not request the user to enter the operating cost, operating time and procurement cost.

Knowledge Base

The knowledge base consists of *declarative knowledge* and *procedural knowledge*. Frames are used to represent the declarative knowledge concerning machines and parts. The entire procedural knowledge is stored in the form of production rules.

The data in the database are used to create declarative knowledge. There are three main types of declarative knowledge in KBSES, namely knowledge concerning:

- Parts
- Operations
- Machines

A sample frame belonging to each type of declarative knowledge is shown in Figs 18.7–9.

The procedural knowledge in KBSES is divided into the following five classes of rules:

- Class 1 rules for collecting data from the user
- Class 2 rules for formulating a model for the problem specified by the user
- Class 3 rules for generating required data in the form of matrices and vectors

```
(Part (Part_name) (Oper_reqd (Oper_1 Oper_2 ... Oper_N))
                  (No_of_oper (n_1 n_2 ... n_n))
                  (mhs_avail (mhs_1 mhs_2 ... mhs_n))
                  (mhs_oper_cost (moc_1 moc_2 ... moc_n))
                  (mhs_oper_time (mot_1 mot_2 ... mot_N)))
```

Fig. 18.7 A sample frame for representing a part.

```
(Operation (Oper_name) (M_reqd (M_1 M_2 ... M_n))
                       (Oper_cost (cost_1 cost_2 ... cost_n))
                       (Oper_time (time_1 time_2 ... time_n)))
```

Fig. 18.8 A sample frame for representing an operation.

```
(Machine (M_name) (Oper(Oper_1 Oper_2 ... Oper_n))
                  (Oper_cost (cost_1 cost_2 ... cost_n))
                  (Oper_time (time_1 time_2 ... time_n)))
```

Fig. 18.9 A sample frame for representing a machine.

- Class 4 rules for selecting the algorithm required to solve the model formulated
- Class 5 rules for incorporating qualitative factors (not considered in the model) into the solution produced by the algorithm

A sample rule belonging to each class is shown below.

Class 1 rule:
    IF   code of a machine (provided by user) does not match with that of a machine already existing in the knowledge base
    THEN request user to provide: code of machine, code of each operation it can perform, corresponding operating cost, time and machine purchasing cost.

Class 2 rule:
    IF   information on available budget is not provided by user
    THEN create model M1 which consists of objective function (5) and constraints (6)–(9), (11)–(14).

Class 3 rule:
    IF   time available on each machine is provided by user
    THEN generate a time vector which indicates time available on each machine.

Class 4 rule:
    IF   model M1 has been selected and the number of integer variables in the model is > 25
    THEN apply algorithm A2
    ELSE apply algorithm A1.

(Basic_model (Obj_func) ((Constraint_ 6) (Constraint_ 7)
(Constraint_ 8) (Constraint_ 9)
(Constraint_10) (Constraint_11)
(Constraint_12) (Constraint_13)
(Constraint_14)))

Fig. 18.10 Frame representation of model M1 in KBSES.

Note that algorithm A1 refers to the branch-and-bound algorithm (see for example, Schrage [16]). Algorithm A2 is a heuristic algorithm and is presented later in this section.

Class 5 rule:

IF   material-handling carrier H_1 is not suitable to tend machine M_1

AND material-handling carrier H_1 is selected to tend machine M_1

THEN select material-handling carrier H_2 that can perform the same functions as material-handling carrier H_1 and that is suitable for machine M_1.

Model and Algorithm Base

Model M1 is stored in the model and algorithm base of KBSES in the form of a frame (Fig. 18.10). It is used to formulate the equipment selection problem that is provided by the user. The specific model formulated depends upon the nature of the problem, which is reflected in the data provided by the user. For example, if the user does not provide budget data, then the system creates a new model which does not involve the corresponding constraint, i.e., constraint (10) in model M1. If the material-handling carriers have already been selected, then model M1a presented earlier is created, and so on.

As already mentioned, the knowledge required to formulate a specific model for the problem statement is stored in the knowledge base of KBSES in the form of production rules. The branch-and-bound algorithm used to solve model M1 or its variants, is stored in the model and algorithm base.

In addition to the above, the model and algorithm base includes a heuristic algorithm to solve model M1. The heuristic algorithm involves two phases. In the first phase, model M1a without constraint (10a) is solved; this results in selection of the required machines. In the second phase, model M1b which includes objective function (5b) and constraints (8), (9), (10b), (12) and (14), is solved. The resulting solution indicates the required number and type of material-handling carriers to be purchased. The advantage of using heuristic algorithm A2 is that its computational time requirement is lower compared with that of algorithm A1.

Inference engine

The inference engine employs the forward-chaining strategy. It sequentially applies the five classes of rules beginning from class 1 rules. Its features are similar to the inference engine described in Heragu and Kusiak [9].

Illustrative Example

In this section, an example problem is solved using KBSES. The problem is to determine the machines and the material-handling carriers to be purchased for the following data:

- 20 units of part P_1, 15 units of part P_2, 18 units of part P_3, 10 units of part P_4, and 12 units of part P_5 to be manufactured
- Available machines: M_1, M_2, M_3, M_4
- Available material-handling carriers: H_1, H_2
- Available time on each machine and material-handling carrier: 5000
- Cost of machines, M_1, M_2, M_3, M_4 are: \$130 000; \$190 000; \$80 000; and \$255,000 respectively
- Cost of material-handling carriers H_1, H_2, are: \$130 000 and \$200 000 respectively
- Available budget is \$800 000

As mentioned before, KBSES acquires the above data in an interactive mode. In addition, it also obtains information on:

- Machine capability, i.e., operations each machine can perform, cost and time required to perform each operation
- Part requirements, i.e., operations to be performed on each part, alternative material-handling carriers on which it can be transported and the corresponding handling cost and time

The information on machine capabilities and part requirements is obtained for each machine and each part. This knowledge is then represented using frames. For the above problem, the frames for representing parts, operations and machines are:

```
(Part (P₁)      (Oper_reqd (o₁ o₂ o₃))
                (No_of_oper (1 1 1))
                (mhs_avail (H₁ H₂))
                (mhs_oper_cost (50 60))
                (mhs_oper_time (50 60)))
(Part (P₂)      (Oper_reqd (o₂ o₃ o₅ o₆))
                (No_of_oper (1 1 1 1))
                (mhs_avail (H₁H₂))
```

```
                         (mhs_oper_cost (70 70))
                         (mhs_oper_time (70 70)))
(Part (P_3)              (Oper_reqd (o_4 o_7 o_8))
                         (No_of_oper (1 1 1))
                         (mhs_avail (H_1))
                         (mhs_oper_cost (70))
                         (mhs_oper_time (70)))
(Part (P_4)              (Oper_reqd (o_1 o_6 o_8))
                         (No_of_oper (1 1 1))
                         (mhs_avail (H_2))
                         (mhs_oper_cost (60))
                         (mhs_oper_time (60)))
(Part (P_5)              (Oper_reqd (o_2 o_4 o_5))
                         (No_of_oper (1 1 1))
                         (mhs_avail (H_1 H_2))
                         (mhs_oper_cost (50 50))
                         (mhs_oper_time (50 50)))
(Operation (o_1) (M_reqd (M_1 M_2 M_3))
                         (Oper_cost (20 15 40))
                         (Oper_time (20 15 40)))
(Operation (o_2) (M_reqd (M_1))
                         (Oper_cost (20))
                         (Oper_time (20)))
(Operation (o_3) (M_reqd (M_1 M_3))
                         (Oper_cost (10 30))
                         (Oper_time (10 30)))
(Operation (o_4) (M_reqd (M_1 M_3))
                         (Oper_cost (8 20))
                         (Oper_time (8 20)))
(Operation (o_5) (M_reqd (M_3 M_4))
                         (Oper_cost (10 8))
                         (Oper_time (10 8)))
(Operation (o_6) (M_reqd (M_2 M_4))
                         (Oper_cost (8 5))
                         (Oper_time (8 5)))
(Operation (o_7) (M_reqd (M_2))
                         (Oper_cost (12))
                         (Oper_time (12)))
(Operation (o_8) (M_reqd (M_2 M_4))
                         (Oper_cost (15 8))
                         (Oper_time (15 8)))
(Machine (M_1)   (Oper (o_1 o_2 o_3 o_4))
                         (Oper_cost (20 20 10 8))
                         (Oper_time (20 20 10 8)))
(Machine (M_2)   (Oper (o_1 o_6 o_7 o_8))
                         (Oper_cost (15 8 12 15))
                         (Oper_time (15 8 12 15)))
```

(Machine (M_3) (Oper (o_1 o_3 o_4 o_5))
　　　　　　　　(Oper_cost (40 30 20 10))
　　　　　　　　(Oper_time (40 30 20 10)))
(Machine (M_4) (Oper (o_5 o_6 o_8))
　　　　　　　　(Oper_cost (8 5 8))
　　　　　　　　(Oper_time (8 5 8))))

Since all the data required by model M1 are provided by the user, the system selects model M1 from the model and algorithm base. If the user had not provided all the data required by model M1, the system would have created a different model. The data provided by the user indicate that the cost of performing operation o_i on machine M_j is equal to the time required to perform operation o_i on machine M_j. Similarly, the cost of handling part P_i on material-handling carrier H_j is equal to the time required to transport part P_i on material-handling carrier H_j. For the above problem the system does not attempt to eliminate any machine from consideration, because the problem size is small. The following data are then generated by the system:

	Machine			
	M_1	M_2	M_3	M_4
o_1	20	15	40	—
o_2	20	—	—	—
o_3	10	—	30	—
o_4	8	—	20	—
o_5	—	—	10	8
o_6	—	8	—	5
o_7	—	12	—	—
o_8	—	15	—	8

$[c_{ij}] = [r_{ij}] = $ (above matrix), Operation

	Material-handling carrier	
	H_1	H_2
P_1	50	60
P_2	70	70
P_3	70	—
P_4	—	60
P_5	50	50

$[h_{ij}] = [s_{ij}] = $ (above matrix), Part

Vector of operations to be performed = $[N_i] = [30\ 47\ 35\ 30\ 27\ 25\ 18\ 28]^T$

In addition to the above, the system generates vectors which indicate the time available on each machine and material handling carrier, their corresponding procurement costs, and the number of units of each part to be manufactured. Since the number of integer variables is greater than 25, algorithm A2 is invoked to solve model M1. The resulting solution is provided below:

$$[x_{ij}] = \begin{array}{c} \\ o_1 \\ o_2 \\ o_3 \\ o_4 \\ o_5 \\ o_6 \\ o_7 \\ o_8 \end{array}
\begin{array}{cccc}
\overset{\text{Machine}}{} \\
M_1 & M_2 & M_3 & M_4 \\
\left[\begin{array}{cccc}
0 & 30 & 0 & - \\
47 & - & - & - \\
35 & - & 0 & - \\
30 & - & 0 & - \\
- & - & 27 & 0 \\
- & 25 & - & 0 \\
- & 18 & - & - \\
- & 28 & - & 0
\end{array}\right] \text{Operation}
\end{array}$$

$$[y_{ij}] = \begin{array}{c} P_1 \\ P_2 \\ P_3 \\ P_4 \\ P_5 \end{array}
\begin{array}{cc}
\overset{\text{Material-}}{\underset{\text{carrier}}{\text{handling}}} \\
H_1 \quad H_2 \\
\left[\begin{array}{cc}
20 & 0 \\
15 & 0 \\
18 & - \\
- & 10 \\
12 & 0
\end{array}\right] \text{Part}
\end{array}$$

Machines M_1, M_2, M_3 and material-handling carriers H_1, H_2 at a cost of $730 000 are selected. The operating and handling costs are $3086 and $4510, respectively. The numbers of times each operation is to be performed on the machines are given in matrix $[x_{ij}]$. The number of units of each part to be transported on the selected material-handling carriers are shown in matrix $[y_{ij}]$. Since the solution is satisfactory, it is presented to the user.

Conclusion

The use of knowledge-based programs for designing manufacturing systems has been discussed. A new approach for solving the manufacturing equipment selection problem has been presented. It involves combining optimization and knowledge-based techniques. This approach is adopted in KBSES, a program for selecting the machines and material-handling carriers required for an automated manufacturing system. The models for the selection of manufacturing equipment available in the literature are not adequate as they do not consider the integration of machines and material-handling carriers. The approach presented allows one to incorporate other models and algorithms.

Acknowledgement. The research presented in this chapter was done while both authors were at the University of Manitoba. The support received from the Natural Sciences and Engineering Research Council of Canada and the University of Manitoba is greatly appreciated.

References

1. Miller DM, Davis RP. The machine requirements problem. Int J Prod Res 1977; 15: 219–231
2. Hayes GM, Davis RP, Wysk RA. A dynamic programming approach to machine requirements planning. AIIE Trans 1981; 13: 175–181
3. Kusiak A. The production equipment requirements problem. Int J Prod Res 1987; 25: 319–325
4. Heragu SS, Kusiak A. Analysis of expert systems in manufacturing design. IEEE Trans Syst Man Cybernet 1987; SMC-17: 898–912
5. Kusiak A. Intelligent manufacturing systems. Prentice Hall, Englewood Cliffs, NJ, 1990
6. Matsushima K, Okada N, Sata T. The integration of CAD and CAM by application of artificial intelligence techniques. Ann CIRP 1982; 31: 329–332
7. Chang TC, Wysk RA. An introduction to automated process planning systems. Prentice Hall, Englewood Cliffs, NJ, 1985
8. Balas E, Ho A. Set covering algorithm using cutting planes, heuristics and subgradient optimization. Math Prog 1980; 12: 37–60
9. Heragu SS, Kusiak A. KBML: a knowledge-based system for machine layout. In: Proceedings 1988 international industrial engineering conference, Orlando, FL
10. Kanet JJ, Adelsberger HH. Expert systems in production scheduling. Eur J Operat Res 1987; 29: 51–59
11. Dolk DR, Konsynski BR. Knowledge representation for model management systems. IEEE Trans Software Eng 1984; SE-10: 619–628
12. Evans JR, Dennis DR, Shafer SM. An intelligent system for production planning modelling and optimization. In: Proceedings IXth international conference on production research, Cincinnati, OH, 1987, pp 1689–1693
13. Miller DM, Davis RP. A dynamic resource allocation model for a machine requirements problem. IIE Trans 1978; 10: 237–243
14. Murty KG. Linear programming. Wiley, New York, 1983
15. Behnezhad AR, Khoshnevis B. The effects of manufacturing progress function on machine requirements and aggregate planning. Int J Prod Res 1988; 26: 309–326
16. Schrage LE. Linear, integer, and quadratic programming with LINDO. Scientific Press, Palo Alto, CA, 1984

Design of Intelligent Manufacturing Systems: Critical Decision Structures and Performance Metrics

S. Parthasarathy and S.H. Kim

Introduction

The systematic design of intelligent manufacturing systems (IMSs) is a complex task that requires an understanding of the nature and structure of manufacturing systems, as well as the models used to represent such systems. This chapter discusses some critical issues in making precise the representation and utilization of knowledge for intelligent manufacturing design. The representational issues include the analysis of the structure of manufacturing plants and the representation of these systems as automata. The utilization issues relate to two conceptually separable clusters: decision rules and performance metrics.

Figure 19.1 identifies the components of IMSs which serve as raw material for knowledge representation. This article describes the strategy for utilizing knowledge in the design of IMSs through the application of decision rules and performance metrics.

The final organization and operation of an IMS depend to a large extent on decisions made at the early stages of the design process. A systematic approach to the synthesis of IMSs requires the identification and classification of general design principles. The ultimate objective of this approach is to develop a set of generalized decision rules that can be encoded into an advisor to assist in the synthesis of an IMS. The decision rules may be validated by observing the operation of the manufacturing system model under various simulated conditions, using a set of performance metrics or functions.

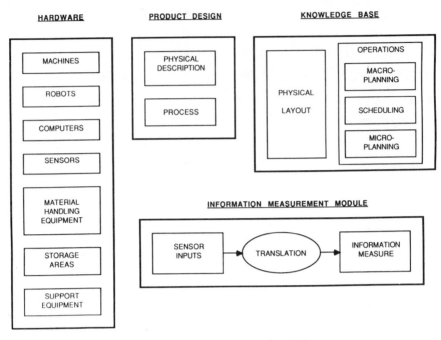

Fig. 19.1 Some components of an IMS.

Attributes of Intelligent Manufacturing Systems

The behaviour of an engineered system is *acceptable* if it fulfils the functional requirements of the system. The acceptability of the system's response is determined by a performance metric or function. If the value of the evaluated response falls within a predefined range, then the system is considered acceptable.

An IMS is one that is capable of operating automatically and efficiently in a changing environment, and in the face of unreliable system components. In other words, the IMS is capable of adaptation, learning, self-repair and self-organization in a dynamic environment that is constantly changing over time. At this point it is appropriate to define the concepts of adaptation, learning, self-repair and self-organization. The following definitions draw on previous perspectives [1].

- A system is called *adaptive* if its response is acceptable in the face of a changing environment or internal structure.
- A *learning* system is an adaptive system whose behaviour space changes over time in response to changes in the environment or within the system itself.

A learning system is one that incorporates some internal representation of changes in the system or the environment. A change in the behaviour of the system could cause a temporary decrease in the performance level. However, an effective learning system will tend to exhibit monotonically increasing levels of performance; if a new behaviour would result in decreased performance, then the system would simply revert to its original behaviour. Moreover, the learning ability of a system is orthogonal to its self-repairing and self-organizing abilities.

- A *self-repairing* system is an adaptive system which redresses disturbances in its internal structure.
- A system is *self-organizing* if, given an initial medley of components, the system unifies the parts to fulfil some purpose.

A self-organizing system is at a higher evolutionary stage than learning and self-repairing systems. As such, it is inherently capable of learning and self-repair.

Finally, a system is *intelligent* if it is adaptive, learning, self-repairing or self-organizing.

A flexible manufacturing system (FMS) is an IMS with little or no capabilities in terms of learning, self-repair and self-organization. However, an FMS is still capable of adaptive behaviour. The flexibility of an FMS lies in its ability to produce a variety of products efficiently on the same set of machines. The subsequent sections regard the FMS as a special case of intelligent production systems.

Flexible Manufacturing Systems

A basic characteristic of a flexible manufacturing system is versatility, the ability to produce a large variety of products using the same equipment. A second factor is the automation of operations, resulting in high productivity levels and resource utilization. FMSs were first introduced to respond quickly to demands in the marketplace. They are particularly suited in situations where [2]:

- The products have fairly short life cycles.
- The rate of new product introduction is high.
- The market is fragmented.
- The market demand is fluctuating.
- The competition is severe.
- Technological change is rapid.

These factors relate to a dynamic environment which demands a timely, flexible response. An FMS allows for an increase in the variety of products, and a reduction in the human effort.

However, researchers have found a marked inflexibility over the years in FMSs. In some cases, the FMS performed more poorly than the conventional systems that they replaced. In a number of cases, this was due to ineffectual management [3], but in others it was the design of the system itself that was suspect.

System designers often follow an informal design process whereby they match an FMS design to a set of products chosen by management. Here the flexibility of the system is limited to the initial set of products and others having similar characteristics. But these systems, for the costs involved, should be more flexible than they are. On the other hand, it is unrealistic and uneconomical to design systems to produce arbitrary products. The solution may lie somewhere in between: an integrated product and process view, rather than a strictly product or process perspective. The FMS should be designed to produce the existing range of products efficiently, with a view towards accommodating future products. The modular design of the system, and the arrangement of the cells to maximize flexibility, are key to achieving a truly flexible system.

The first step in developing the knowledge to synthesize IMSs systematically is to develop a conceptual and formal model of an IMS. A number of preliminary steps in this direction will be taken after a brief review of some related work.

Some Related Work

Despite the growing importance of intelligent manufacturing systems in industrial society, little systematic research has been conducted in this domain. Our knowledge of the fundamental nature of production operations is sketchy at best. However, the design of intelligent factories requires a systematic understanding of information parameters and metrics for performance evaluation. This section highlights some of the previous work performed in these areas.

Information accounts for a large fraction of the total cost of production plants [4]. Yet, in spite of the central position that information occupies in manufacturing, little has been written about the nature and characteristics of information as they relate to system performance and the fulfilment of functional requirements.

In a related context, students of management science have investigated some of the parameters of information processing in organizations. An example is the observation that the information needed by a subsystem to perform a given task is equal to the difference between the information required for the task, and that which is already available [5].

A promising source of existing knowledge lies in information theory, a field dealing with the transmission of messages. The mainstay of this field is the problem of conveying signals reliably through a conduit. However, any attempt to apply information-theoretic concepts to the wider class of issues arising in factories, requires a broadening of the key concepts in the classical approach to information. To this end, classical information theory may be interpreted in the context of purposive behaviour [6–8]. This information-theoretic model has been used to study the implications for the design of intelligent systems in different classes of applications, ranging from mobile robots to production process control. The increasing trend towards computerized control and intelligent devices implies that a proper understanding of information characteristics will become even more important in the future [9].

Models of physical systems, being idealized constructs, will often represent reality imprecisely [10]. Therefore, real-time data must be monitored and reconciled. Because a realistic situation is not static, the system may change and therefore require new information to cope with disturbances. It is vital that this information be relevant, complete and timely. The problem of timeliness has led to the proposition that one way to categorize time delays is to classify them into different functional stages in terms of sensing, reasoning and acting [11, 12]. Since the information processing needs differ for each stage, the design issues must be handled in corresponding fashion.

A better understanding of production operations and intelligent mechanisms will also enable us to address the theoretical limitations to the processing of information and the extent to which activities can be automated. Some research in this area indicates that, in general, an automated agent cannot determine the function or purpose of a system merely by inspecting its defining specifications [13, 14]. The growing recognition of the need to develop a scientific basis for manufacturing systems, and intelligent machines in general, suggests that significant strides in the study of flexible plants will be made in the years ahead.

Conceptual Framework for IMSs: the Manufacturing Machine

It may be convenient to consider the manufacturing system as a "black box" or a purely "input–output machine" as shown in Fig. 19.2(a) [6, 15]. A more detailed model of the manufacturing system, shown in Fig. 19.2b, distinguishes between the physical and information systems. The physical system takes in raw materials such as steel, wood or chemicals, and produces a set of final products. The software component takes in information such as the product specifications, and produces output information such as the level of system performance.

Both the physical and informational systems transform or change the inputs to obtain the outputs. The changes that take place within the physical system can be classified into three basic types of transformation: geometric, internal and union transformations [15]. These changes are effected using a set of machines, robots and other support equipment. Within the information system the inputs are processed by humans and

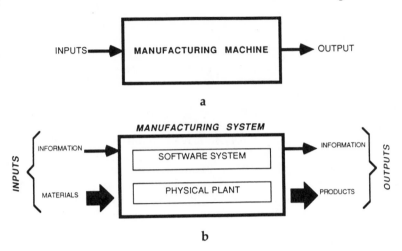

Fig. 19.2 Conceptual framework.

computers. This information processing is performed with reference to a suitable representation of the actual physical system, along with a set of objectives and constraints. Thus a representation or model of the physical system is crucial to the processing of information within the plant.

Figure 19.3a illustrates the input to the information system. This input includes the functional requirements of the products. The functional requirements or specifications of a product depend on a number of factors including market demand and supply; management objectives and constraints such as cost, time, space and quality; and governmental and environmental factors. To illustrate, the input to an automobile factory may be "Produce 1000 Cougars and 2000 Lynx per month". Here the numbers 1000 and 2000 reflect market information, and thereby production requirements. This input is very general and must be decomposed into the lower levels, as illustrated in Fig. 19.3b. The requirement of 1000 Cougars per month (f_1) is decomposed into design, planning, scheduling and other aspects, each of which may in turn be broken down further.

Perspective on a Formal Framework for IMSs

A consistent representational framework for intelligent manufacturing systems is necessary for their systematic analysis and synthesis. In the past, analytical models of the manufacturing system have fared poorly in representing discontinuous or nonlinear systems. Another major drawback of analytical models is their inability to represent and manipulate qualitative information.

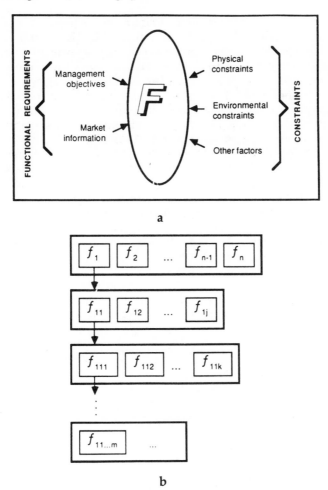

Fig. 19.3 a Input specifications. **b** Hierarchical structure of functional requirements.

A representation of an IMS must facilitate analysis and synthesis of both quantitative and qualitative information. The representation must also capture the knowledge hierarchy within the manufacturing system. A convenient structure is a hierarchy corresponding to the structure of a business organization. This type of structure is characterized by the policy- and decision-making level of top management, the planning and coordination level of middle management and the execution level of the production supervisor. A formal model of the manufacturing system will allow us to study the limits and develop guidelines for its construction and operation.

A formal framework for IMS is proposed, incorporating the principles of information theory, transformation theory of production processes [15], automata theory and computability theory. Figure 19.4 depicts the formalization of manufacturing knowledge through these theories. This approach represents the manufacturing system and its components as automata, whether of the deterministic or stochastic type. Transformation theory, in conjunction with automata and computability theories, can be used to study the nature and limits of automation in a manufacturing system and to develop guidelines for automation. Information theory and transformation theory are used to develop a set of performance metrics to evaluate the performance of a manufacturing system. The purpose of developing these metrics is to monitor and compare alternatives. A kernel model of the manufacturing system results from the distillation of the theoretical model, the performance metrics and the results on automatability. The theoretical models along with the results on automatability and the performance metrics can then be used to validate existing design rules for the synthesis of manufacturing systems and to generate novel design rules for IMS.

Decision rules and performance metrics are two critical knowledge components of an IMS. Table 1 illustrates the overall framework for the nature and structure of these two components. Structure relates to a

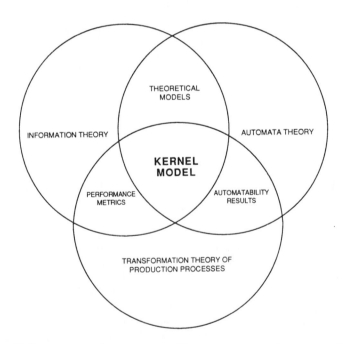

Fig. 19.4 Formalization of manufacturing knowledge through transformation, information and automata theories.

Table 1. Framework for the nature and structure of manufacturing knowledge

Structure	Nature
Decision rules	
Facts or commands	Alternatives and complementary rules
Conditional or simple rule	Cooperating and competing rules
	Conflicting rules
Performance metrics	
Classical information metric	Alternative metrics
Entropy metric	Complementary metrics
	Competing metrics
	Conflicting metrics

property of an individual decision rule or metric, while nature is a set attribute. In other words, *structure* reflects the composition of the individual elements while *nature* denotes the interdependencies that exist among a collection of elements. This framework is elaborated further below.

Formal Model of an IMS

A simple model for adaptive systems, suggested by Glorioso and Colon Osorio [1], has been incorporated into a system model for intelligent manufacturing systems. The model consists of a physical system and a regulating system; the latter module may be further partitioned into a performance evaluation module and an activator module.

Figure 19.5 illustrates the model configuration. The manufacturing system maps the input I, into an output Ω. The input I, is a function of the product specifications and time t. The index set of the input to the system is denoted by $J = \{1, 2, ..., m\}$. In other words, the inputs to the system are denoted by a set $I = \{I_1, I_2, ..., I_m\}$. The output Ω, is a function of the input I, the state Q of the manufacturing system, and time t.

The performance evaluation element (PEEL) evaluates the system with respect to some metric. It embodies the goals and constraints of the system, serving to map the input and output in terms of a metric Π. The activating element (ACTEL) interprets or transforms the measure Π into specific instructions that guide the manufacturing system to acceptable levels of performance. The PEEL and ACTEL together account for the intelligence of the system and constitute the learning or teaching module.

In general, the performance measure Π is a vector or list of attributes rather than a scalar. For example, a microprocessor fabrication plant may have the performance measures Π_1 = daily rate of chip production and Π_2 = yield rate. Then the performance of the plant is given by $\Pi = \langle \Pi_1, \Pi_2 \rangle$.

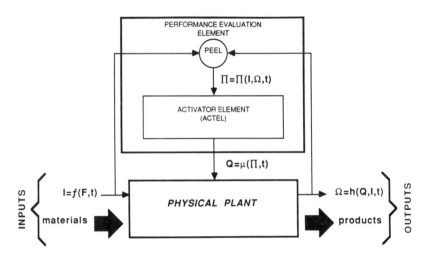

Fig. 19.5 Model of an IMS.

Each performance metric Π_i may be quantitative or qualitative. An example of the former is the yield rate, which may take on any real value between 0 and 1; an example of the latter may relate to the expansion flexibility of a production line which might be classified as $\Pi_i = \langle$poor, fair, good$\rangle$.

The concept of acceptable behaviour may now be defined. Let Π_i be the set of all feasible values of Π. Then A_i is the subset of Π_i, which represents all the acceptable values in Π_i. For example, the acceptable values of performance for the yield rate may be the range $A_2 = (80\%, 100\%)$. The list $A = \langle A_1, A_2, ..., A_m \rangle$ then denotes the space of acceptable behaviour.

The IMS can then be formally defined as a six-tuple of the form:

$$\langle I, \Omega, A, \Pi, \mu, Q \rangle$$

where the components are as follows:

I input set which is a function of the product specifications, the system environment and time

Ω output set or product that is a function of the input set and the state of the manufacturing system

A function which defines the range of acceptable system behaviour

Π performance metric or function that evaluates the performance of the system with respect to the goals and constraints of the manufacturing system

μ   function that interprets or transforms the performance measures into a set of instructions. This function ensures that system performance improves over time

Q state set that describes the condition of the manufacturing system and its components

The acceptable performance of a system may be defined as follows. The performance of a system S is said to be *acceptable* if it lies within the space of acceptable behaviour. More precisely, we may write

$$\forall\, i \in J\,(\Pi_i\,(S) \in A_i)$$

This definition will serve as the basis for the formulation of adaptive systems. A system S is said to be *adaptive* if, after a change in the environment or in the system itself, S responds in such a way as to maintain acceptable behaviour, or to restore acceptable behaviour if it was disturbed.

The notion of relative intelligence may now be defined. Let S^* and S_* be two systems whose performances are evaluated by $\Pi = \langle \Pi_1, \Pi_2, ..., \Pi_n \rangle$. In other words:

$$\Pi\,(S^*) = \langle \Pi_1^*, \Pi_2^*, ..., \Pi_n^* \rangle$$
$$\Pi\,(S_*) = \langle \Pi_{1*}, \Pi_{2*}, ..., \Pi_{n*} \rangle$$

Then S^* is said to be *more intelligent* than S_* if the following conditions hold:

$$\exists\, i\,(\Pi_i^* > \Pi_{i*})$$
$$\forall\, j\,(\Pi_j^* \geq \Pi_{j*})$$

This definition incorporates the concept of dominance; a system *dominates* another if it is superior in at least one aspect and not inferior in any way.

To illustrate, let Π_1 be a measure of flexibility and suppose that S^* and S_* are two plants that are comparable in every respect except that S_* is more flexible than S_*. Then $\Pi_1^* > \Pi_{1*}$ while $\Pi_i^* = \Pi_{i*}$ for all other values of i. In this case S^* dominates S_*, and is therefore *more intelligent*.

The simple model discussed above may be extended to accommodate the layered and hierarchical structure of manufacturing systems. Figure 19.6 illustrates the configuration for a multi-level manufacturing system associated with a hierarchical organization, common to both natural and man-made systems. Glorioso and Colon Osorio [1] recommend an interconnection scheme for intelligent multi-goal systems in which

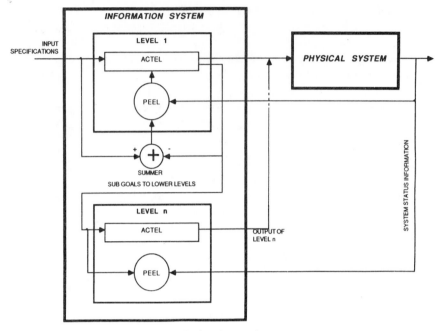

Fig. 19.6 Layered control structure.

control is distributed throughout the system. Figure 19.7 illustrates the configuration of such a distributed control structure for a hierarchical manufacturing system with three levels. For example, the three levels illustrated in the figure may correspond to system layout, system planning and system operation respectively. In this system, the overall objective is input to the highest level, which then determines what tasks it will perform and which others to delegate to lower levels. At each level, performance is evaluated with respect to the tasks that are to be performed and then compared at the preceding level.

Decision Rules

The ultimate structure and operation of an IMS depend heavily on decisions made at the early stages of the design process. Traditional design processes for the design of an FMS involve matching equipment specifications with the tasks required to produce a given set of parts. Such a design methodology limits the capabilities of the final system. A

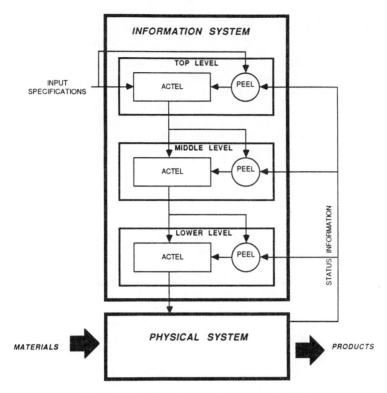

Fig. 19.7 Distributed control structure for IMS.

systematic procedure for synthesizing an IMS requires the identification and classification of generic design principles for factories.

A framework for decision making within a manufacturing system leads to a better understanding of the decision process involved in the design of such systems. Decision rules can be classified into categories based on the structure of the manufacturing system and its objectives. Finally, the identification of the structure and nature of the decisions themselves, simplifies the formalization and encoding of the rules. Kim and Suh [16, 17] discuss the formalization of design rules and their subsequent encoding into PROLOG.

Nature and Structure of Decision Rules

The decision rules may be classified on the basis of their nature and structure, taken individually as well as in sets. Table 2 summarizes the types of decisions and their symbolic representation. A manufacturing system must satisfy certain requirements which may be stipulated as

Table 2. Nature and representation of decision rules

Category	Type	Symbolic representation
Individual structure $\{D\}$	Fact or command Conditional or simple rule	$D: Q$ $D: P \rightarrow Q$
Set property $\{D_1, D_2\}$	Alternatives and complementary rules Cooperating and competing rules Conflicting rules	$D_1: P \rightarrow Q_1$ $D_2: P \rightarrow Q_2$ $D_1: P_1 \rightarrow Q$ $D_2: P_2 \rightarrow Q$ $D_1: P_1 \rightarrow Q$ $D_2: P_2 \rightarrow \neg Q$

facts or assertions. These commands have the simplest structure. Rules D_1 and D_2 in Table 3 illustrate the structure of facts or commands. The rules address the safety requirements that must be maintained in every system.

However, the majority of decision rules are of the IF...THEN type and the general representational structure of such rules is: $P \rightarrow Q$. All the rules in Table 3, except for rules D_1 and D_2, are of this type. D_1 and D_2 also illustrate the general nature of some rules which can then be interpreted or translated into more specific rules. D_2 also illustrates the heuristic nature of some decision rules.

In general, decisions made at any level of the manufacturing system affect and are affected by those at other levels. For example, flexibility decisions often affect more than one aspect of system versatility. Therefore, it is relevant to bring out the nature of the dependence of decisions within a manufacturing system. Decision rules may be of different types: · cooperating and competing, alternative and complementary, or conflicting in nature. The dependency is a set property; a decision D_i by itself cannot be considered cooperating or conflicting, in contrast to a set of decisions $D = \{D_1, D_2, \ldots D_n\}$. Table 2 summarizes the types of decision interdependencies and the related symbolic representations. Table 3 illustrates these types of interdependencies through sample decision rules.

Alternative and complementary rules have the same representational structure, as do cooperating and competing rules. The following explanation can be considered informal definitions of the four types of rules. A rule $D_i: P \rightarrow Q_1$ is said to be an alternative to a rule $D_j: P \rightarrow Q_2$ if both the rules work towards maximizing or improving the same objective (P). A rule $D_i: P_1 \rightarrow Q$ is said to cooperate with a rule $D_j: P_2 \rightarrow Q$ if the action results in the improvement or maximization of the different premises P_1 and P_2.

Decisions can compete in two ways: for selection and for resources. Two decisions $D_i: P \rightarrow Q_1$ and $D_j: P \rightarrow Q_2$ may compete for selection because they both lead to an improvement in the objective P, and the

Table 3. Sample decision rules

Facts or commands
D_1:   Safety requirements must be following in manufacturing systems
D_2:   In a manufacturing cell, safety requires a minimum distance of 6 feet between adjacent machines.

Conditional or simple rule
D_3:   IF
              a cell contains two lathes manned by a robot
      THEN
              arrange the lathes such that their spindle axes are perpendicular

Alternatives and complementary rules
D_4:   IF
              expansion flexibility is the priority
      THEN
              a modular layout has a higher preference rating than an unstructured layout
D_5:   IF
              expansion flexibility is the priority
      THEN
              a rectangular or straight line layout has a higher preference rating than a circular or looped layout

Cooperating and competing rules
D_6:   IF
              routing flexibility is the priority
      THEN
              use wire-guided AGVs over rail-guided AGVs or conveyors
D_7:   IF
              expansion flexibility is the priority
      THEN
              use wire-guided AGVs over rail-guided AGVs or conveyors

Conflicting rules
D_8:   IF
              high volume production is the priority
      THEN
              use a conveyor system as the material transportation system
D_9:   IF
              expansion flexibility is the priority
      THEN
              do not use a conveyor system as the material transportation system

---

application of one may be followed by the other in order to maximize the objective. These two rules are said to be complementary in nature. When two rules D_i: $P_1 \rightarrow Q$ and D_j: $P_2 \rightarrow Q$ compete for the same resources to improve different objectives, they are said to be competing rules.

Classification by System Structure

Decision rules apply to three main aspects of the manufacturing system: *layout*, *planning* and *operation*. The layout aspect includes the physical and communicational structure of the system. The planning aspect includes process planning and scheduling. Finally, the operational aspect deals with the actual execution of tasks within the system. In Table 3, D_4 is primarily a layout rule, while D_6 relates to planning and D_8 to operation.

A correspondence exists between the decision hierarchy within manufacturing systems and the hierarchical nature of business organizations. Figure 19.8 illustrates this correspondence; the top-level decision making involves the physical and communicational layout of the IMS, while the middle-level decision making affects planning and scheduling of required tasks, and low-level decision making influences the operational aspects of the system. Typically, top-level decisions involve a time horizon ranging from months to years, and middle-level decisions involve a range from days to weeks while low-level decisions deal with durations of the order of seconds and minutes. In short, decisions can be classified into long-range, medium-range and short-range.

As previously stated, the decisions at each level are not independent of the other levels. Top-level decisions influence a large number of decisions at lower levels. For example, top management's decision to employ different robots for loading and unloading a machine will increase the complexity of planning at the middle level, as well as the complexity of programming and coordinating robot actions at the lower level. Decisions are tempered by constraints on the resources available, and the decisions made at each level attempt to maximize their objectives by using the resources available.

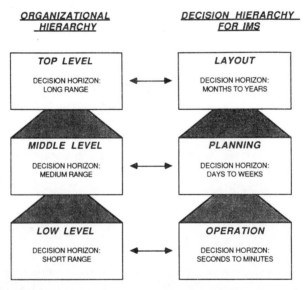

Fig. 19.8 Decision hierarchies in business organizations and in the synthesis of IMS.

Classification by System Objectives

An important objective of any IMS, and the primary objective of an FMS, is flexibility. Flexibility can be related to different aspects of the manufacturing system. For example, the ability to perform a large number of operations on a single machine relates to machine flexibility, whereas the ability to add new machines to an existing manufacturing system with minimum interference relates to expansion flexibility. The different types of flexibility in the context of performance metrics are listed below. The decisions relating to flexibility of IMSs can be classified by the type of flexibility they influence. For example, D_6 tends to improve the routing flexibility of a system whereas D_5 will tend to augment the expansion flexibility.

Applications of Framework to Design Level Issues

The classification of decision rules improves the computational efficiency of an IMS design system, by providing a basis for the selection and execution of actions. The classification of decision rules based on their structure may be used to determine the *order of selection* of rules. In the synthesis of an IMS, facts that are true for all systems are applied first, followed by alternatives and complementary rules, and then cooperating and competing rules. Conflicting rules are left until the end, when more involved conflict resolution methods are used to determine the order of selection.

The use of *meta-rules* guides the selection of rules in order to minimize design time needed to achieve their goals. The meta-rules form the basis of an intelligent advisory system for the synthesis of an IMS. Consider the case of two rules D_i: $P_1 \to Q$ and D_j: $P_2 \to Q$. If management objectives dictate that P_1 must be maximized and P_2 minimized, then action Q must be executed only as a last resort. Or consider this set of cooperating rules:

$$D_m = \{D_{1m}: P_1 \to Q_m,$$
$$D_{2m}: P_2 \to Q_m,$$
$$D_{3m}: P_3 \to Q_m,$$
$$D_{4m}: P_4 \to Q_m\}$$

and the set:

$$D_n = \{D_{1n}: P_1 \to Q_n,$$
$$D_{2n}: P_2 \to Q_n,$$
$$D_{3n}: P_3 \to Q_n\}$$

In this situation, the firing of action Q_m satisfies more goals than the activation of Q_n. Both of these high level observations can be encoded into meta-rules.

The classification of rules also aids in the generation of novel decision rules through abstraction. The general strategy used in the abstraction of new decision rules is an experimental one and can be summarized as follows. Consider the case of conflicting rules D_i: $P_1 \rightarrow Q$ and D_j: $P_2 \rightarrow \neg Q$. For a given scenario, execute action Q, and measure the system performance. Similarly, measure system performance under the same conditions when $\neg Q$ is executed. Repeat the evaluations for different scenarios. Then any consistent observations can be abstracted into decision rules. This abstraction of decision rules helps in the evolution of the prototype system into a fully-fledged one.

Performance Metrics

The objective of developing a theoretical model of IMS is not limited to the analysis of such systems. The ultimate goal is to develop a set of generalized decision rules for IMS that can be encoded into an advisor to assist in the synthesis of factories. The most effective decision rules can be determined by the performance of the system model under different simulated conditions. For such evaluation, a set of metrics is necessary.

A performance metric of particular interest is that of the flexibility of a manufacturing system, which arises when organizational complexity is incorporated into an information-theoretic framework [6]. In general, the flexibility of a manufacturing system may be defined as the ability to produce a variety of products efficiently using the same set of machines. Flexibility relates to several issues in the design and operation of FMSs, including the following: the flexibility of the physical and communicational layout of the manufacturing system, the rearrangement of and addition to the layouts, the planning and scheduling of operations for a given task and the number and composition of cells and their functions for a given set of products. The literature identifies several issues in deriving a metric of flexibility [2]:

- machine flexibility
- product flexibility
- process flexibility
- operation flexibility
- routing flexibility
- volume flexibility
- expansion flexibility

Such a composite metric of flexibility could be used, for example, to demonstrate the tradeoff between a system with highly flexible cells that can perform many tasks, versus one in which the cells are more specialized.

In addition to a measure of flexibility, a metric of system performance with respect to time is of interest in the evaluation of manufacturing plants. From the definitions of learning, adaptive, self-repairing and self-organizing systems, it is clear that the flexibility and temporal metrics will together constitute a measure of system intelligence. Thus, an intelligence metric for manufacturing systems must combine attributes such as flexibility and timeliness.

Nature and Structure of Performance Metrics

As with decision rules, sets of performance metrics can be classified into alternative, complementary, competing and conflicting metrics. Table 4 illustrates the types of interdependencies with some examples.

The classifications based on the nature of metrics are set properties. For example, a metric M_0 can conflict with M_1 but M_0 by itself cannot be classified as conflicting. Alternative metrics are those which are equivalent and serve equally well. For example, consider a job shop whose performance is to be evaluated. The performance may be evaluated by the measurement of the throughput of the shop (activity focus), or the average turnaround period for workpieces to progress through the shop (event focus).

Competing metrics are those that lead to the evolution of different designs. To illustrate this, consider a metric that evaluates the reliability of an IMS, say mean time between failures (MTBF), versus a metric that evaluates the availability of an IMS, say percent uptime. These metrics

Table 4. Nature and structure of performance metrics

Category	Type	Example
Individual structure (M)	Classical information metric Entropy metric	M: $-p \ln p$ M: $-\Sigma\, p \ln p$
Set property ($S_m = \{M_1, M_2\}$)	Alternative metrics	M_1: Throughput metric M_2: Turnaround time metric
	Complementary metrics	M_1: Machine utilization metric M_2: Throughput metric
	Competing metrics	M_1: Reliability metric (MTBF) M_2: Availability metric (%-uptime)
	Conflicting metrics	M_1: Machine utilization factor M_2: Reliability metric (MTBF)

compete, since a system that operates continuously returns a high value for percent uptime; but such continuous operation may result in frequent machine failures and hence a low value for MTBF.

Complementary metrics are those that may be applied successively to evaluate an IMS. For example, successive application of metrics for machine utilization and throughput results in a system that exhibits a high machine utilization value and a high throughput. Conflicting metrics correspond to conflicting objectives. For example, a machine utilization metric is a measure of the successful operation of the system, whereas a reliability metric, such as MTBF, is a measure of the failures in a system.

The metrics can be quantitative, qualitative, or both. For example, a metric that evaluates the flowrate is quantitative in nature, while a metric that evaluates expansion flexibility in terms of ⟨good, fair, poor⟩ is qualitative. The structure of the quantitative performance metrics is based on information theory [6, 18, 19]. The performance metrics are based on the classical information metric and the entropy measure. The general formula for the entropy of a system is given by

$$H(S) = - \sum p_i \ln p_i$$
$$\text{such that} \sum p_i = 1$$

where $i = 1$ to m, for m denoting the number of components under consideration, and p_i a ratio reflecting the characteristic of the system being measured. The merits and limitations of this expression are discussed by Kumar [19]. The application of this general expression to the measurement of system performance is illustrated below.

In evaluating the expansion flexibility of a manufacturing system, it is not possible to arrive at a numerical answer unless we arbitrarily assign numerical values to different types of layouts and then define an algorithm to compute the total flexibility. Since such a method is arbitrary, it may be more appropriate to use a qualitative evaluation as illustrated by the following example.

Consider the case where an FMS has to be designed for a new product. Further, management expects an increased demand for the product and other similar items in the future. Thus, the design priority is a system structure which can be readily augmented and rearranged. In this case a rectangular layout would have a higher priority than a circular layout, and a modular system structure would have a higher priority than an unstructured layout; the expansion flexibility of a rectangular layout is greater than that of a circular layout. This fact cannot be expressed numerically unless arbitrary values are assigned to each of the alternatives. Therefore, it may be better to encode the impact of layout on flexibility as follows:

IF

the priority is expansion flexibility

THEN
>a modular layout has a higher flexibility rating than an unstructured layout

AND
>a rectangular or straight line layout has a higher flexibility rating than a circular or looped layout.

The need to have qualitative metrics in addition to quantitative measures for flexibility and timeliness is obvious. A comprehensive knowledge-based advisory system must incorporate both algorithmic and heuristic performance functions.

General Performance Metrics

The generalized measure of entropy can be used to measure system performance by choosing a suitable ratio p to substitute in the formula for $H(S)$. However, this metric does not take into account factors such as the versatility and complexity of cells, nor the composition and arrangement of the cells in the system.

In order to account for these effects, a suitable weighting system is necessary. Such a weighting system assigns weights to system components depending on the contribution of the elements to system performance. For example, if t_i is the output of the ith cell measured in units/hour and T_{total} is the total output of m cells, then the weighting assigned to the ith cell is (t_i/T_{total}). Such a weighting factor can be expressed in general as follows:

$$\text{weighting factor} = w_i = t_i/T_{total}$$

where t_i is the nominal output of a component and T_{total} is the corresponding nominal output of the aggregation of components.

For example t_i may be the cost of equipment in the ith cell and T_{total} the total cost of the system; or t_i may be the machine-to-robot ratio in the ith cell and T_{total} the machine-to-robot ratio of the system. The performance of the system may then be expressed as the sum of the between-cell entropy and the weighted average of the within-cell entropies.

The foregoing interpretation of the information metric allows us to draw on related work in the field of economics. Theil [20] provides a detailed analysis of the measurement of entropy in aggregated systems. The formula for the entropy of a system may be expressed as:

$$H(S) = \sum w_i \ln (1/w_i) + \sum w_i H_i(S), \text{ for } i = 1 \text{ to } m$$

where $H(S)$ is the system entropy, and $H_i(S)$ is the individual cell entropy. The first term on the right-hand side of the expression is the between-cell entropy, while the second term is the weighted average of the within-cell entropies $H_i(S)$.

Another measure of system performance suggested by Theil is a measure of system inequality obtained by subtracting $H(S)$ from its own maximum value. System inequality is a measure of the inequalities in performance among the individual components of the systems. Theil's general formula for system inequality can be used to identify flow bottlenecks within a system and thus identify points of inflexibility. Such a measure can be applied to a more precise analysis of system performance. The measure of inequality can also be used to analyse the effects on performance of a change in the configuration or composition of a system or cell.

Sample Application: Modularity Metric

This section discusses a metric for modularity based on the expression for the entropy of a two-component system. The modularity metric proposes a slightly different interpretation for p in the general expression for entropy. This serves to illustrate the versatility of the general entropy measure in evaluating system performance. The approach illustrates the use of information-theoretic arguments in developing a set of performance metrics for IMSs. This metric for modularity does not account for the complexity and versatility of the cells, nor individual machines, nor the distribution and arrangement of the machines in the cells. This simplification serves to illustrate the concepts involved more clearly.

A modular system is one in which machines and other equipment are arranged in rational groups or cells with a variety of aims in mind. For example, a modular layout efficiently utilizes space and reduces the complexity of planning and routing. The metric for modularity is a measure of the degree to which the layout reduces the complexity involved in the planning and operation of the system.

Consider a system S that consists of n machines arranged into m cells, where n^* is the number of different types of machines, and m^* the number of different cells. The ratio of the number of cells (m) to the total number of machines (n) is defined as the degree of modularity (Δ). The ratio of the number of different cells (m^*) to the total number of cells (m) is defined as the degree of variety within the system (V_s). Finally, the ratio of the number of different machines (n^*) to the total number of cells can be considered as the degree of variety within the cells (V_c). The performance metric for system modularity (Π_m) is given by the following formula:

$$\Pi_m = \Pi_\Delta + \Pi_{V_s} + \Pi_{V_c}$$

where

$$\Pi_\Delta = \text{entropy of modularity} = -(p_1 \ln p_1 + q_1 \ln q_1)$$
$$p_1 = \Delta = m/n \text{ and } q_1 = 1-p_1$$
$$\Pi_{V_s} = \text{entropy of system variety} = -(p_2 \ln p_2 + q_2 \ln q_2)$$
$$p_2 = V_S = m^*/m \text{ and } q_2 = 1-p_2$$
$$\Pi_{V_c} = \text{entropy of cell variety} = -(p_3 \ln p_3 + q_3 \ln q_3)$$
$$p_3 = V_c = n^*/m \text{ and } q_3 = 1-p_3$$

Since Π_m is given as a simple sum of three terms, maximizing it is tantamount to maximizing each of Π_Δ, Π_{V_s} and Π_{V_c}.

Table 5 and its graphic representation in Fig. 19.9 illustrate the application of Π_Δ to a nine-machine system. The graph illustrates the following aspects of the modularity metric: for a low value of m, the degree of modularity is low. The value of Π_Δ illustrates the fact that the complexity associated with programming and coordinating operations within a cell is high. For a high value of m, the degree of modularity is high, and the corresponding value of Π_Δ illustrates the fact that the

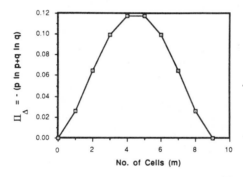

Fig. 19.9 Degree of modularity: cells versus Π_Δ.

Table 5. Data for cells versus Π_Δ

Cells (m) for $n = 9$	$p = m/n$	$q = 1-p$	Π_Δ
0	0	1	0
1	0.111	0.889	0.026
2	0.222	0.778	0.065
3	0.333	0.667	0.099
4	0.444	0.556	0.118
5	0.556	0.444	0.118
6	0.667	0.333	0.099
7	0.778	0.222	0.065
8	0.889	0.111	0.026
9	1	0	0

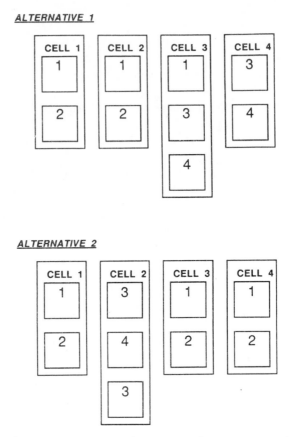

Fig. 19.10 Alternative arrangements for a nine-machine system: result of Π_M.

complexity associated with interfacing or coordinating the large number of cells is high. According to the graph, the best choice would be to have four or five cells for a total of nine machines. By a similar reasoning process, Π_{V_s} is maximized for four or five different machines, while Π_{V_c} is optimized for two or three different cells. Figure 19.10 illustrates two alternative optimal configurations that exhibit equally high values for Π_m.

Discussion

The framework for decision rules and metrics provides a basis for designing intelligent factories. Performance metrics enable a factory to evaluate its performance and make improvements if necessary. The decision rules help in guiding the selection of appropriate responses to a dynamic environment, and in selecting the appropriate metric to evaluate the responses. The classification of metrics helps an intelligent system in choosing the proper performance metrics.

The weighted entropy measures and the measure of inequality within a system provide a framework that can be used to analyse several system issues. For example, inequality measures can be used to study the balance between the versatility of system components and their complexity. Consider a system consisting of m cells. If cell i is highly versatile, and cell j is very inflexible, then the inequality measure will bring out this difference in versatility through a higher value for the measure. The capabilities of cell j may be augmented in order to achieve greater equality. The inequality of the system will decrease to the point where cells i and j are equally versatile (assuming that the other cells are unchanged and that the sum of the capabilities of cells i and j remains a constant) and increase when cell j is made more versatile than cell i. The primary objective is to keep the inequality at a minimum.

If the flowrate of cell i is much greater than that of cell j, then the inequality measure will bring out this flow bottleneck. By simulating different physical and communicational configurations of the system, the inequality measure can be used to reduce and ultimately to eliminate flow bottlenecks. Simulations can be used to determine a favourable machine-to-robot ratio (R_{mr}) for the cells as well as for the system. Then the weighting factors used in the measurement of system entropy are given by the ratios of the R_{mr} of the cells to that of the system. Changing the ratios within the cells and simulating the different alternatives, can help determine the optimum ratios for the cells and the system.

The limits to automation, the effects of simple automation, and the tradeoffs between comprehensive automation and complexity can be studied by calculating the maximum value of system entropy under simulated conditions. Similarly, the effects of standardizing IMS layout and components can be determined using the inequality measure.

Future Work

A simulation model of a three-cell flexible manufacturing system is under development. The applications of the models and frameworks discussed in the previous sections, will be applied to this FMS simulation model. This study will assist in the validation of the proposed framework and in developing newer and better methods for designing factories. The study will consider cost and the difficulty of designing new equipment and interfaces, as well as the efficiency implications. The simulation model along with the design rules and performance metrics will be incorporated into a knowledge-based advisor for the design of intelligent systems. Two issues critical to the development of a fully-fledged advisory system are a temporal reasoner and an intelligent decision maker. The following sections discuss these two aspects in greater detail.

Temporal Model for Manufacturing Systems

An IMS is defined as an adaptive manufacturing system capable of learning and self-organization. These attributes are dynamic qualities which reflect changes over time. Hence a key component of system intelligence is temporal reasoning or the ability to represent and reason about time. Temporal reasoning ability is crucial to the planning and scheduling of tasks within a manufacturing system.

Allen [21, 22] addresses the requirements of a general temporal model. A temporal framework must allow for significant imprecision and uncertainty of information, since much temporal information is relative in nature and bears no relation to absolute dates or times. Consider two events, A and B; event A could occur before, after or during event B, or may overlap B. The temporal model should facilitate representation and manipulation of such information. The model should allow for variation in the grain of reasoning so as to consider durations of days and months in addition to seconds and minutes. Above all, the model should facilitate easy manipulation, since in a rapidly changing environment the IMS should respond in time to adapt to the changes.

The ability to reason about time is crucial to the planning and scheduling of operations in a system. Hence, a temporal metric is needed to evaluate the flowrates, recovery times, changeover times, start-up times and the timeliness of the system's actions. For maximum utilization of a system's resources and efficiency of system operation, it is often desirable to perform operations simultaneously where possible. A temporal metric could be used to demonstrate the tradeoff between simultaneous and sequential arrangement of tasks for a set of products. The temporal metric will be especially useful in the generation and evaluation of alternative plans, schedules and routings for a set of specified tasks, as well as in robot planning and implementation.

Intelligent Decision Making

A simulation model of FMS as a special case of IMS is being developed at our laboratory based on the proposed framework for IMS. The classification of decision rules, together with the set of performance metrics, can be applied to the simulation model to develop a system that learns how to make intelligent decisions. This may be achieved by developing learning systems [23] and assigning certainty factors to each decision rule. The certainty factors will reflect the criticality of each rule, and will be determined by iterated trials. For example, consider the following rule.

D: The machine-to-robot ratio in a cell must not exceed 2

The prototype model will initially assign a certainty factor of 1 to this rule. Every time this rule is applied, the certainty factor will be modified depending on the performance of the system as evaluated by the metrics. Through simulation of different scenarios, a realistic value for the certainty factor may be extracted. The certainty factor is tracked at every step to identify consistent patterns and possible exceptions to the rule. The final result of this experimentation might be expressed in a form such as:

D: The machine robot ratio in a cell must not exceed 2,
except when $P_e = \{P_{1e}, P_{2e}, \ldots P_{ne}\}$

Further, the procedure adopted to generate the final decision rule can be encoded as meta-knowledge. In this way, the simulation system can learn to formulate its own control rules as a result of computational experiments. These may then be checked against a physical factor to determine whether they should be accepted, discarded or further refined.

References

1. Glorioso GM, Colon Osorio FC. Engineering intelligent systems: concepts, theory and applications. Digital Press, Digital Equipment Corporation, Bedford, MA, 1980
2. McDougall GHG, Noori HA. Manufacturing-marketing strategic interface: the impact of flexible manufacturing systems. In: Kusiak A (ed) Modelling and design of flexible manufacturing systems. Elsevier Science Publishers, Amsterdam, 1986, pp 189–205
3. Jaikumar R. Postindustrial manufacturing. Harvard Bus Rev 1986; no. 6: 69–76
4. Suh NP. The future of the factory. Robotics Comput Integr Manuf 1984; 1: 39–49
5. Galbraith J. Designing complex organizations. Addison-Wesley, Reading, MA; 1973
6. Kim SH. Mathematical foundations of manufacturing science: theory and implications. PhD thesis, MIT, 1985
7. Kim SH. A generalized information-theoretic framework for intelligent systems. Technical Report, Laboratory for Manufacturing and Productivity, MIT, Cambridge, MA, 1987
8. Nakazawa H, Suh NP. Process planning based on information concept. Robotics Comput Integr Manuf 1984; 1: 115–123
9. Kuo BC. Digital control systems. Holt, Rinehart and Winston, New York, 1980, p 644
10. Tsypkin YZ. Adaptation and learning automatic systems. Nikolic ZJ (trans) Academic Press, New York, 1971
11. Kim SH. A mathematical framework for intelligent manufacturing systems. Proceedings symposium on integrated and intelligent manufacturing systems, ASME, Anaheim, CA, 1986, pp 1–8
12. Lawson JS Jr. The role of time in a command and control system. In: Athans AM et al., Proceedings fourth MIT/ONR workshop on distributed information and decision systems motivated by command-control-communications problems, vol 4, Laboratory for Information and Decision Systems, Report LIDS-R-1159, MIT, 1981, pp 19–60

13. Kim SH. Frameworks for a science of manufacturing. Proceedings North American manufacturing research conference, Minneapolis, MN, 1986, pp 552–557
14. Kim SH, Suh NP. Mathematical foundations for manufacturing. Trans ASME J Eng Indust 1987; 109: 213–218
15. Parthasarathy S, Kim SH. Formal models of manufacturing systems. Technical Report, Laboratory for Manufacturing and Productivity, MIT, Cambridge, MA, 1987
16. Kim SH, Suh NP. Application of symbolic logic to the design axioms. Robotics Comput Integr Manuf 1985; 2: 55–64
17. Kim SH, Suh NP. Formalizing decision rules for engineering design. Technical Report, Laboratory for Manufacturing and Productivity, MIT, Cambridge, MA, 1987
18. Hagler C, Kim SH. Information and its effect on the performance of a robotic assembly process. Proceedings symposium on intelligent and integrated manufacturing: analysis and synthesis, Boston, MA, December 1987, pp 349–356
19. Kumar V. Entropic measures of flexibility. Int J Prod Res 1987; 25: 957–966
20. Theil H. Economics and information theory. North-Holland Publishing Company, Amsterdam, 1967
21. Allen JF. An interval-based representation of temporal knowledge. In Proceedings IJCAI, Vancouver, BC, 1981, pp 221–226
22. Allen JF, Koomen JA. Planning using a temporal world model. Proceedings IJCAI, Karlsruhe, W. Germany, 1983, pp 741–747
23. Kim SH. A unified framework for self-learning systems. Proceedings manufacturing international '88, Vol III: symposium on manufacturing systems, Atlanta, GA, 1988, 165–170

Subject Index